W0042432

# RECENT DEVELOPMENTS IN ION EXCHANGE

# 2

Proceedings of the International Conference, ION-EX '90, held at the North East Wales Institute, Wrexham, Clwyd, UK, 9–11 July 1990.

## Members of the Organising Committee

| | |
|---|---|
| Mr E. R. Adlard | Royal Society of Chemistry (NW Region) |
| Mr W. J. Beesley (*Treasurer*) | North West Water Authority |
| Mr B. Crooks | ICI Pharmaceuticals |
| Dr A. Davies | Alfred H. Knight Ltd |
| Dr A. Dyer (*Chairman*) | University of Salford |
| Dr H. Eccles | SCI Solvent Extraction and Ion Exchange Group |
| Mr R. C. George | Dionex (UK) Ltd |
| Dr J. Greene | Nuclear Electric |
| Dr M. J. Hudson | University of Reading |
| Mr H. Hughes (*Secretariat*) | The North East Wales Institute |
| Mr B. W. King | Phase Separations Ltd |
| Mr M. Parker | Beckman Instruments (UK) Ltd |
| Mr M. Parker | Waters Division of Millipore |
| Mr D. Ryder | Ciba-Geigy plc |
| Mr P. Wall | BDH Ltd |
| Dr P. A. Williams (*Secretariat*) | The North East Wales Institute |

Supported by the Royal Society of Chemistry North Wales Section and North West Region (Analytical Division) and the Society of Chemical Industry.

# RECENT DEVELOPMENTS IN ION EXCHANGE

# 2

*Edited by*

## P. A. WILLIAMS

*Research Division, The North East Wales Institute,
Deeside, Clwyd, Wales, UK*

and

## M. J. HUDSON

*Chemistry Department, University of Reading,
Whiteknights, Reading, UK*

ELSEVIER APPLIED SCIENCE
LONDON and NEW YORK

ELSEVIER SCIENCE PUBLISHERS LTD
Crown House, Linton Road, Barking, Essex IG11 8JU, England

*Sole Distributor in the USA and Canada*
ELSEVIER SCIENCE PUBLISHING CO., INC.
655 Avenue of the Americas, New York, NY 10010, USA

WITH 45 TABLES AND 171 ILLUSTRATIONS

© 1990 ELSEVIER SCIENCE PUBLISHERS LTD
© 1990 UNITED KINGDOM ATOMIC ENERGY AUTHORITY—pp. 213–218
Softcover reprint of the hardcover 1st edition 1990

**British Library Cataloguing in Publication Data**

Recent developments in ion exchange 2.
  1. Ion exchange
  I. Williams, Peter A. (Peter Anthony)   II. Hudson, M. J.
  (Michael James), *1940–*
  541.3723
  ISBN-13:978-94-010-6836-9      e-ISBN-13:978-94-009-0777-5
  DOI: 10.1007/978-94-009-0777-5

**Library of Congress CIP data applied for**

No responsibility is assumed by the Publisher for any injury and/or damage to persons or property as a matter of products liability, negligence or otherwise, or from any use or operation of any methods, products, instructions or ideas contained in the material herein.

**Special regulations for readers in the USA**

This publication has been registered with the Copyright Clearance Center Inc. (CCC), Salem, Massachusetts. Information can be obtained from the CCC about conditions under which photocopies of parts of this publication may be made in the USA. All other copyright questions, including photocopying outside the USA, should be referred to the publisher.

All rights reserved. No part of this publication may be reproduced, stored in a retrieval system, or transmitted in any form or by any means, electronic, mechanical, photocopying, recording, or otherwise, without the prior written permission of the publisher.

# PREFACE

These Conference Proceedings deal with the papers presented at the International Conference on Ion Exchange Processes (ION-EX '90) which was held at The North East Wales Institute of Higher Education, 9–11 July 1990. The camera-ready paper format was chosen so that delegates could receive their copy on arrival at the Conference. The Proceedings include reviews of biological materials, inorganic ion exchangers, the nuclear industry, theoretical aspects and new advances. In addition, there are research papers dealing with industrial ion exchange procedures and new materials. The Proceedings should therefore be of interest to those who need to be brought up to date in the various aspects of processes which involve ion exchange and ion chromatography which are now accepted as important in analysis, separation processes and process control. In each of these areas there have been important developments which are herein described.

As Editors we should like to express our thanks to the individual authors for preparing their manuscripts in the required format and to Haydn Hughes and Linda Sneddon for their invaluable assistance in compiling these Proceedings.

PETER A. WILLIAMS
MICHAEL J. HUDSON

# CONTENTS

## Part 3: Nuclear Industry

## Part 4: Theoretical Aspects and New Advances

**Part 5: New Materials**

**Part 6: Industrial Applications**

# Part 1

# BIOLOGICAL MATERIALS

# THE IMPORTANCE OF ION EXCHANGE PROCESSES IN LIVING SYSTEMS

ROBERT J.P. WILLIAMS
Inorganic Chemistry Laboratory, University of Oxford
South Parks Road, Oxford OX1 3QR, UK

## ABSTRACT

In biology there is a great variety of charged surfaces formed from poly-saccharides, lipids, proteins and polynucleic acids. For the most part they are negatively charged and therefore act as cation exchangers. The exact chemical anionic groups, their concentrations and their whereabouts in the organism are important since the concentration levels of free cations is very variable from one part of the system to another. Anion binding is to amino-groups of proteins. In the simplest biological systems, direct comparisons can be made with man-made ion-exchangers. However, ion-exchange is also used in biology to make current carrying and trigger devices.

## INTRODUCTION

Ion exchange in biological systems is governed at first sight by the same properties of the free ions and of the binding matrices as are found in man-made exchangers or in such non-living materials as soils. We write the simplest expression for equilibrium binding of a single ion as

$$\text{Binding} = K_{SM}[M]$$

where K, the binding constant, characterises the surface (S) interaction with the metal cation or anion (M) and [M] are the metal ion or anion concentrations although activity should be used if greater accuracy is required. Binding will show saturation of course. Before elaboration of the equation and consideration of more complex cases we need to look at the nature of biological surfaces for binding and at the concentrations of free ions in biological fluids. It is here that we meet the first major problem. Biological surfaces are very diverse and from one locality to another in a biological system [M] can be very variable. I shall take examples of the nature of biological surfaces after a brief description of the concentration ranges of cations and anions in biology.

## Concentrations in Biology

In so far as external solutions are concerned biology is very diverse since life exists in the sea which has a high concentration of salt - the worst type of sea is 3 Molar in salt - and in the very dilute solutions of fresh water where the total salt is less than one hundreth molar. This external situation is poised opposite an internal total anion plus cation concentration of about 0.2 molar. The value is almost independent of the life form and almost independent of the site in the organism whether it be intracellular or in circulating fluids. This internal salt is made up of simple 1:1 electrolytes largely, i.e. NaCl or KCl, plus a variety of other substances. In the body fluids NaCl dominates but inside cells the variety includes much more KCl and less NaCl together with much more phosphate bound and free and a huge variety of organic anions and cations both small and very large. The whole system is therefore in an energised steady state of electrolyte distribution due to ion pumping across membranes and selective syntheses. The syntheses also produce the larger cooperative units, membranes and particles, which are the ion exchange resins for the purposes of this paper.

Perhaps the most important features of the salt distribution in biological systems from the ion-exchange point of view are given in Table 1 which is self explanatory. The distribution of ions in biology is very far from homogeneous. We turn to the very hetereogeneous surfaces, Fig.1.

### TABLE 1
Concentration of the main cations in living cells and their environment
- mM/mM $dm^{-3}$

| System | $Na^+$ | $K^+$ | $Ca^{2+}$ | $Mg^{2+}$ |
|--------|--------|-------|-----------|-----------|
| Sea water | 460 | 10 | 10 | 52 |
| Valonia | 80 | 400 | 1.5 | 5.0 |
| Red-blood cells | 11 | 92 | $10^{-3}$ | 2.5 |
| Blood plasma | 160 | 10 | 2 | 2 |

## External Surfaces of Unicellular and Multicellular Systems

The outer surface of many, if not most, biological systems is composed of rather strong acid anion centres, often belonging to sugars. The surfaces are therefore ionised. The inorganic metal ions then interact with these O(oxygen)-donors amongst which are $-OSO_3^-$, $-CO_2^-$ and $-OPO_3^{2-}$ units.

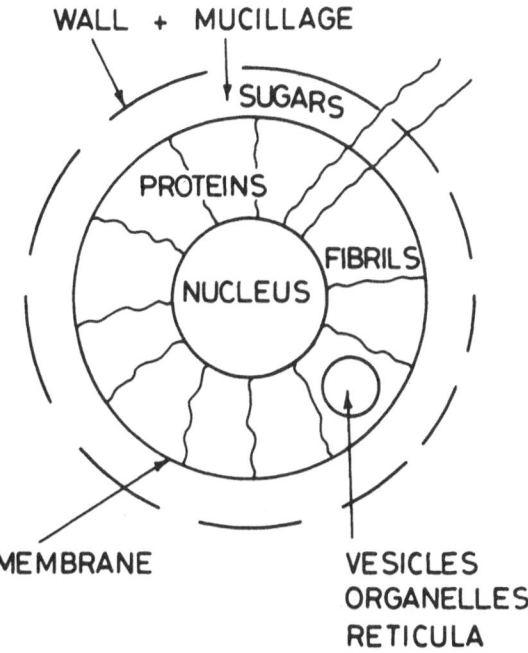

FIGURE 1. A simple diagram of a cell showing the different structures and compartments. The variety of surfaces become the ion-exchange beds of biology but they are deformable.

While such centres bind weakly to monovalent cations they bind to divalent ions quite respectably but in a not very discriminatory manner, Table 2. The common divalent cations in biology are $Mg^{2+}$ and $Ca^{2+}$ and they overwhelm in concentration the concentrations of all other metal ions except $Na^+$ and $K^+$ both in the sea and in fresh water, Table 1. Using equation (1) it is clear that these wet surfaces are $Mg^{2+}/Ca^{2+}$ ion-exchange beds to which the cations bind quite well and so stabilise the anionic network. They also equilibrate quite rapidly with their surroundings. As we all know if we introduce other elements such as $Zn^{2+}$, $Cd^{2+}$, $Pb^{2+}$ even in small amounts into these solutions the surfaces will take up these foreign ions to a roughly tenfold greater extent than the concentration ratio to $Ca^{2+}$ would suggest.

Many biological surfaces are exposed to air rather than to aqueous solutions. It is now a common observation that the air borne elements in different forms can exchange with the elements, often $H^+$, relatively weakly attached to the surface. In this way lichens act as detectors of air pollution.

TABLE 2

Stability constants ($\log K_M$) of alkaline-earth and first series transition metal ions with some carboxylate, phosphate and sulphate ligands (T = 20° or 25°C).

$$\log K_M$$

|  | $Mg^{2+}$ | $Ca^{2+}$ | $Sr^{2+}$ | $Ba^{2+}$ | $Mn^{2+}$ | $Fe^{2+}$ | $Co^{2+}$ | $Ni^{2+}$ | $Cu^{2+}$ | $Zn^{2+}$ |
|---|---|---|---|---|---|---|---|---|---|---|
| Acetate | 1.25 | 1.24 | 1.19 | 1.15 | 1.40 | - | 1.46 | 1.43 | 2.23 | 1.57 |
| Malonate | 1.95 | 1.85 | (1.25) | 1.34 | (2.3) | (2.5) | 2.98 | 3.30 | 5.55 | 2.97 |
| Phosphate | 1.60 | 1.33 | 1.0 | - | 2.58 | - | 2.18 | 2.08 | 3.2 | 2.4 |
| Pyrophosphate | 5.4 | 4.9 | (4.7) | (4.6) | - | - | 6.1 | 6.98 | 7.3 | (5.1) |
| Methyl-phosphate | 1.52 | 1.49 | - | - | 2.19 |  | 2.00 | 1.91 | - | 2.16 |
| Sulphate | 2.20 | 2.31 | 2.3 | 2.3 | 2.0 | 2.3 | 2.47 | 2.40 | 2.4 | 2.3 |
| Thiosulphate | 1.84 | 1.98 | 2.04 | 2.21 | 1.95 | 2.17 | 2.05 | 2.06 | - | 2.30 |

Source: Stability constants. The Chemical Society Special Publications No. 17 and 25. London 1964.

## Internal Surfaces of Multicellular Organisms

Some parts of the interior surfaces of many multicellular organisms, i.e. the external surfaces of cells which are in the circulating fluids, are not too different from the exposed surfaces of cells of all kinds. One example is the cell surfaces exposed to human body fluids. We can treat the problem in the same way as we analyse ion-exchange resin columns; the blood or other body fluid being the liquid in the column, the input solutions are from the digestive system and the output is the various excretary mechanisms. The fact that input has to go through a variety of membranes before it reaches the flowing circulatory liquids of the body, or that the output is filtered through the kidney is of no consequence to us here. Nearly all the surface of the cells over which the fluids pass are made of glycosylated proteins (saccharide-protein compounds). Some of the glycosyl units carry charged sulphate, carboxylate and phosphate groups and densely packed hydroxyls and act as ion-exchanger beds. The selectivity for cations is very similar to that of the outside surfaces so that we find that the exterior of an internal cell is a magnesium/calcium cation exchanger with a surplus of dispersed negative change neutralised as necessary in a Gouy-Chapman layer by sodium rather than potassium. A very important feature is that the surfaces are now rather loosely structured so that the structure of the ion-exchanger bed is cation dependent. Equation (1) no longer applies since K varies with degree of saturation. The implication is that it is necessary to have very strictly controlled ion concentrations in blood for example so as to maintain a well defined organisation. This is a very interesting use of ion exchange.

There is a further new feature; many of the glycoproteins of connective tissue are extensively sulphated and carry no other anions. There is a big difference between carboxylated/phosphorylated resins and sulphated resins. The first are based on anions derived from weaker acids and have affinity orders

$$Li > Na > K > Rb > Cs$$
and $$Mg > Ca > Sr > Ba$$

while the reverse order is found for sulphated surfaces which bind very weakly especially to $Na^+$, $Mg^{2+}$ and $Ca^{2+}$. In biology which is full of these cations but at low concentrations this means that charge neutralisation does not occur for the sulphated resins. The negative, bound, sulphate ions then repel one another forming an open swollen gel, where there are few bound cations, rather than a tight surface. This open gel is required around cells in order that food, small organic molecules, can diffuse to the cells. The outer-most network needs to be a less swollen cross-linked (by $Ca^{2+}$ and $Mg^{2+}$) mesh to protect the system. The risk with the sulphated system is that it may retain heavy large ions but even this can be turned to an advantage by biology, see below. All in all we see that sulphated, carboxylated and phosphated ion exchangers are used to generate ion binding exactly parallel to that seen in the ion exchanger industry, to a first level of approximation, but that the structure of the exchanger is now involved in the total function.

So far we have described the biological (resin) surfaces which are made from polymerised sugars, and may be compared with sulphonated polystyrene. We cannot leave the discussion of outer surfaces until we have described the crystalline biominerals such as calcite (shell) and hydroxyapatite (bone). These minerals have very high surface areas since they are made from

microcrystals. Once again we observe that they are composed of
oxyanions of carbon and phosphorus. They are           avid collectors
of all kinds of divalent ions and these calcium salts always contain $Mg^{2+}$,
$Sr^{2+}$ and even lanthanides on their surfaces. The effect of $Mg^{2+}$ on these
growing (living), crystalline, ion exchange resinsis that it not only
affects the morphology of the ion-exchange 'beads' but can affect even the
allotropic form. In the sea animals calcium carbonate is normally
aragonite while in land animals it is calcite. Curiously this difference
is found even in the balance organ of the brain which in man and land
animals is made of calcite but in marine animals is made from aragonite
with a heavy magnesium content. We do not need to stress the problems
of ion exchange on all these minerals should water become contaminated
with radio-active elements. The problem deepens because the ion
exchanger bed is constantly forming and dissolving so that,in the re-
modelling,ions absorbed on the surfaces get trapped in the interior when
chelation therapy aimed at removing unwanted ions will fail. Chelation
therapy is an obvious example of ion-exchanger elution.

Before re-examining the validity of the only equation, equation (1),
which has been used in this paper we should look at the other surfaces of
cells and cell systems.

### Lipids

The membranes of biology are liquid ion-exchange beds, Fig. 2, rather
unlike the solids or semisolids discussed so far. The fatty acid or ether
tails form either bilayers or monolayers in which all the molecules can
flow. These layers carry headgroups which again resemble the layers on
the outside of cells in that they are anionic by virtue of carboxylate and
phosphate groups. The phosphates are frequently diester phosphates, to
which we return, but where there are carboxylates and mono-ester
phosphates the binding of ions is selective very much as in the Stern
(complexing) ion exchange layer of the outer cell surfaces. A new
principle which these systems introduce is the variable curvature of the
surfaces, Fig. 3. The curvature is not independent of the bound ions, it
is a surface tension property, so that ion-exchange reactions control the
shape of membranes. The interested reader is referred to a review when it
is pointed out that as cations bind the different lipids will phase-
separate causing the membranes to curve, Williams (1987).

In the description of surfaces so far we have not considered very closely the concentration term in equation (1) but have kept our minds on binding constants. In the discussion of lipids this is no longer possible since lipid membranes separate the inside from the outside of cells. Cells expend very considerable energy in the movement of cations and anions in and out of their cytoplasm (internal fluid) and they do so in a selective manner. The main exported ions are $Na^+$, $Ca^{2+}$ and $Cl^-$ and the imported ions are $K^+$, $Mg^{2+}$ and $HPO_4^{2-}$ plus many organic ions. As a consequence the inner and outer lipid surfaces of the cell membrane differ in exposure to ions. But the surface energy is interactive with the solution composition in a dynamic manner and the molecules in the liquid resin (lipid bilayer) rearrange both transversely and laterally within the membranes to minimise the total free energy of the system. The whole of the morphology of a cell is then under some ion-exchange control. These are peculiar ion-exchangers indeed. Equation (1) has only a formal value for presenting discussion.

$$R-\overset{\overset{O}{\|}}{C}-O-CH_2$$
$$R'-\overset{}{\underset{\overset{\|}{O}}{C}}-O-\overset{}{C}H-CH_2-O$$
$$\overset{}{\underset{}{P}O_2^-}$$
$$O-CH_2-CH_2-N^+(CH_3)_3$$

(a)

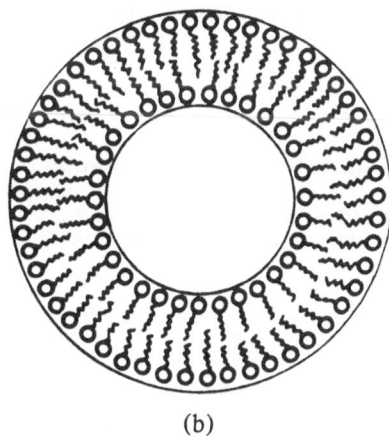

(b)

FIGURE 2. An artificial vesicle or liposome which is the simplest object which can be likened to the lipids of a cell membrane. The circles in (b) are the headgroups and the wobbly lines of (b) are the fatty acid chains, R, of (a).

### Inside the Cell

Apart from the inside of the outer membrane of a cell there are a variety of bodies inside the cell cytoplasm. Some of these, called organelles, are similar to small primitive whole cells trapped in the large cell e.g. mitochondria and chloroplasts. They have surrounding membranes. There are also a variety of membrane structures forming vesicles, vacuoles, Golgi apparatus and reticula. All of these structures have membranes made from lipids so that their surfaces are rather like the surfaces of the outer membrane of the cell already discussed. It is time however to draw attention to the proteins which are in all these membranes and in some

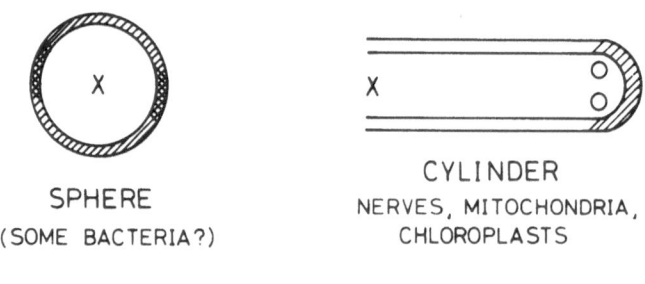

SPHERE
(SOME BACTERIA?)

CYLINDER
NERVES, MITOCHONDRIA,
CHLOROPLASTS

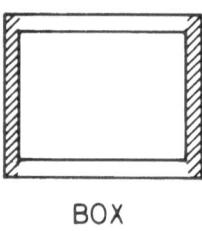

BOX
PLANTS, MUSCLE, FIBROBLASTS

FIGURE 3. Cells have shapes imposed on them by the disposition of molecules in structures. All have polarity. The shape means that ion-exchange properties are localised but it must be remembered that ion-binding also affects the shape of these fluid beds.

membranes account for 50% of the structures. These proteins fan out into space around the lipid bilayer·giving a second more hydrophilic layer extending 100Å both on the inward and outward facing surfaces of the lipid bilayer. They are also quite mobile but are more highly structured than the lipids. It is worth a paragraph or two to consider them as ion-exchange surfaces which are partially liquid.

**Proteins as Ion-Exchangers**

While describing the surfaces of cells there was little need to discuss
anion binding since the surface was treated as a negatively charged zone in
large part. The surface is largely a cation-exchanger. At the next level
of detail this is not true since many biopolymers carry a small percentage
of positively charged groups. Frequently in lipids these are contained in
choline-like, $-N(CH_3)_3^+$ centres and do not interact strongly with the
anions in solution. Only when we come to look at proteins do we find
local surface areas which have a large number of $-NH_3^+$ and guanidinium
headgroups. Thus proteins can bind anions as well as cations and we need
to turn to the nature of the free anions in biological fluids.

The major small anions are chloride, phosphate, many small carboxylates and some sulphate. Amongst these phosphate is the obvious
candidate for binding to protein surfaces. There is clearly a great
significance to the fact that many small molecules e.g. sugars are handled
by biology as sugar phosphates. The great advantage of the proteins
looked upon as anion-exchange resins is their selectivity. The protein
surface is constructed to match in shape and chemical side-chains a
specific molecular ion be this $Ca^{2+}$, acetate, or some phosphated
compound. Here we would need a very detailed analysis in order to show
how to construct an ion-exchange resin of extreme selectivity.

This is not the place to elaborate the analysis of the interaction
between proteins and small molecules. One point needs making.
The protein molecules are not rigid, though they are not disordered liquids
like the membrane lipids. They alter their shapes and structures to
varying degrees but specifically on binding anions or cations. Where
structural changes are small we can liken the binding to simple ion
exchange but where there is a larger structural response the ion exchanger
becomes a signalling device, Fig. 4. Can we learn something from this?
The electrical circuits of biology are electrolytic. It follows that the
only localised wires must be diffusion on ion-exchanger surfaces. It also
follows that condensers are ion-exchange beads of some kind e.g.
calciquestrin. It takes little imagination to build circuits from ion
exchange resins and ions. Man knows these devices in ion-exchange storage
batteries and in ion conductors. They are a major feature of biology.

Proteins introduce an extra new feature. The binding centres they
present for cations now have great variety such as is met in chelating
analytical reagents. Selectivity as a chelating resin is now apparent e.g.

the Irving-Williams series of binding to nitrogen and sulphur donors becomes clear. There is no space to go into detail but much can be learnt from the study of metal ion binding to proteins for the advancement of ion exchange studies.

METAL ION CONCENTRATION

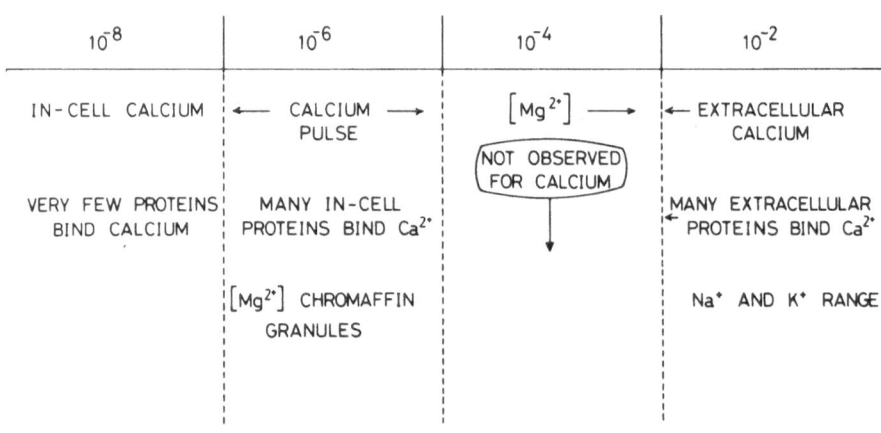

FIGURE 4. The ion concentrations in cells are very different from those outside cells. A wide variety of binding centres are differentially distributed. Switches of calcium concentrations, usually as a pulse of calcium into a cell, is part of the electrolytic circuitry of biology which depends critically on ion-exchange properties.

**DNA and RNA**

Inside cells there is another kind of ion-exchange polymer, the polynucleotides, which build the nucleus, DNA, and the ribosomes, RNA. These are polyesters based upon doubly substituted phosphate. As centres for the binding of cations such as $K^+$ and $Mg^{2+}$ these centres are weak, comparable in fact to the headgroups of many lipids. However they can become powerful ion-exchange centres in two ways. If there are possible states of the fold of the RNA or DNA which throw together several of these diester groups then the local concentration of negative charges induced can be ,and is found to be, stabilised by $Mg^{2+}$ ions in particular, see the references for details.

A second point is that,even when spread out,the negative charges can bind a series of positive charges, amines, which are on a chain - a polyamine. Thus ion-exchange competition here as elsewhere is governed by the relative density of charge on the cation and the anion. We have investigated this problem for solution equilibria using series of small and increasingly larger anions and cations, analysing for the most favourable interactions. The general conclusions are:

(1) High density anion charges such as $RPO_4^{2-}$, small polyphosphates and multicarboxylates such as $EDTA^{4-}$ interact best with high density cation charge as seen on ions such as $Mg^{2+}$ and $[Co(NH_3)_6]^{3+}$

(2) Lower density but still localised anion charge such as that on $SO_4^{2-}$ interacts but with larger cations of high charge density such as $Ba^{2+}$. Barium sulphate is insoluble and so is magnesium hydroxide but the opposite pairings give soluble salts.

(3) Regularly spaced anion charges of relative low charge density $(R_2PO_4^-)_n$, $(RCO_2^-)_n$ etc. interact with the above cations only if the polymer has a tendency to fold so that the local charge density becomes high. In other circumstances, i.e. stretched out polyelectrolytes, polyamines (especially propylene-based amine chains) interact very effectively with the spread out charges of this kind of polyanion.

These equilibria are all based on ion-exchange to a large degree, although some hydrophobic component is important in any interaction between organic cations and anions. They are the basis of the control of DNA transcription and RNA translation. Protein ion-exchangers are more selective and have more specific interactions. The surfaces of cells are different again in that the chemical groups are more restricted and there is no regular repeat as in DNA or RNA.

## Cooperativity between Ions and Exchangers

Equation (1) was written in the simplest form so as to lead to a discussion of the primary features (a) the concentration of the ions (2) the nature of the ion exchanger surfaces. It is clear that many factors alter the binding equation in more concentrated solutions, e.g. in mixed solutions, at different pH, $T°C$ and pressure. All these factors are common in the treatment of ion exchangers *in vitro*. The big distinction between *in vitro* and *in vivo* systems is that as well as competitive phenomena there are many cooperative features *in vivo*,Fig.5. The cooperativity or anticooperativity means that K in equation (1) is dependent on conditions.

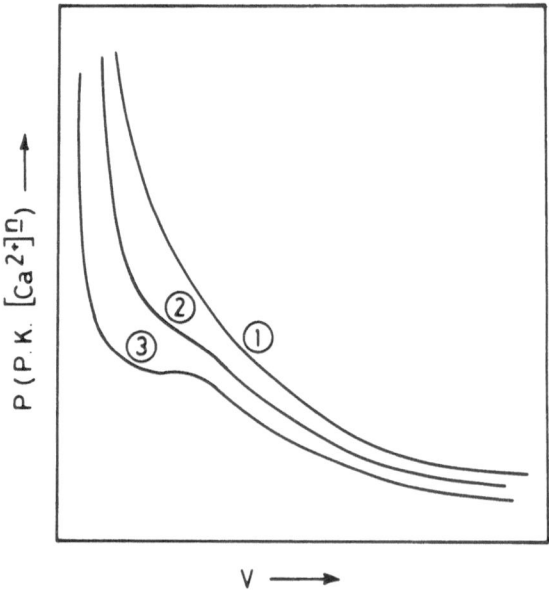

Fig.5 The plot of pressure, P, against volume, V, for a perfect gas follows line 1 while increasing cooperativity is shown in curve 2 until in the case of curve 3 there is phase (liquid) condensation. The same critical phenomena are seen for fluid ion-exchange particles in biological liquids under the influence of calcium ion concentrations. Their volumes, initially, say a random coil, can follow simple binding, 1, or collapse to a condensed state, at critical calcium occupancy values, 3.

Moreover the value of K depends on the condition of the exchanger which is here totally dependent upon the occupancy. The nature of a biological surface is interactive with the nature and number of ions bound. This flexibility means that while ion-exchange is of immense concern for all biological organisation it is extremely difficult to describe. We have to write an equation of the kind

$$\text{Binding } (M_n X_m) = \Sigma_o^{n,m} K(M_n X_m)[M]^n [X]^m$$

which is intended to show that there are concentration terms in M, anion or cation, and in the exchanger, X, which is fluid. The concentration terms are cooperative in M and X so that K takes on different values for different occupancies. Here we do not need to state that occupancy and states of the biopolymers regulate the synthesis and concentration of M and X so that the system is self synthetizing and self-assembling. It is enough to note that the biological systems can point to some clever ways of looking at ion-exchange while experience with ion-exchange can help to understand some features of biological systems.

## REFERENCES

1. Williams, R.J.P. The structural roles of amphiphilic molecules, Biochem. Soc. Trans. 1987, **15**, 36-47.

2. Williams, R.J.P. On First Looking into Nature's Chemistry, Chem. Soc. (London) Revs. 1980, **9**, 281-325.

# THE USE OF CHEMICALLY SUPPRESSED ION CHROMATOGRAPHY IN ELEMENTAL ANALYSIS

JOHN P SENIOR
SmithKline Beecham Pharmaceuticals
Physical Organic Chemistry, The Frythe, Welwyn, Herts, England

## ABSTRACT

At the Frythe Chemically Suppressed Ion Chromatography (CSIC) is employed
in Pharmaceutical Research.  Currently the technique is being developed
for application to elemental analysis.
The term Combustion Ion Chromatography (CIC) has been used to describe
the combination of a combustion method and Ion Chromatography (IC).  The
Schoniger combustion is widely used in analytical chemistry and provides
a means for the extraction of an element from its matrix and subsequent
quantitative conversion to a measureable ionic derivative.  CSIC is
capable of the high degree of instrumental precision required to perform
elemental analysis and careful consideration of instrumental calibration
can produce the required accuracy.
A method for the determination of elemental sulphur has been developed
and been available as a service for over two and a half years.  The
validation data suggest the method is reliable and robust.  Methods for
the determination of elemental fluorine and elemental phosphorus are
currently being developed.  Data from initial validation exercises are
encouraging.

## INTRODUCTION

Historically the chromatography of ions is associated with the

development of the ion-exchange mechanism of chromatographic separation.

However, it was not until publication of the paper by Small et al. [1] in

1975 and the subsequent production of viable commercial instruments in the late 1970's that the term IC (defined simply as the chromatography of ions) came into common usage. Traditionally the technique involved an ion-exchange separation followed by conductivity detection. Today two different modes of IC are available which differ in their treatment of the baseline signal prior to detection via a suppressor device while Electronically Suppressed Ion Chromatography (ESIC) is essentially electronic back-off against the baseline signal.

At SK&F Research Chemically Suppressed Ion Chromatography (CSIC) is used in a variety of applications. Its principal role is assisting in the characterisation of organic compounds of pharmaceutical interest. This involves elemental analysis, trace analysis and studies of counter-ions [2].

In recent years we have devoted much time to the development of CSIC for use in elemental analysis. Initially, the objective was to replace the time consuming wet chemical method used for determination of elemental sulphur. This has been achieved and we are now developing methods for the determination of elemental fluorine and phosphorus. In addition we are investigating the potential offered by the technique for a simultaneous multi-elemental method, whereby more than one element can be determined using the same sample solution, and in some cases in the same chromatographic run.

Elemental analysis requires liberation of the analyte from its sample matrix and its subsequent measurement. These requirements can be fulfilled by a suitable combustion procedure and IC, a combination sometimes referred to as Combustion Ion Chromatography.

Method development is separated into three phases. Phase I involves active development involving optimisation of operating parameters and culminates in validation exercises using analytical standards. Phase II involves extended validation and assessment of method robustness by analysis of a wide variety of 'real' samples. Phase III refers to long term validation when the method is available as a service to customers. The strategy adopted involves production of a method to determine a suitable ionic derivative of the element and then development of a procedure which provides a quantitative conversion to this derivative from combustion of the corresponding element. Much of the methodology is common to all three procedures and is presented in the following sections. Operating conditions specific to each separate procedure are presented in Appendix 1.

## METHODS AND MATERIALS

### Sample Preparation

The necessity to liberate heteroatoms from organic matrices can be satisfied by employing Schoniger Combustion. This essentially involves combustion of sample in an oxygen rich atmosphere and subsequent absorption of the combustion products into solution in a form which is amenable to measurement by IC. The flexibility of the technique is in the choice of the absorption solution in order to promote reduction or further oxidation of the initial combustion products.

The Schoniger apparatus consists of a modified Erlenmeyer flask typically made of glass and a sample basket holder made of platinum or silica. The absorption solution (10ml) is added to the flask which is then purged with oxygen for five minutes. The sample (approximately 10mg) is weighed by difference into a one inch square piece of filter paper and placed in a basket holder which is then secured. The flask is placed inside a safety cabinet and combustion is triggered. The flask is then left standing to allow absorption of the combustion products which can be facilitated by shaking. The solution is made to volume (50ml) inside the flask and injected directly into the chromatograph.

## The Determination of Anions by CSIC

The requirement for precise and accurate measurement of the analyte as an ionic derivative can be satisfied by the use of CSIC. The instrument we use is a Dionex 2020i system incorporating a Dionex APM module. Chromatographic separation is achieved by employing a Dionex AG4A guard column and a Dionex AS4A separator column with a 2.2mM sodium bicarbonate/ 2.8mM sodium carbonate mixture as the mobile phase at a flow rate of 2.0ml $min^{-1}$.

Detection is achieved by conductivity measurement using a Dionex CDM1 with detector output at 100µs and temperature compensation at 1.7. Chemical suppression is achieved using Dionex AFS with a 25mM sulphuric acid suppressant at a flow rate of 2.0ml $min^{-1}$.

Calibration is performed via an external standard bracketing method using a single calibration solution. Calibration solutions are prepared by using known amounts of the appropriate sodium salts and the mobile phase. The use of the mobile phase as extraction medium is strongly recommended in order to reduce the production of physical and chemical changes after presentation to the instrument.

Linearity in terms of concentration vs response indicate linear concentration ranges, however experience suggests that matching of analyte concentration in the calibration and sample solutions is desirable.

Data collection in terms of peak height measurement is achieved using a HP3390A electronic integrator and data processing is via our in-house software package. The principle behind the calculation of unknown amounts of analyte is the use of a time weighted factor derived from a response factor vs time calibration plot, produced from the calibration data.

Validation of analytical procedures is performed within the framework of the three phases of method development outlined previously. Analytical accuracy and precision are assessed by the validation exercises of Phase I and monitored in Phases II and III via the inclusion of analytical standards in each batch of samples analysed.

## RESULTS

Data from Phase I validation exercises, performed using analytical standards (Appendix 2) are presented below (Table 1).

TABLE 1

PHASE I VALIDATION DATA

| ELEMENT | S | F | P |
|---|---|---|---|
| THEORY % | 15.82 | 8.49 | 11.80 |
| FOUND % | 15.82 | 8.65 | 11.67 |
|  | 15.48 | 8.22 | 12.03 |
|  | 15.66 | 8.54 | 11.54 |
| MEAN % | 15.65 | 8.47 | 11.75 |
| MAXIMUM SPREAD ±% | 0.17 | 0.22 | 0.25 |

## CONCLUSION

The combination of Schoniger Combustion and Chemically Suppressed Ion Chromatography provides a generalised procedure which can be applied to elemental analysis and is advantageous to traditional wet chemical methods. Modifications to this generalised procedure provides methods for the determination of elemental sulphur, fluorine and phosphorus. The sulphur method has been available since February 1987 and is regarded as robust and reliable. Data collected in initial validation exercises of the fluorine and phosphorus methods are encouraging. The robustness of these methods is currently being assessed and validation extended by analysis of the analyte in a variety of chemical matrices.

APPENDIX

1. Specific Parameters for Elmental Analyses

| ANALYTE | SULPHUR | FLUORINE | PHOSPHORUS |
|---|---|---|---|
| ABSORPTION SOLUTION | 4% $H_2O_2$ | ELUENT | 4% $H_2O_2$ |
| IONIC DERIVATIVE | SULPHATE | FLUORIDE | ORTHOPHOSPHATE |
| MEASURING VALUE | 90ppm | 12ppm | 45ppm |
| MEASURING RANGE | 50-120ppm | 5-25ppm | 30-60ppm |

2. Method Development - Phase I - Analytical Standards

| ELEMENT | ANALYTICAL STANDARD |
|---|---|
| SULPHUR | S-BENZYLTHIURONIUM CHLORIDE |
| FLUORINE | SK&F 96022 |
| PHOSPHORUS | TRIPHENYLPHOSPHINE |

ACKNOWLEDGEMENTS

I would like to thank the following for their significant contribution to the work presented in this paper; M. J. Graham, J. C. Tribe, A. J. Lowe and D. Middleton.

REFERENCES

1. H. Small, T. Stevens and W. Baumann. Novel Ion Exchange Chromatographic method using Conductimetric Detection. Analytical Chemistry, 1801, vol. 47, 1975.

2. J. P. Senior. Ion Chromatography and Pharmaceutical Research - A Study of Counter-ions. Recent Developments in Ion Exchange (eds. P.A.Williams and M.J.Hudson). Elsevier Applied Science Publishers, London, 1987, p60-66.

# BIOLOGICAL SEPARATIONS USING LATEX-BASED PELLICULAR RESINS

Dr Ken Cook
Dionex (UK) Limited
Albany Park, Camberley, Surrey

## INTRODUCTION

Traditional macroporous resins contain their ion exchange sites within large pores. These deep pores impede mass transfer of molecules between the resin and the mobile phase. The result is broader peaks, poor resolution and lower recovery. In addition large molecules can be excluded or trapped by the pores, limiting the ability to separate large protein and DNA Fragments.

Pellicular resins on the other hand concentrate a vast number of ion exchange sites into a very narrow microbead layer on the surface of the main non-porous resin bead [Figures 1a & 1b]. This configuration gives a high loading capacity while maintaining very short diffusion paths. These features create high efficiencies through rapid mass transport. It is also an ideal resin for efficient chromatography of large molecules with high yields. Other distinct advantages for the biological field include high mechanical stability and total pH compatibility, allowing harsh column clean-up procedures with 1M acid and 1M base.

The performance of traditional ion exchange resins has always been disappointing compared to the efficiency of reverse - phase chromatography. However, the introduction of high performance pellicular ion exchange resins has renewed interests in ion exchange and giving the biochemist a much needed additional tool in biological separations.

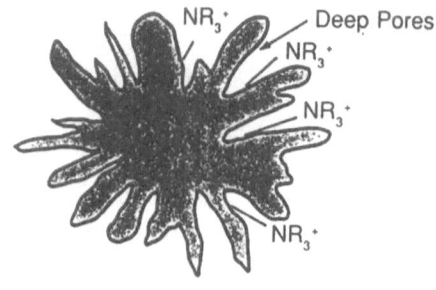

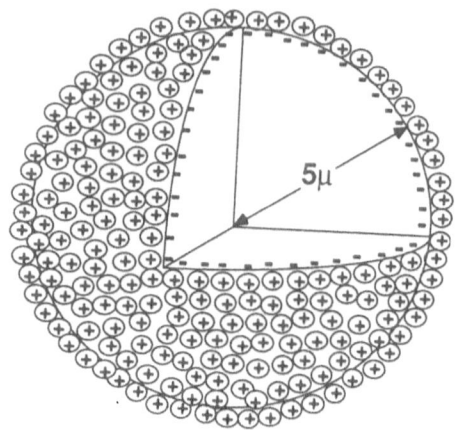

- Slow Diffusion
- Low Efficiency
- Non-Rigid

Figures 1a & 1b

## RESULTS AND DISCUSSION

The most significant advance using pellicular resins has
undoubtedly been in the separation of carbohydrates. Sugars
give the chromatographer some difficult problems in separation
and detection. They are non-chromophoric, so they cannot be
detected easily using UV detectors. They are non volatile,
making them unsuitable for direct analysis by GLC. The
biggest problem however, lies in their similar structures.
Sugars such as glucose and galactose are almost identical in
structure, giving very little ground in which to force a
separation. Traditional methods for liquid chromatographic
analysis of carbohydrates have generally used silica based
resins, with refractive index (RI) detection. The columns
used suffer from poor selectivity and efficiency with limited
column lifetimes. In addition gradient elution becomes
unrealistic because of the RI detection.

The pellicular strong anion exchange resin takes
advantage of the wealthy acidic nature of carbohydrates to
give highly selective separations. Neutral monosaccharides
have pka values between 12 and 14. At pH values above 12 they
effectively exist as multivalent anions which can be separated
by strong anion exchange mechanisms. The eluents used are
usually sodium hydroxide based. This approach cannot be used
with silica based columns due to their poor stability at high
pH. However, polystyrene-divinylbenzene resin based columns
are ideally suited to this separation mechanism.

The CarboPac PA1 column has been particularly developed for carbohydrate separations. Alditols and monosaccharides elute first on this column with larger carbohydrates being more strongly retained, eluting later. An easy way to change the column selectivity for smaller carbohydrates is to adjust the pH using sodium hydroxide gradients. A good example of sodium hydroxide gradient elution is shown in figure 2 where 18 commonly occurring sugars are separated in under 40 minutes. This demonstrates a considerable advantage over other commercially available columns.

## Carbohydrates by Gradient Elution

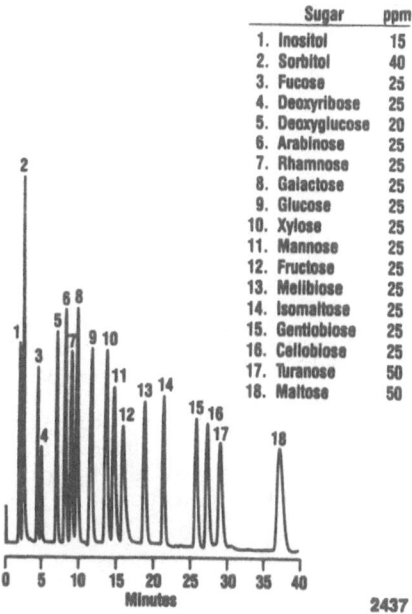

| | Sugar | ppm |
|---|---|---|
| 1. | Inositol | 15 |
| 2. | Sorbitol | 40 |
| 3. | Fucose | 25 |
| 4. | Deoxyribose | 25 |
| 5. | Deoxyglucose | 20 |
| 6. | Arabinose | 25 |
| 7. | Rhamnose | 25 |
| 8. | Galactose | 25 |
| 9. | Glucose | 25 |
| 10. | Xylose | 25 |
| 11. | Mannose | 25 |
| 12. | Fructose | 25 |
| 13. | Melibiose | 25 |
| 14. | Isomaltose | 25 |
| 15. | Gentiobiose | 25 |
| 16. | Cellobiose | 25 |
| 17. | Turanose | 50 |
| 18. | Maltose | 50 |

2437

Figure 2

The pulsed amperometric detector used in this case allows the use of gradients, it is specific for carbohydrates and is more sensitive than RI detectors.

In order to rapidly elute strongly retained components, ionic gradients are used. This generally involves sodium acetate gradients in a sodium hydroxide based eluent. An example of polysaccharide separation can be seen with Amylose [Figure 3], where amylose polysaccharides are eluted in order of chain length. In the chromatogram in figure 4 identical amylose chains going over 40 units in length are separated on the basis of single additional carbohydrate units.

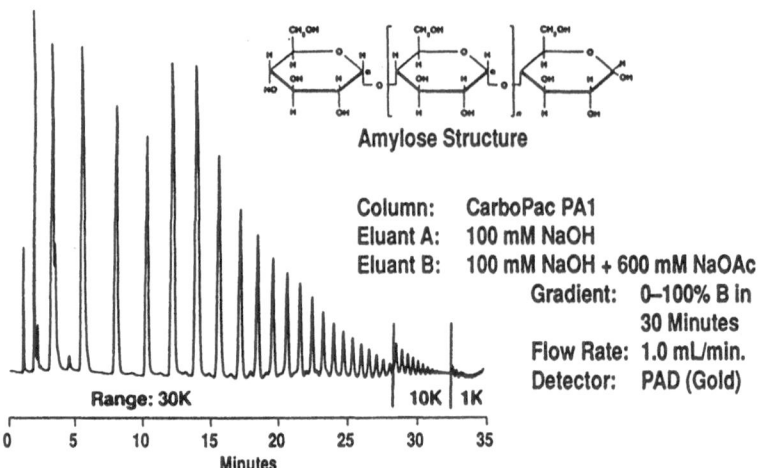

Amylose Structure

Column:     CarboPac PA1
Eluant A:   100 mM NaOH
Eluant B:   100 mM NaOH + 600 mM NaOAc
            Gradient:   0–100% B in
                        30 Minutes
            Flow Rate: 1.0 mL/min.
            Detector:   PAD (Gold)

Range: 30K                              10K | 1K

0      5      10     15     20     25     30     35
                      Minutes

Figure 3.

A further example of the resolving power of pellicular microbead resins is the unique ability to resolve structural isomers of oligosaccharides derived from biological samples.

The structure and separation of some Fetuin-derived glycopeptides is shown in Figure 4a and 4b.  The glycopeptides in this chromatogram are well resolved, even though they only differ by a single linkage position of a terminal galactose.

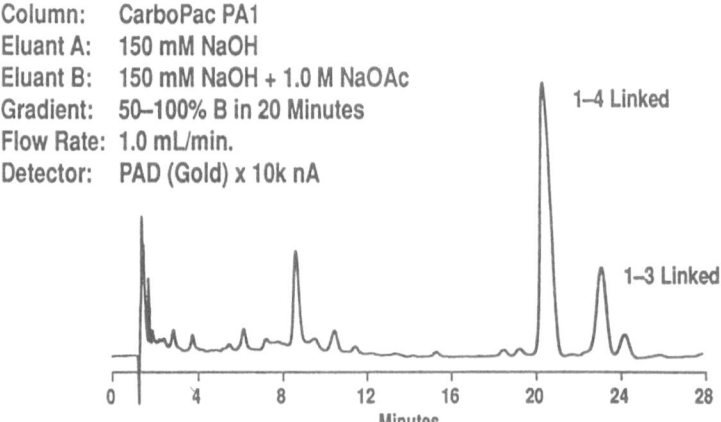

Column:     CarboPac PA1
Eluant A:   150 mM NaOH
Eluant B:   150 mM NaOH + 1.0 M NaOAc
Gradient:   50–100% B in 20 Minutes
Flow Rate: 1.0 mL/min.
Detector:   PAD (Gold) x 10k nA

1–4 Linked

1–3 Linked

0      4      8      12     16     20     24     28
                      Minutes

Figure 4a.

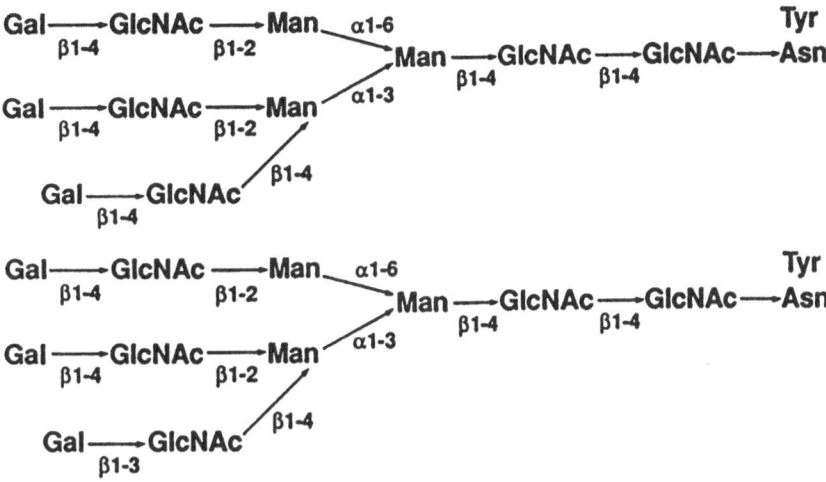

Figure 4b.

The importance of glycosylation in biological systems is becoming more widely recognised. The introduction of this new separation and detection system for carbohydrates has had a large impact on this growing research area.

Pellicular resins have also proved useful in protein and DNA separations. These larger molecules can be separated more efficiently with higher yield on  pellicular resins.  The ProPac PA1 column has been developed for these applications. The separation of DNA restriction fragments gives a very good indication of the improved resolution and recovery for macromolecules.  The microbeads do not exclude large molecules from the exchange sites, enabling resolution of DNA fragments from 8 to 587 base pairs long [Figure 5].

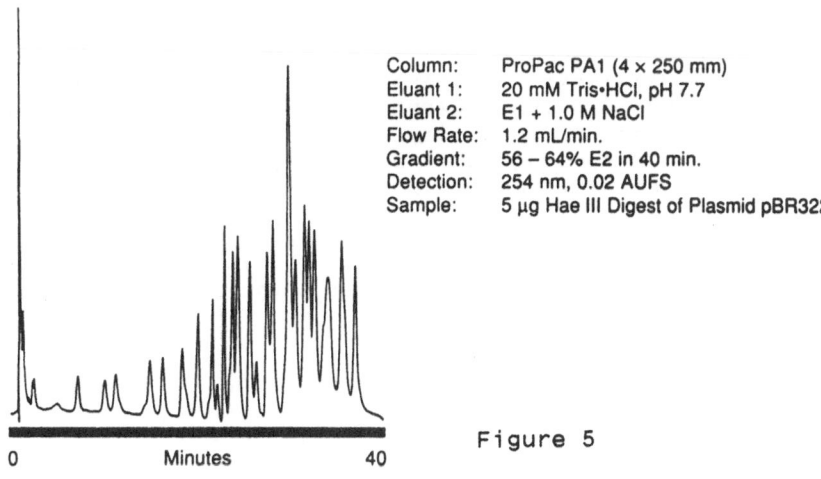

| | |
|---|---|
| Column: | ProPac PA1 (4 × 250 mm) |
| Eluant 1: | 20 mM Tris•HCl, pH 7.7 |
| Eluant 2: | E1 + 1.0 M NaCl |
| Flow Rate: | 1.2 mL/min. |
| Gradient: | 56 – 64% E2 in 40 min. |
| Detection: | 254 nm, 0.02 AUFS |
| Sample: | 5 µg Hae III Digest of Plasmid pBR322 |

Figure 5

Futhermore,this separation is achieved in 40 minutes
which could take up to 10 times longer on a comparable
macroporous resin.  The non-porous structure of the resin
increases the speed of column equilibrium during gradient
analysis.  This ensures all the exchange sites rapidly achieve
equilibration with the new elution buffer.  This fast
equilibration and mass transfer results in sharp peaks from
gradient runs, usually obtained only with reverse phase
columns [Figure 7].

Tryptic digests give particularly good high resolution
fingerprints using sodium chloride gradients.  Anion exchange
has proved particularly useful in protein and DNA separations
with macroporous resins.  The use of higher resolution
pellicular resins should enhance this technique even further.

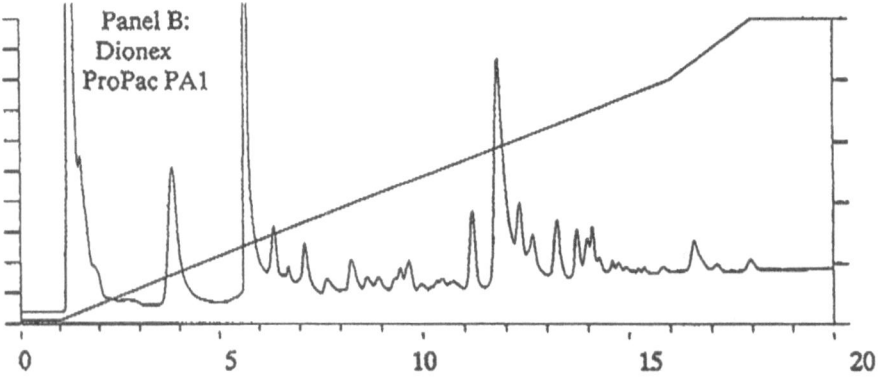

Figure 6.


Anion exchange on pellicular resins can also be used for
amino acid analysis, a separation traditionally done using
cation exchange.  The separation gives a very different
elution pattern to traditional resins with arginine eluting
first.  The chromatography is carried out with high pH eluent
to encourage the formation of anionic charge on the amino
acids.  This separation can be very useful in the analysis of
such important analytes as the negatively charged
phosphorylated amino acids which can elute as unresolved early
peaks from cation exchange resins [Figure 7]

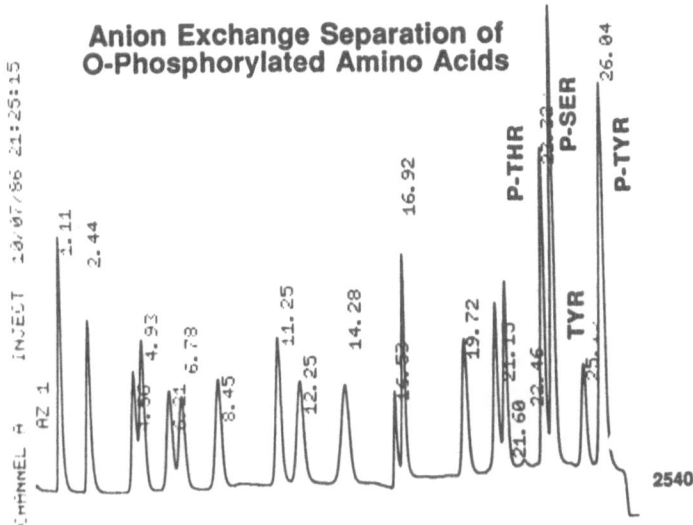

Figure 7

Nucleic acids, organic acids and other ionic biomolecules can also be efficiently resolved on pellicular resins, giving a wide application base for these resins in biological fields. The resins are packed in polymeric column bodies which will withstand pressures of up to 4,000 psi. These columns are also well suited to corrosive biological salt buffers and will not contaminate samples with metal ions. In short these resins provide a family of columns well suited to biological analysis.

## SUMMARY

Microbead ™ ion exchange columns offer unique ion exchange selectivities with speed and resolution unobtainable with comparable macroporous resins.

Pellicular resins are formed by binding sub-micron ion exchange Microbeads ™ to non-porous resin particles. This results in a resin with extremely rapid mass transport, high efficiencies, high mechanical stability and total pH compatibility. Furthermore, column equilibration is much faster than for macroporous resins.

These unique pellicular resins have found immediate applications in many biological fields. High resolution and improved recovery of proteins from complex mixtures can be achieved in shorter separation times. The separation time for DNA fragment analysis can be reduced significantly to that required by traditional macroporous resins. Microbead ™ anion exchange resins combined with pulsed amperometric detection now provides the most sensitive and specific system available for the separation and detection of carbohydrates. Other important application areas include nucleic acids, amino acids and ionic biological molecules.

# Simultaneous determination of alkali and alkaline earth metal ions with isocratic ion chromatography using the SUPER-SEP cation column (according to Schomburg)

Dr. Markus W. Läubli
Development Chemist
METROHM Ltd.
CH-9101 Herisau
Switzerland

## ABSTRACT

The fast and efficient isocratic separation of mono- and divalent cations on a polymer-coated silica material has been introduced by Schomburg and coworkers [1]. The separation takes place on a crosslinked layer of poly(butadiene-maleic acid) with weak organic acids as eluents. Ion chromatography with electronic background suppression employing this type of column has been used successfully in the analysis of water, foodstuffs and beverages [2]. The field of application of this column is growing fast. With acidified standard and sample solutions (pH = 3) the linear range of concentration is about three orders of magnitude.

## INTRODUCTION

Ion chromatography with electronic suppression of the background conductivity has proven an excellent tool for the determination of anions or cations in different samples. The development of a polymer coated stationary phase based on poly-(butadiene-maleic acid) on a spherical silica material [1] allows the analysis of alkali and alkaline earth metal cations in a single run with no need to apply gradient elution or column switching. The separation takes place in about 16 minutes with perfect separation even for the monovalent cations.

## EXPERIMENTAL

All the experimental work was performed on a METROHM Ion Chromatography System (Figure 1).

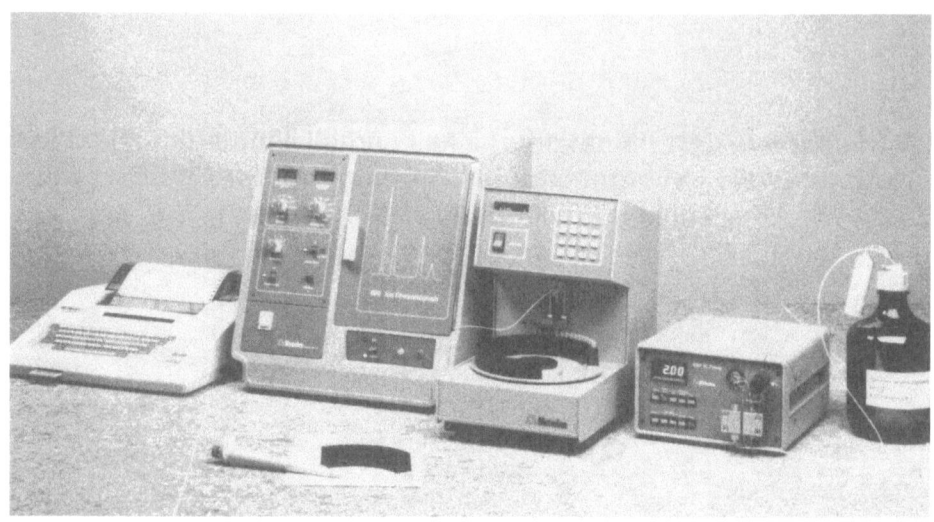

Figure 1:  METROHM Ion Chromatograph

| Instrument: | – 697 IC Pump (Metrohm Ltd., CH-9101 Herisau, Switzerland)<br>– 690 Ion Chromatograph (Metrohm Ltd.)<br>– 586 Labograph  (Metrohm Ltd.)<br>– PE Nelson PC Integrator |
|---|---|
| Column: | SUPER-SEP cation column (acc. to Schomburg), 125 × 4 mm |
| Eluent: | 5 mmol/L citric acid,<br>0.75 mmol/L pyridine-2,6-dicarboxylic acid |
| Flow rate: | 1.0 mL/min |
| Pressure: | 4.2 MPa |

## RESULTS AND DISCUSSION

The separation of alkali and alkaline earth metal cations is performed on a cross-linked layer of poly(butadiene-maleic acid). The two carboxylic groups of the maleic acid units may function as cation exchanger sites as well as complexing sites for divalent cations. Therefore the separation mechanism can be understood as a mixture of cation exchange with $H^+$ as eluting ion and a complexing mechanism with citric and pyridine-2,6-dicarboxylic acid as complexing agents in the eluent. The elution order with citric or tartaric acid as eluent is $Li^+ < Na^+ < NH_4^+ < K^+ \ll Mg^{2+} < Ca^{2+} < Sr^{2+} < Ba^{2+}$. If added to the eluent, pyridine-2,6-dicarboxylic acid changes the elution order of the divalent cations ($Ca^{2+} < Mg^{2+} < Sr^{2+} < Ba^{2+}$) due to its complexing strength [3]. Figure 2 a/b shows the influence on the retention times of citric acid and pyridine-2,6-dicarboxylic acid, each in a constant back-

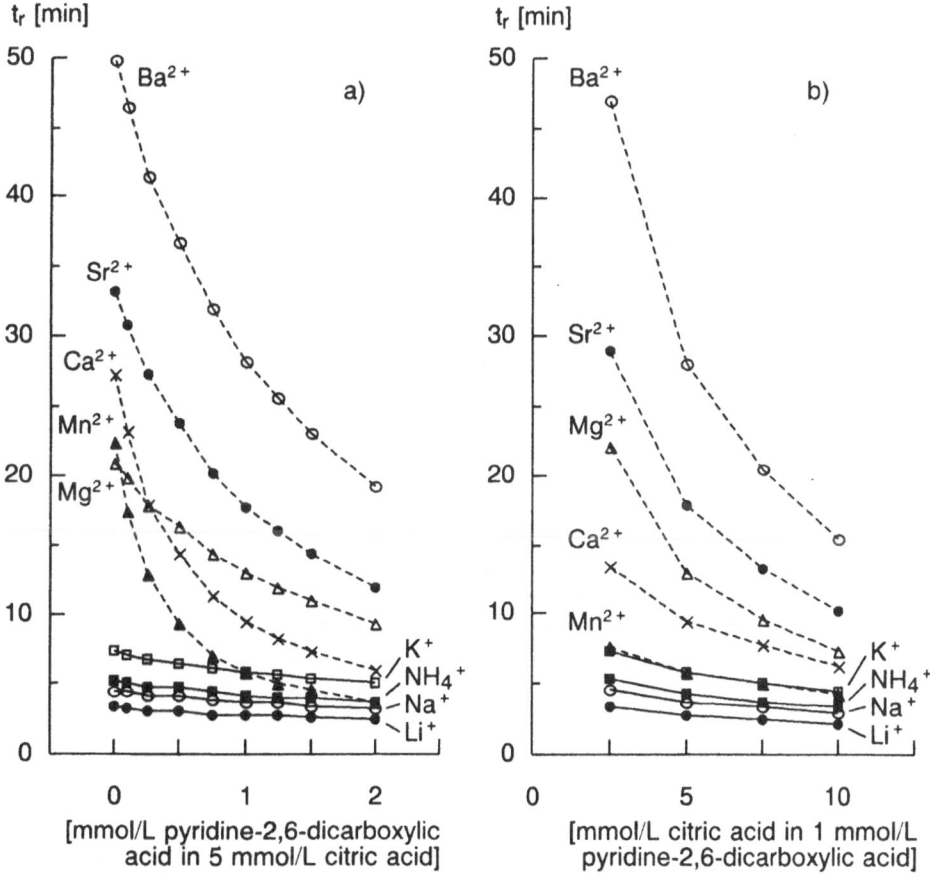

Figure 2: Influence of pyridine-2,6-dicarboxylic acid (a) and citric acid (b) on the retention behaviour of cations separated on the SUPER-SEP cation column

ground of the other. Transition metals are usually eluted in the front peak and do not interfere. Under standard conditions manganese appears between potassium and calcium [2].

The packing of the SUPER-SEP cation column is stable in a range between pH = 2 and pH = 7; no organic modifier should be used in the eluent. Under these conditions the column has a lifetime comparable to other HPLC-columns. Figure 3 a shows a standard chromatogram of a new SUPER-SEP cation column. Figure 3 b shows a standard chromatogram of the same column after more than 1500 injections.

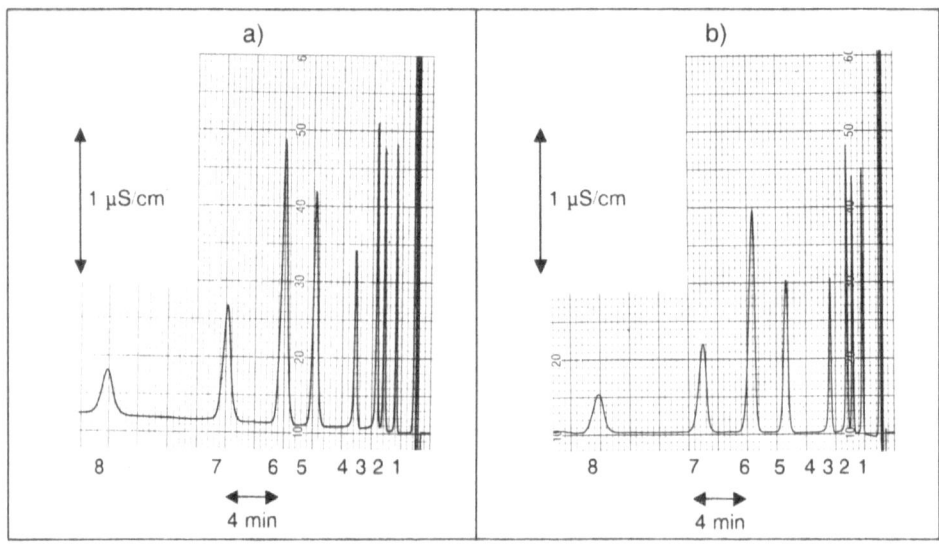

Figure 3: Standard chromatogram,  Eluent:  5 mmol/L citric acid, 0.75 mmol/L
pyridine-2,6-dicarboxylic acid
a) first injection
b) after more than 1500 injections

| | |
|---|---|
| 1: 1 mg/L $Li^+$ | 5: 10 mg/L $Ca^{2+}$ |
| 2: 5 mg/L $Na^+$ | 6: 10 mg/L $Mg^{2+}$ |
| 3: 5 mg/L $NH_4^+$ | 7: 20 mg/L $Sr^{2+}$ |
| 4: 10 mg/L $K^+$ | 8: 20 mg/L $Ba^{2+}$ |

## Linearity

The linearity of the system was tested by injection of mixed standard solutions with varying dilution ratios. The concentration varied by a factor of 100. In Figure 4 a/b the calibration curves based on area or on height are given for sodium and calcium. With area-based calibration the linear range goes up to about 100 mg/L for sodium, ammonium, potassium, calcium and magnesium (injection volume 10 µL). The detection limits under these conditions are 30 µg/L for sodium and ammonium, 50 to 100 µg/L for potassium, calcium and magnesium [2]. Calculation based on height measurement is not recommended. The calibration curves, especially for the divalent cations, show a deviation from linearity already at quite low concentrations (Figure 4 a/b).

The pH of samples and standard solutions has to be adjusted to a value of 3 to achieve a maximum range of linearity.

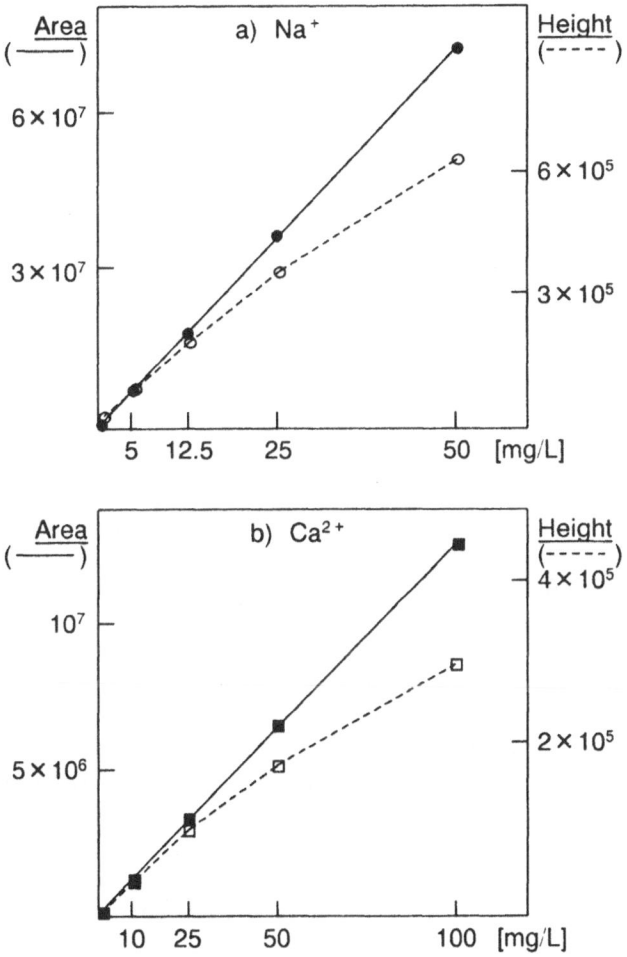

Figure 4: Calibration curves based on area and height respectively
a) for sodium (0.5 to 50 mg/L)
b) for calcium (1 to 100 mg/L)
(standard solutions in 2 mmol/L HNO₃)

## Applications

The sample solutions to be used for analysis have to be aqueous and pH should be adjusted to about 3. Above this pH the concentration of the divalent cations tend to be overestimated. Microfiltration (0.45 µm) is recommended. Figure 5 shows the chromatogram of Herisau tapwater under these conditions. The results for calcium and magnesium are the same as those obtained with potentiometric titration.

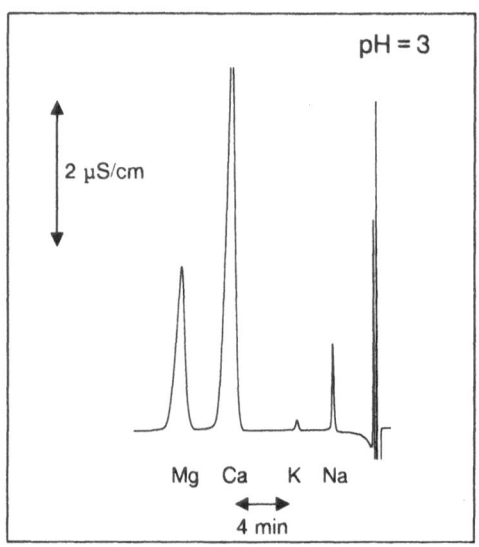

Figure 5: Cation determination in tap water from Herisau
pH adjusted to 3.0 with 1 mol/L $HNO_3$;
3.9 mg/L $Na^+$; 1.4 mg/L $K^+$; 83.8 mg/l $Ca^{2+}$; 18.3 mg/L $Mg^{2+}$
(external standard in 2 mmol/L $HNO_3$).

The concentration of the different cations in the injected solution should be below 100 mg/L each for direct determination. If minor components are of interest, the main component may have a higher concentration. The determination of cations in a spiked sea salt is such an example (Figure 6). 1 g of the salt sample is dissolved in 100 mL of 2 mmol/L $HNO_3$. This solution is further diluted for the determination of $Na^+$ (1:100; 315 mg/g) and for $K^+$, $Ca^{2+}$ and $Mg^{2+}$ (1:25; 2.6 mg/g, 1.5 mg/g and 0.2 mg/g, respectively). The concentration of $NH_4^+$ is < 0.3 mg/g, which is the limit of detection under these conditions.

A fertilizer is analysed for cation content. 1 g of fertilizer powder is mixed with 1 mL concentrated $HNO_3$ and diluted with deionised water to 100 mL. 1 mL of this solution is diluted again to 100 mL. This solution is injected into the chromatograph after filtration through a 0.45 µm disposable syringe filter (Figure 7). The fol-

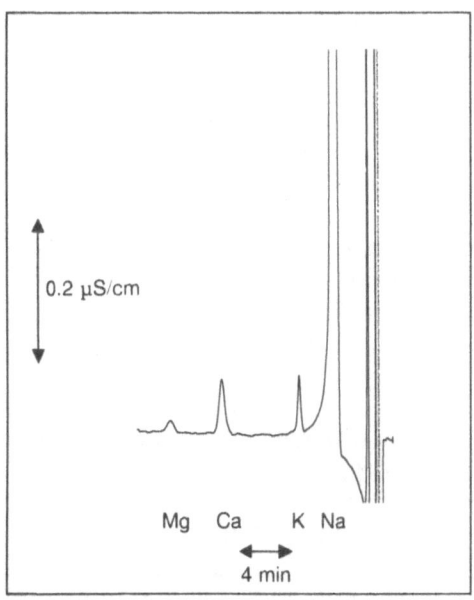

Figure 6: Cation determination in a spiked sea salt
315 mg/g $Na^+$; 2.6 mg/g $K^+$; 1.5 mg/g $Ca^{2+}$; 0.2 mg/g $Mg^{2+}$

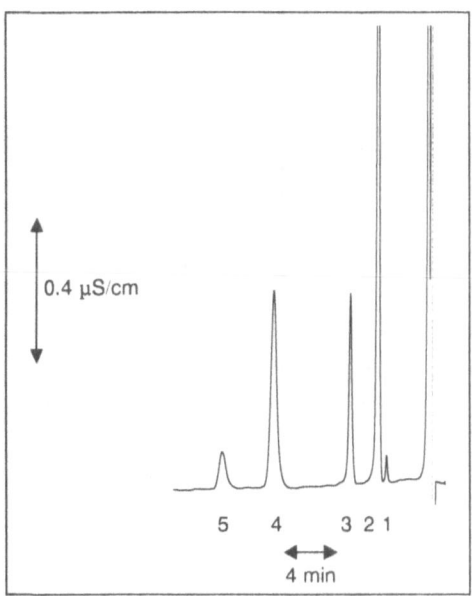

Figure 7: Cation determination in a fertilizer
1: 2.4 mg/g $Na^+$; 2: 85.6 mg/g $NH_4^+$; 3: 56.9 mg/g $K^+$;
4: 59.9 mg/g $Ca^{2+}$; 5: 6.9 mg/g $Mg^{2+}$

lowing concentrations were determined: 2.4 mg/g sodium, 85.6 mg/g ammonium, 56.9 mg/g potassium, 59.9 mg/g calcium and 6.9 mg/g magnesium.

If necessary different sample preparation steps may be used, such as solid phase extraction of organic components, extraction of cations from an organic phase into an aqueous phase, etc.

Apart from inorganic cations, substituted ammonium ions can be determined as well. Table 1 gives the retention times for a number of different ions determined on the SUPER-SEP cation column.

Table 1: Retention times of different organic and inorganic cations on the SUPER-SEP cation column

| Cation | tr/min | Cation | tr/min |
|---|---|---|---|
| Li$^+$ | 2.8 | Di(methyl)amine | 7.8 |
| Hydroxylamine | 3.5 | Tri(ethanol)amine | 8.5 |
| Tris-(hydroximethyl)-aminomethane | 3.8 | Morpholine | 9.3 |
| Na$^+$ | 3.8 | Ca$^{2+}$ | 11.3 |
| NH$_4$$^+$ | 4.4 | Tri(methyl)amine | 14.2 |
| 2-Amino-2-methyl-1,3-propandiol | 4.7 | Mg$^{2+}$ | 14.3 |
| Ethanolamine | 5.0 | Sr$^{2+}$ | 20.2 |
| Methylamine | 5.4 | 2-Diethylaminoethanol | 29.4 |
| K$^+$ | 6.1 | Tetra(methyl)-ammonium | 30.1 |
| Di(ethanol) amine | 6.3 | Ba$^{2+}$ | 32.0 |
| Mn$^{2+}$ | 7.1 | | |

# REFERENCES

1. Kolla, P., Köhler, J., Schomburg, G.,
   Polymer-coated cation-exchange stationary phase on the basis of silica,
   Chromatographia, 1987, **23**(7), 465-72.

2. Schomburg, G., Kolla, P., Läubli, M.W.,
   Ion chromatography of alkali and alkaline earth metal ions using a silica-
   based polymer-coated stationary phase,
   American Laboratory, 1989, **19**(3), 40-5.

3. Kondratjonok, B., Schwedt, G.,
   Isokratische, simultane Ionen-Chromatographie von Alkali-, Erdalkali- und an-
   deren Metallionen,
   Fres. Z. Anal. Chem., 1988, **332**, 333-7.

# Part 2

# INORGANIC ION EXCHANGERS

# RECENT ADVANCES IN INORGANIC ION EXCHANGERS

ALAN DYER
Cockcroft Building, Department of Pure and Applied Chemistry,
University of Salford,
Salford M5 4WT, UK.

## ABSTRACT

Progress in aspects of inorganic ion exchangers is reviewed. Recent developments in the theoretical treatment of ion exchange phenomena is related especially to zeolites. The synthesis of new materials with cation and/or anion exchange capabilities are described and specific examples of analytical and industrial importance cited. Advances in the more well established exchangers (eg zeolites, hydrous oxides, and phosphates) are considered.

## INTRODUCTION

In his closing remarks to the International Conference on Ion Exchange (IEX '88) held at Cambridge in 1988, Dr Robert Kunin expressed the firm opinion that, had he the opportunity to start his scientific life again, he would devote his career to the inorganic ion exchange materials.

This comment, coming from one of the most eminent pioneers of resin technology, was an acknowledgement of the potential offered by inorganic materials to provide robust, highly selective exchange media at costs competitive to the relatively unselective and less stable resins.

The last two decades have seen a clear revival in inorganic ion exchange studies. A major area has been in the study of zeolites prompted by their essential industrial applications as catalysts[1], high performance sorbents, molecular sieves and detergent builders[2]. In all these cases their successful utilization depends heavily on a detailed knowledge of their ion exchange characteristics.

The need for accurate ion exchange data has spawned significant advances in the understanding of the fundamentals of the ion-exchange process. Here zeolites have proved to be highly suitable models for the testing of ion-exchange theories[3] - only studies on clay minerals have kept pace with these advances[4].

So far as non zeolitic inorganic ion exchangers (clays excepted) are concerned far less is known of their synthesis, structures and properties and this hampers equivalent progress in their uses and commercial availability. There can be no doubt as to the potential offered by the possible creation of layer and framework structures based upon linkages of coordination polyhedra. It has been computed that there are a possible 6 million three dimensional linkages of the $[SiO_4]^{4-}$ and $[AlO_4]^{5-}$ tetrahedra alone, so reference to a standard inorganic textbook gives an insight into the wide variety of other coordination polyhedra which might be persuaded to create open structures having ion exchange properties.

This prospect has extended zeolite chemistry in that zeotypes can be synthesized having coordination polyhedra such as $[PO_4]^{3-}$ and $[Co O_4]^{5-}$ forming part of a framework structure[5]. This progress has been prompted by the search for new catalysts and sorbents and, as yet, very little is known of their ion exchange nature.

It is only a matter of time before the inventive synthetic routes to zeolites (such as by using large organic molecules as templates, or void filters, in an aqueous reaction medium[6]) can be extended to systems containing other polyhedra entities.

The growth in structure determining techniques which can be applied to fine powders should also help to solve the structures of well characterized inorganic ion exchangers (non-zeolitic as well as zeolitic). Examples of these techniques are neutron diffraction, magic angle spinning NMR and the use of synchrotron radiation.

To conclude this introduction some comments about the current usage of inorganic exchangers are needed. Probably the most well known use is that of the treatment of aqueous nuclear waste streams but this is outside the brief of this review (See Bibler[7] and Ruvarac and Clearfield[8]). Other areas arise in environmental clean-up, analytical separations and those uses requiring the selective removal of an ionic species. These will be included in this review which will now survey the range of inorganic materials available and recent progress in the study of their properties.


## CATION EXCHANGERS

### Zeolites

Introduction: These highly crystalline hydrated aluminosilicates are characterized by framework (3-dimensional) structures. The isomorphous substitution of Al into the framework creates a net negative charge which is balanced by cations present inside the structure. Many (but not all) the zeolites have open structures so that cations in their water environments can be easily replaced by ion exchange.

Although the zeolites were originally defined as a class of naturally occurring minerals they are relatively easy to synthesize and this has produced many novel zeolites unknown in nature. The most well known of these synthetic species is zeolite A which is widely used as a detergent builder. This uses the zeolite ion exchange as a water softener to replace the noxious phosphate builders. The structure of zeolite A is illustrated in Fig.1.

Zeolite A is also widely used as a molecular sieve and desiccant (ie Linde Molecular Sieves 3A, 4A and 5A). Another commonly synthesized zeolite has the structure of the mineral faujasite (Fig.2)and this can be formed with a range of Si/Al compositions. When $Si/Al \approx 1$ the zeolite is known as "X zeolite" but when the $Si/Al \to 2$ and above the material is known as 'Y-zeolite' and is the major component of the most widely used catalyst to crack crude oil.

A more complete description of zeolites can be found in references 9 and 10. Several reviews of earlier work on zeolite ion exchange are available[3, 11].

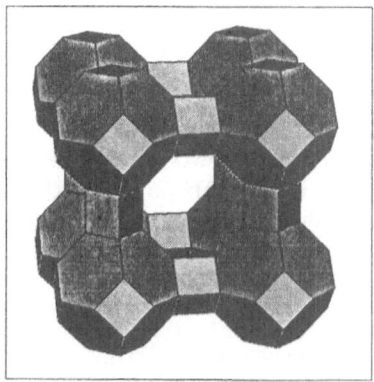

Fig. 1   Structure of synthetic zeolite A

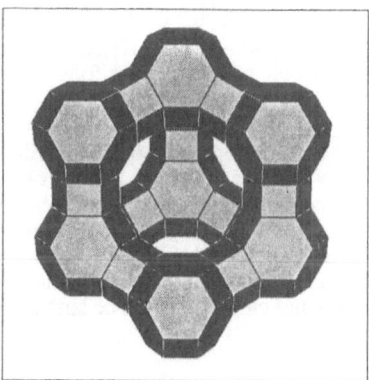

Fig. 2   Faujasite structure - isostructural with
synthetic zeolites X and Y

Recent progress in zeolite ion exchange theory: As mentioned earlier zeolites have been preferred models for theoretical studies of the ion exchange process. Their well described structures, reproducibility and purity (especially the synthetic products), lack of swelling and high exchange capacities are attractive to test and extend theoretical postulates.

No comprehensive treatment has yet been published, aside from the reviews cited above, but a review devoted specifically to the modern theories of ion exchange in a zeolite context is in press[12].

One major problem which has been progressed well comes from the consideration of ternary exchange systems. Townsend and coworkers have demonstrated that earlier models[13] can be applied to the exchanges of $Na^+/K^+/Cd^{2+}$ in zeolite X[14] and $Na^+/NH_4^+/Mg^{2+}$ in zeolite Y[15]. A simplification of their own approach to the elucidation of ternary systems has been provided by Franklin and Townsend[16] which is based upon comparisons to equivalent work on resin exchangers.

A different approach has been made by Triay and Rundberg[17], they describe a numerical technique of regularization with non negativity constraints based upon earlierwork[18]. The method analyses isotherm data by a computational technique to generate "recovered selectivity coefficients" which can then be refit to the experimental uni-univalent isotherm. The intent of this technique is to give a method of determining selectivity coefficient distributions which correspond to distinct heteroenergetic cation sites within the exchanger.

Selectivities also have been modelled in chabazite[19] giving special regard to the effect of ion solvation whereas Tao and Yang have proposed equations suitable for application to incomplete exchange[20].

So far as kinetic studies of ion exchange have been concerned, Dyer and Yusof[21,22] have used a Carman-Haul approach to follow $Na^+$, $K^+$ and $H_2O$ self diffusion in a synthetic analcime. They observed that isotropic pathways were followed in the three channels which comprise the analcime structure.

Thompson and Tassopoulos[23] proposed a semi-analytical interpretation of two staged diffusional processes observed in zeolite cation exchangers. Good fits were obtained but needed the invokation of the concept of a "threshold" concentration. This "threshold" must be overcome in the larger cages of a zeolite (when, say, 2 cages are available to accommodate cations and water molecules as in zeolite A - see Fig.1) prior to transport between the larger and smaller cage. This implies that the initial transfer from the bulk liquid to the large cage and the subsequent progress to the smaller cavities cannot be simply diffusion or kinetically controlled.

Other Studies: Some workers have completed zeolite cation exchange under novel conditions. Beyer et al[24] continue earlier work on solid phase ion-exchange to demonstrate that the ZSM-5 structure remains unchanged when the exchanges are from solid salts (eg $Li^+$, $Na^+$, $K^+$, $Rb^+$, and $Cs^+$ chlorides). Fyfe and coworkers[25] also report solid phase exchange which they relate to mechanical mixing, water content and crystallite size.

Uyama et al[26] have performed exchanges of $Na^+$ by $Li^+$, $K^+$, $Rb^+$, and $Cs^+$ into zeolites A and X using nitrates dissolved in non-aqueous ammonia solutions. They noted an excess of exchange over equivalent results in aqueous media which they ascribe to inclusion of ammoniated metal ions and their nitrates.

Work on the measurement of the heats of cation exchange continues. Roques-Malherbi and coworkers[27] report microcalorimetric measurements for alkali metal uptake in heulandite and norderite. Mordenite (with A and X) was the subject of a similar study for $Zn^+$ and $Mn^{2+}$ exchange[28] and heats of exchange for alkali metal and alkaline earth ion into a calcium rich clinoptilolite have recently been measured[29].

A successful exchange of $Cu^{2+}$ into zeolite Y controlled electrochemically has been reported by Maisuradze et al[30] and the effect of $Ag^+$ and $NH_4^+$ exchange on the stability of the synthetic gismondine like phases NaP-1, NaP-2 noted[31].

Applications: The way in which the use of the fundamentals of exchange kinetics and equilibria can be related to the solution of practical problems is exemplified in the studies by Rees[32] in a recent conference report[2]. The same publication contains several other papers which take a very practical view of the current status of zeolites as detergent builders[33,34,35,36].

There has been a continued growth in environmentally orientated studies designed to use zeolites (especially natural species) for the selective removal of ions from aqueous solutions. Work related to this has been summarized in table I. New processes for sewage treatment using clinoptilolite rich volcanic tuffs have been patented[47]. Capacities of 50 - 2000 $m^3$/day have been obtained at 5 different Hungarian sewage plants.

Associated work describes the use of ion exchange capacity measurement for the quantitative determination of the zeolite content of tuffs from the Tokaj Hills[48].

## Other silicates

Nernst-Planck equations have been used to follow $Na^+/H^+$ and $K^+/M^+$ exchanges in an antimony$^v$ silicate[49]. Forward and reverse exchanges were described under conditions where ion exchange diffusion in the solid phase was the rate determining step. A recent US patent[50] reports the synthesis of crystalline titanosilicates with ion exchange properties and hydrophilite has been suggested for Cd uptake[51]. In another silicate study Bergk et al[52] found that the presence of methyl alcohol increases the ion exchange capacity of magadiite.

## Hydrous oxides

Introduction: These materials have been well studied because of their ease of preparation. They exhibit special selectivities of ion sieve nature and are radiation stable. There are comprehensive reviews[53] available describing earlier work on their preparation, properties and characterization.

Hydrous titanium oxides: Interest in this ion exchange media continues. Venkataramani and Gupta[54] have examined the water sorption and ion exchange properties of 5 different hydrated titanium oxides (as well as hydrated bismuth, ferric and thorium oxides.)

Abe and coworkers[55] describe $Pb^{2+}$, $Mg^{2+}$, $Ca^{2+}$, $Cd^{2+}$ and $Hg^{2+}$ selectivity reliant on the anion present and also separations of $Cd^{2+}$, $Hg^{2+}$, and $Pb^{2+}$ applicable to low concentrations of these ions in natural waters.

New preparations of hydrous $Ti^{iv}$ and $Sn^{iv}$ oxides with good stability to alkali have been reported by Yamazaki et al[56] and reports of composite materials based upon titanium come from Weller[57] (for cation and anion removal from simulated brines) and Onorin et al[58] ($Li^+$, $Sr^{2+}$, $Fe^{2+}$ selective forms).

TABLE 1
Waste water treatment and related applications using zeolites

| Zeolite | Cations | Substrate | Reference | |
|---|---|---|---|---|
| Clinoptilolite | $NH_4^+$, $Pb^{2+}$, $Mg^{2+}$ $Cd^{2+}$, $Zn^{2+}$, $Co^{2+}$, $Ni^{2+}$ | Simulated Wastewater | Horvathova | (37) |
| Clinoptilolite | $Hg^{2+}$, $Pb^{2+}$, $Zn^{2+}$ $Ca^{2+}$, $Cd^{2+}$ | Aqueous Solution | Iacomi, et al | (38) |
| Clinoptilolite Mordenite | $Mg^{2+}$, $Ca^{2+}$ $K^+$, $NH_4^+$ | Aqueous Solution | Kang, et al | (39) |
| Clinoptilolite | Various | Sea water | Khamizov, et al | (40) |
| Clinoptilolite Ferrierite Mordenite | $Cd^{2+}$, $Pb^{2+}$ | Aqueous Solution | Loizidou & Townsend | (41,42) |
| 13X, 10X, 10A 5A, Z-12, Z-20 | $Cu^{2+}$ | Simulated Brines | Lopez Ruiz & Gil Montero | (43) |
| Mordenite (natural & synthetic) | $Rb^+$, $Cs^+$ | Geothermal water supplies from geothermal power plants | Onedera, et al | (44) |
| Clinoptilolite | $Ag^+$, $[Ni(NH_3)_6]^{3+}$ | Waste waters from photo-graphic materials processing and Mg production | Rustamov & Makhmadov | (45) |
| Natural | $Ca^{2+}$, $Mg^{2+}$ | Hard water | Wang, et al | (46) |

A layered hydrous titanium oxide has been tested for $Rb^+$ and $Cs^+$ uptake[59]. It showed a complex stepwise exchange, creating three intermediate phases each having a different degree of swelling which could be related to cation hydrations.

Hydrous ferric oxide: A detailed study of the porous nature, acidity and structural characteristics (IR and XRD) have been described for different ferric oxides shown to be variously composed of $\alpha FeOOH$, $\alpha Fe_2O_3$ and $\gamma Fe_2O_3$. The study includes $Na^+$ exchange capacity measurements ($0.07 \rightarrow 0.28$ meq $g^{-1}$) at different pK values[60].

Hydrous manganese oxide: Ooi and coworkers[61] present a detailed study of hydrous manganese oxide designed to test its ability to remove $Na^+$ and $K^+$ from sea water. Selectivity series are given for the $Li^+$, $Na^+$, and $K^+$ exchanged forms of this oxide in relationship to some alkali and alkaline earth metal ions. Misak and Mikhail[62] however report irreversibility of the $Na^+/H^+$ exchange, and also measure $Na^+$, $Cs^+$ diffusion coefficients on a material quoted as $Na_4 Mn_{17} O_{22} . 13H_2O$.

Other studies: When hydrous oxides have been investigated for direct applications the following publications can be cited. Hydrated antimony pentoxide (HAP) is well known for its use to scavenge 24-Na interference from neutron activated samples prior to gamma spectroscopic analysis. A recent report suggests that this function can be improved if the HAP is dispersed in a phenol-formaldehyde resin[63].

Both $SnO_2$ (hydrated) and $TiO_2$ (hydrated) can be used to selectively remove $Pb^{2+}$ from zinc electroplating[64], and Mo loaded onto both alumina and $SnO_2$, in their hydrated forms can be used as a 99-Mo generator[65]. The freeze drying of hydrated Ti, Zr and Nb oxide improves their ability to take up $U^{4+}$ [66].

## Phosphates
Again the earlier work on these interesting ion exchangers has been comprehensively reviewed[67,68,69,70].

Work has continued on titanium phosphates (TIP) by Llavona et al[71] who have suggested an ion exchange mechanism in the TIP layer structure and quote free energies, enthalpies and entropies for $Li^+$, $Na^+$, $K^+$, and $Cs^+$ exchanges from acid solution. Smirnov et al[72] suggest the use of TIP to separate $Na^+$ and $K^+$ from a solution of their iodides.

Less well examined phosphates have been prepared. A tin$^{iv}$ antimonophosphate has been synthesized in both crystalline and amorphous phases, and selectivities for 22 metal ions obtained. Useful selectivities were found for $Pb^{2+}$, $Ce^{3+}$ $\cdot Sm^3$ and $La^{3+}$ and quantitiative chromatographic separations made of the ion pairs $Mg^{2+}/Pb^{2+}$, $Ca^{2+}/Pb^{2+}$ and $Sr^{2+}/Pb^{2+}$ [73].

A stannic hexametaphosphate proves to have a usable capacity and good chemical and thermal stabilities. It is selective for $Ag^+$ and $Pb^{2+}$ and can separate $Fe^{2+}$ from $Fe^{3+}$[74,75].

## Other inorganic cation exchangers
Janardanan and coworkers[76] have prepared and characterized a cerium$^{iv}$ arsenite exchanger ($Ce O_2$, $As_2O_3$, $5H_2O$) and describe the following binary separations which can be executed on this novel material; $Mg^{2+}/Pb^{2+}$, $Zn^{2+}/Pb^{2+}$, $Zn^{2+}/Hg^{2+}$, $Ca^{2+}/Hg^{2+}$.

Zirconium tungstate[77] can be prepared in a granular form appropriate to column use and so used to separate the following ion pairs $Co^{2+}/Fe^{3+}$, $Mn^{2+}/Fe^{3+}$, $Cu^{2+}/Fe^{3+}$, $Zn^{2+}/Fe^{3+}$, $Al^{3+}/Fe^{3+}$, $Hg^{2+}/Zn^{2+}$, and $Hg^{2+}/Cd^{2+}$. Another tungstate with useful capacities is a ceric tungstate which can achieve carrier free separations of 95-Nb from 95-Zr and 113m-In from 113-Sn[78]. Yields are 100% and separations are complete after 15 mins. The same publication describes uptakes of 21 other cations, at low solution concentration, including the multivalent species $Zr^{4+}$, $Hf^{4+}$, $Sn^{4+}$, $Nb^{5+}$, $Se^{6+}$, $Mo^{6+}$ and $W^{6+}$.

The selectivities of this type of exchanger can be seen further in the reports of separations of $Mn^{2+}$ from $Al^{3+}$, $La^{3+}$, $Fe^{3+}$ and $Cu^{2+}$ on zinc antimonate[70] and a study of zirconium$^{iv}$ molybdate[80] quoting distribution coefficients for 17 metal ions and the column separation of $Hg^{2+}$ from various other ions.

To conclude this report of progress in inorganic cation exchangers it is appropriate to mention the fact that several insoluble ferricyanides have cation exchange capabilities. Their application[81] has been to aqueous nuclear waste treatment so will not be discussed here.


## Anion Exchangers

Inorganic solids for anion exchange have received much less attention than those with cation capacities. Very few exhibit reversible uptake/release of anion species so true inorganic anion exchangers are rare. Nevertheless anion removal from aqueous solution still is useful so this review will not confine itself to formally defined anion exchangers. Dyer and Jamil[82] have recently listed anion exchange materials which can be regarded as commercially available. They also describe the irreversibility of exchange noted for most of the substances listed.

One compound listed is the layer compound hydrotalcite and this has been the subject of a recent review[83].

Most of the hydrous oxides, discussed earlier, are, of course, amphoteric and Dabral et al[84] make use of this when paper impregnated with tin oxide is used for the ion exchange chromatography of iodide, thiocyanide, sulphide, phosphate, dichromate, vanadate thiosulphate, arsenite, ferric and ferrocyanide species.

Similarly Paterson and Smith[85] describe the ion sieve behaviour of $\beta$ FeOOH for the halide ions and a reversible fluoride exchange on monoclinic zirconia[86].

A novel anion exchanger of gel type, and suitable for column work, is the stannic telluride described by Srivastava et al[87]. The following affinity sequence was observed;

$$Br^- < Cl^- < VO_3^- < SCN^- < AsO_4^{3-} < CrO_4^{2-} < MoO_4^{2-} < Fe(CN)_6^{4-} < PO_4^{3-} < SO_4^{2-}$$

and separation of $SO_4^{2-}$ from $CrO_4^{2-}$, $MoO_4^{2-}$, $SCN^-$ and $Br^-$ attained. The material also exhibited low affinities for some metals.

A new zeotype has been claimed[88] in a recent patent having the composition of

$$M_{x/m}^{m+} \quad (AlO_2)_{1-y} \quad (PO_2)_{1-x} \quad (SiO_2)_{x+y} \quad N_{y/n}^{n-}$$

This is known as MCM-2 and N refers to an anion exchangeable moiety.

The concept whereby a convenient solid surface can be "customized" to confer specific ion exchange properties has been used by Papava et al[89]. They describe a method whereby the $H^+$ form of the natural zeolite clinoptilolite can be successively reacted with epichlorohydrin, glycidyl ether, triethylamine and sodium hydroxide to produce exchangeable $OH^-$ groups. Capacities of $2.2 \rightarrow 2.5$ meq g$^{-1}$ for $Cl^-$ and $SO_4^=$ are quoted and the modified material is resistant to alkali and hydroxyacids.

## Conclusion

This review describes progress in inorganic ion exchangers during the last three years. Clearly new materials continue to be synthesized and there is a slow movement to analytical and industrial uses for these materials and the more well established compounds.

Perhaps the recent paper by Kato et al[9] reviewing their use to remove metal and halide ions from electronic materials pressages the move to satisfy Dr Kunin's dernier vu.

## REFERENCES

1.  Dwyer,J., Chem. Ind., 1984, 258.

2.  "Zeolites as Catalysts, Sorbents and Detergent Builders", Karge, H.G. and Weitkamp, J. (eds.), Studies in Surface Science and Catalysis 46, Elsevier, 1989.

3.  Townsend, R.P., in "New Developments in Zeolite Science and Technology", Murakami, Y., Iijima, H. and Ward, J.W. (Eds.), Studies in Surface Science and Catalysis 28, Elsevier, 1986, 273.

4.  Chu, S.Y. and Sposito, G., Soil Sci. Soc. Am. J., 45, 1981, 1084.

5.  Flanigen, E.M., Patton, R.L. and Wilson, S.T. in "Innovation in Zeolite Material Science", Grobet, P.J., Mortier, W.J., Vansant, E.F. and Schulz-Ekloff, G., (Eds.), Studies in Surface Science and Catalysis 37, Elsevier, 1988, 13.

6.  Szostak, R., "Molecular Sieves Principles of Synthesis and Identification", Van Nostrand Reinhold, 1989.

7.  See Bibler, J., loc. cit.

8.  Ruvarac, A.L. and Clearfield, A., J.Serb. Chem. Soc., 53, 1988, 283.

9.  Dyer, A., "An Introduction to Zeolite Molecular Sieves", J. Wiley, 1988.

10. Breck, D.W., "Zeolite Molecular Sieves", J. Wiley, 1974.

11. Barrer, R.M. in "Natural Zeolites, Occurrence, Properties, Use", Sand, L.B., and Mumpton, F. (Eds.) Pergamon Press, 1978, 385.

12. Dyer, A. in "Inorganic Ion Exchangers in Chemical Analysis", Qureshi, M., and Varshney, K.G. (Eds.) CRC Press, to be published later 1990.

13. Lutz, W., Seidel, R. Chem. Techn., submitted to the editor.

14. Franklin, K.R. and Townsend, R.P., J. Chem. Soc., Faraday Trans., 1, **84**, 1988, 687.

15. Franklin, K.R. and Townsend, R.P., J. Chem. Soc., Faraday Trans. 1, **84**, 1988, 2755.

16. Franklin, K.R., Townsend, R.P., Whelan, S.J. and Adams, C.J. in "New Developments in Zeolite Science and Technology", Murakami, Y., Iijima, A. and Ward, J.W. (Eds.), Studies in Surface Science and Catalysis 28, Elsevier, 1986, 289.

17. Triay, I.R. and Rundberg, R.S., Zeolites , **9**, 1989, 217.

18. Triay, I.R. and Rundberg, R.S., J. Phys. Chem., 1987, 5269.

19. Barreto, M.C., Ciambelli, P., Del Re, G., and Peluso, A., J. Phys. Chem. Solids, **48**, 1987, 1.

20. Tao, Z. and Yang, G., Gaodeng Xuexiao Huaxue Xuebao, **8**, 1987, 170.

21. Dyer, A., and Yusof, A.M., Zeolites, **7**, 1987, 191.

22. Dyer, A. and Yusof, A.M., Zeolites, **9**, 1989, 129.

23. Thompson, R.W. and Tassopoulos, M., Zeolites, **6**, 1986, 9.

24. Beyer, H.K., Karge, H.G. and Barbely, G., Zeolites, **8**, 1988, 79.

25. Fyfe, C.A., Kokotailo, G.T., Graham, T.D., Browning, G.C., Hyland, M., Kennedy, G.J. and Deschutter, C.I., J. Amer. Chem. Soc., **103**, 1986, 522.

26. Uyama, H., Kanzaki, Y. and Matsumoto, O., Mat. Res. Bull., **22**, 1987, 157.

27. Roque-Malherbe, R., Berazain, A. and Del Rosario, J.A., J. Thermal Anal., **22**, 1987, 949.

28. Bilba, N., Bilba, D., Mihaila, G. and Naum, N., Rev. Roum. Chim., **32**, 1987, 563.

29. Kirov, G., Petrova, N. and Filizova, L., Dokl. Bolg. Akad. Nauk., **42**, 1989, 89.

30. Maisuradze, G.V., Tsitsishvili, G.V., Kekeliya, D.V., Dolidze, L.D. and Meladze, K.G., Izv. Akad. Nauk. Gruz. SSR, Ser. Khim., **13**, 1987, 130.

31. Tsitshishvili, G.V., Krupennikova, A.Y. and Dolaberidze, N.M. Dokl. Akad. Nauk. SSSR, **297**, 1987, 658.

32. Rees, L.V.C., in ref. 2, pg. 661.

33. Roland, E., in ref.2, pg. 645.

34. Schwuger, M.J. and Liphard, M., in ref. 2, pg. 673.

35. Leonhardt, W. and Sax, B.M., in ref. 2, pg. 691.

36. Upadek, H. and Krings, P., in ref.2, pg. 701.

53

37. Horvathova, E., Chem. Prum., **38**, 1988, 351.

38. Iacomi, F., Popovici, E., Alexandrouei, M. and Barbat, A., An. Stiint. Univ." Al.I. Cuza", Iasi, Sect. 1b, **22**, 1986, 99.

39. Kang, S.J., Egashira, K. and Choi, J., Nendu Kagaku, **27**, 1987, 27.

40. Khamizov, R.K., Butenko, T., Bronov, L.V. Skovyra, V.V. and Novikova, V.A., Izv. Akad. Nauk. SSSR. Ser. Khim., **11**, 1988, 2461.

41. Loizidou, M. and Townsend, R.P., J. Chem. Soc., Dalton Trans., **8**, 1987, 1911.

42. Loizidou, M. and Townsend, R.P., Zeolites, **7**, 1987, 153.

43. Lopez Ruiz, J.L. and Gil Montero, A., Afinidad, **45**, 1988, 19.

44. Onedera, Y., Iwasaki, T., Hayashi, H. and Torii, K., Nippon Kogyo Kaishi, **104**, 1988, 277.

45. Rustamov, S.M. and Makhmudov, F.T. Zh. Prikl. Khim. (Leningrad), **6**, 1988, 34.

46. Wang, Z., Ding, Y., Qian, Z. and Zhang, P., Shuichuli Jishu, **41**, 1988, 167.

47. Zeofloc$^R$ process, Hung,Pat.No. 195, 457 H (see also Olah, J., Papp, J., Meszaros-Kiss A., Mucsi, G.Y. and Kallo, D., in ref. 2, pg. 711).

48. Czaran, E., Domokos, E., Meszaros-Kiss, A. and Papp, J., Acta. Chim. Hung., **121**, 1986, 343.

49. Varshney, K.G., Khan, A.A. and Rani, S., Colloids Surf., **25**, 1987, 131.

50. McCullen, S.B., Chen, N.Y. and Han, S., US Pat. 4, 828, 812, 1987.

51. Battacharyya, S. and Banerjee, S.P., J. Indian Counc. Chem., **A**, 1988, 29.

52. Bergk, K.H., Nietzold, G. and Schweiger, W., Z. Chem., **28**, 1988, 78.

53. See articles by Rurarac, A., and Abe, M., in "Inorganic Ion Exchange Materials", Clearfield, A. (Ed.), CRC Press, 1982, pp. 141, 161 resp.

54. Venkataramani, B. and Gupta, A.R., Indian J. Chem., **27A**, 1988, 290.

55. Abe, M., Wang, P., Chitrakar, R. and Tsuji, M., Analyst, **114**, 1988, 435.

56. Yamazaki, M., Kanno, Y. and Inoue, Y., Bull. Chem. Soc. Jpn., **62**, 1989, 1837.

57. Weller, J.P., US Pat. 14, 692, 431, 1987.

58. Onorin, S.A., Vol'khin, V.V. and Khodyashev, N.B., Izv. Vysch. Uchebn. Zaved. Tsvetn. Metall., **3**, 1987, 51.

59. Sasaki, T., Komatsu, Y. and Fujiki, Y., Inor. Chem., **28**, 1989, 2776.

60. Petro, N.S.H., Ghoneimy, H.F. and Misak, N.Z., Colloids Surf., **27**, 1987, 81.

61. Ooi, K., Miyai, Y. and Katoh, S., Sep. Sci. Technol., **22,** 1987, 1779.

62. Misak, N.Z. and Mikhail, E.M., Solvent Extraction Ion Exchange, **5,** 1987, 363, 939.

63. Belewicz, A., Bartos, B., Narbutt, J. and Polkowska-Motsenko, H., Anal. Chem., **59,** 1987, 1737.

64. Kaneko, M., Hori, T., Inoue, M., Itahana, T. and Sakata, T., Jpn. Kokai Tokyo Koho, JP.63, 70,000, 1987.

65. Buerck, J., Ali, S. and Ache, H.J., Radiochem. Acta., **46,** 1989, 151.

66. Pakholkov, V.S. and Zelenin, V.I., Radiokhimiya, **30,** 1988, 518.

67. Clearfield, A., in "Inorganic Ion Exchange Materials" Clearfield, A., (Ed.), CRC Press, 1982, pg.1.

68. Alberti, G., ibid, pg.75.

69. Constantino, U., ibid pg.111.

70. Alberti, G., in "Recent Advances in Ion Exchange", Williams, R. and Hudson, M.J., (Eds.), Elsevier, 1987, pg.233.

71. Llavona, R., Suarez, M., Garcia, J.R. and Rodriguez, J., Inorg. Chem., **28,** 1989, 2863.

72. Smirnov, G.I., Grishina, T.D., Kachur, N.Y. and Chernyak, A.S., Izv. Vyssh. Uchebn. Zaved. Khim. Technol., **31,** 1988, 23.

73. Thind, O.S. and Mittal, S.K., Synth. React. Inorg. Met-Org. Chem., **17,** 1987, 93.

74. Yao, X., Yao, S., Liu, J., Liu, L. and Cheng, J., Gaodeng Xuexiao Huaxue Xuebo, **9,** 1988, 1006.

75. Yao, X, Liu, L., Liu, L. and Cheng, J., Fenxi Huaxue, **17,** 1989, 97.

76. Janardanan, C., Nair, S.M. and Savariar, C.P., Analyst, 1988, 133.

77. Sarkar, B. and Basu, S., Indian J. Chem., **28A,** 1989, 346.

78. Battacharyya, D.K. and De, A., J. Radioanal. Nucl. Chem., **108,** 1987, 109.

79. Ma, F., Yin, B. and Li, Z., Lizi Jiaohuan Yu Xifu, **4,** 1989, 24.

80. Qureshi, S.Z. and Rahman, N., Indian J. Chem., **28A,** 1989, 349.

81. Narbutt, J., Bilewicz, A. and Szeglowski, Z., Eur. Pat. EP, 217, 143, 1987.

82. Dyer, A. and Jamil, M., in "Ion Exchange For Industry", Streat, M. (Ed.), Ellis Horwood, 1988, pg. 494.

83. Sato, T., and Shimado, M., Kagaku, (Kyota), **42,** 1987, 708.

84. Dabral, S.K., Sing, K.P., Muktawat, J. and Rawat, J.P., Indian J. Chem., **27A,** 1988, 745.

85.  Paterson, R. and Smith, P.M., <u>J. Colloid Interface Sci.</u>, **124,** 1988, 581.

86.  Paterson, R., Smith, P.M. and Thomae, M., in <u>"Ion Exchange For Industry"</u>, Streat, M. (Ed.), Ellis Horwood, 1988, pg. 504.

87.  Srivastava, S.K., Singh, A.K. and Renu Khanna, <u>Indian J. Chem.</u>, **26A,** 1987, 534.

88.  Derouane, E.G. and von Ballmoos, R., <u>Braz. Pedido,</u> P.1B 2 85, 04, 000, 1987.

89.  Papava, G. Sh., Khotenashvili, N.Z., Maiskradze, N.A., Dokhturishvili, N.S., Gelashvili, N.S. and Gavashelidze, E. Sh., <u>Plast. Massy.</u>, **7,** 1988, 51.

90.  Kato, H., Yamamoto, N. and Iinuma, T., <u>Denshi Zairyo,</u> **27,** 1988, 111.

# SYNTHESIS AND ION EXCHANGE PROPERTIES OF LITHIUM ION SELECTIVE INORGANIC ION EXCHANGER BY APPLYING ION MEMORY EFFECT

MITSUO ABE, RAMESH CHITRAKAR[*], MASAMICHI TSUJI AND YASUSI KANZAKI
Department of Chemistry, Faculty of Science
Tokyo Institute of Technology, 2-12-12 Ookayama
Meguro-ku, Tokyo-152, JAPAN.

## ABSTRACT

A new crystalline antimonic acid $HSbO_3.12H_2O$ was prepared by $Li^+/H^+$ ion exchange reaction with concentrated nitric acid solution from $LiSbO_3$. The $LiSbO_3$ was obtained by heating $LiSb(OH)_6$ at 900°C. The $LiSb(OH)_6$ was prepared by the addition of LiOH solution to an Sb(V) chloride solution at 60°C. The X-ray diffraction pattern( XRD ) of $HSbO_3.12H_2O$ was indexed to a monoclinic cell (space group $P2_1/m$ or $P2_1$) with a=8.676 Å, b=4.752 Å, c= 5.263 Å and $\beta$=90.75°. The pH titration curves of monoclinic antimonic acid (M-SbA) showed apparently monobasic acid for the systems of alkali metal ions/$H^+$. The uptake order of the metal ions are $K^+ < Rb^+ < Cs^+ < Na^+ < Li^+$ throughout the pH range studied. The low uptakes of $K^+$, $Rb^+$, $Na^+$ and $Cs^+$ at high pH might be due to steric or ion sieve effects for large unhydrated cations on M-SbA. Thermodynamic data were derived for $Li^+/H^+$ exchange on M-SbA from pH titration curve.

## INTRODUCTION

Hydrous oxides of tetravalent and pentavalent ions were known to possess interesting ion exchange properties [1]. Some hydrous oxides of Sn, Zr, Sb, Ta and Nb were also known as inorganic proton conductors [2-3]. Among these oxides, hydrous antimony pentoxide "so-called antimonic acid" shows cation exchange behavior. Systematic studies of preparations of the antimonic acid were carried out by one of the authors [4-6]. Various

[*] Present address: Department of Mines & Geology, Kathmandu, Nepal.

antimonic acid materials in amorphous, glassy and crystalline forms, have been obtained with different chemical composition and ion exchange properties, depending on the method of their preparations as well as aging [4]. The affinity series for alkali metal ions vary depending on the crystalline form of the exchanger as well as on the sorption media. In acid media, the series Li < Na <K < Rb < Cs was found on both amorphous and glassy antimonic acids, while the series Li << K < Cs < Rb << Na was found on the crystalline antimonic acid $Sb_2O_5.4H_2O$ which has a face centered cubic structure (a=10.38 Å) with a space group of Fd3m [4,7]. The size preference selectivity, for cations having about 1.0 Å of crystal ionic radii, was observed on the crystalline antimonic acid by extensive studies of selectivity of various cations [8-10].

Vol'khin et al.[11] have prepared lithium ion sieve manganese oxide by proton exchange from spinel type manganese oxide containing lithium ions. This oxide showed extremely high selectivity for lithium ions [12]. Cryptomelane type manganese oxide prepared from potassium permanganate showed an extremely high selectivity for potassium ions[13]. Almost the same crystal structure was observed before and after exchanging lithium ions or potassium ions in these oxides with $H^+$. It exhibits an ion memory effect where it keeps favorable space and sites for original metal ions even if the metal ion in the materials were exchanged with $H^+$.

It is quite hopeful that hydrous antimony pentoxide having different crystal structures will show different selectivity for various metal ions. Various attempts have been made in order to obtain new types of crystalline materials for hydrous antimony pentoxide by this concept of ion memory effect[14].

This paper describes the successful preparation and some of the characterization, and ion exchange properties of a new crystalline antimonic acid $HSbO_3$ which cannot be found in the literature[15,16].


## EXPERIMENTAL


Preparation of $LiSb(OH)_6$. A $CO_2$ free 1M LiOH solution (1200 mL) prepared with $N_2$ bubbling was added slowly to 4M $SbCl_5$ (40 mL) with stirring. The precipitate ( $LiSb(OH)_6$ ) obtained was kept in mother liquor at 60°C for 2 days, washed with water by decantation, filtered and finally air dried. The $LiSbO_3$ was obtained by heating $LiSb(OH)_6$ at 900°C for 4 hours. Conversion of $LiSbO_3$ to $HSbO_3$. A 10 g of $LiSbO_3$ sieved to 100-200 mesh

size was washed with water to remove any adherent particles.    The slurry
was poured into a glass column and a 11M $HNO_3$  solution was passed
through at a flow rate of 0.1 mL/min.    The acid treated sample was washed
with demineralized water until neutral and then filtered and air dried.
All the samples were humidified in a desiccator containing saturated NaCl.
Chemical analysis. X-ray diffraction( XRD ), thermal analysis and infra red
spectra. Antimony and lithium                        were determined as
described in [11]. The XRD pattern was measured using Phillips Automated
Powder X-ray Diffractometer, model PW-1700 with  $CuK\alpha$ radiation.   Thermal
analysis  was carried out using a Rigaku Denki Thermoflex, model 8001 at a
heating rate of $10^\circ$ C/min with   $\alpha$ $-Al_2O_3$ as the reference material in air.
IR spectra were measured by the KBr  disc  method  with  a JASCO   Infrared
Spectrometer,  model  DS-701G.
pH titration curves.    0.10 g of the exchanger was immersed into a mixed
solution (10.0 mL) of $MOH + MNO_3$ or $MNO_3 + HNO_3$ (M=$Li^+$, $Na^+$, $K^+$, $Rb^+$, $Cs^+$)
with ionic strength 0.1 in varying ratios with intermittent shaking at
$30^\circ$C. Additional measurements were carried out at $30^\circ$, $45^\circ$ and $60^\circ$C for
$Li^+/H^+$ only. Kinetic experiments showed that the system attained
equilibrium within one week.   After pH measurement, the metal ion
concentrations in the supernatant solution was determined as described in [17].
In the reverse pH titration  for $Li^+$ at $30^\circ$C, $Li^+$ exchanged samples (4.3
mequiv/g) at $30^\circ$C were mixed with varying ratios of $LiNO_3+HNO_3$ solution at
$30^\circ$C. After equilibrium was attained, the pH of the solutions as  measured
and $Li^+$ released from the exchanger as  determined.


## RESULTS AND DISCUSSIONS


Preparation of $LiSbO_3$.   In this experiment, the $LiSb(OH)_6$  was prepared by
the addition of LiOH solution in prehydrolyzed  $SbCl_5$ solution. The
formation of $LiSb(OH)_6$ was rapid, being complete in one day when the
temperature of aging was $60^\circ$C.    The $LiSbO_3$ was obtained by heating
$LiSb(OH)_6$ at $900^\circ$C for 4h and showed apparently granular particles suitable
for column operation.  The XRD patterns of the prepared $LiSb(OH)_6$ and
$LiSbO_3$ were identical to that reported in the literatures[18].
Preparation of $HSbO_3$.  In order to obtain $HSbO_3$, a nitric acid solution at
different concentrations was passed through a column of $LiSbO_3$. The $Li^+$ in
$LiSbO_3$ was exchanged to $H^+$ with removal of about 99% $Li^+$ by 11M $HNO_3$ over a month.
The sample in acid form  was washed with demineralized water and then dried
in air   using a fan.   The dried sample was very stable in air over one

year, while wet $HSbO_3$ was gradually transformed to cubic crystalline
antimonic acid (C-SbA) on keeping it in demineralized water over 5 days
at room temperature.

Chemical analysis. The chemical analysis of prepared samples were in good
agreement with the theoretical values of Li and Sb in $LiSb(OH)_6$ and $LiSbO_3$
within 2% of experimental error.

Thermal analysis. Thermal analysis of $LiSb(OH)_6$ were in good agreement
with the results reported by Franck [18]. The amorphous material was
obtained after heating it at the end of these endothermic peaks. The
crystallization to $LiSbO_3$ was started from about 720°C and increased in the
intensities of the XRD pattern. The constant intensities were observed
after a small exothermic peak around 810°C.

Small endothermic peaks at 80°C, 280°C, and 400°C and a large
endothermic peak around 835°C were observed for the acid form.
A very weak endothermic peak at 80°C is due to            free water
(interstitial water). The sample heated up to 200°C for 4 hours did not
show any change in the d values while diffraction intensities were slightly
reduced. The sample heated at around 350°C for 4 hours showed the weak
XRD pattern of rutile type structure (d=3.36 Å, 2.50 Å, 2.34 Å and 1.70 Å).
The slow weight loss with endothermic peak around 400°C is due to the
loss of structure water in the sample which agrees well with results of the
disappearance of the OH band in the IR spectra of the sample heated at 500°C.
The X-ray analysis indicated that the sample heated at 900°C was $\beta$-$Sb_2O_4$.
The weight loss between 780°C and 900°C        corresponded        to those from
$Sb_2O_5$ to $Sb_2O_4$. A composition of $HSbO_3 0.12H_2O$ was established for the
acid form        by assuming that        one equivalent amount of exchangeable $H^+$
to one of        Sb was present.

X-ray diffraction. The XRD patterns of the samples are shown in
FIG. 1. The pattern of $HSbO_3$ was        slightly different        from
$LiSbO_3$ which was orthorhombic in structure with space group Pncn. The pattern
of $HSbO_3$ was fitted to a monoclinic system. Extinctions of the type 0k0 with
k= 2n can be established with a space group as either $P2_1/m$ or $P2_1$. The lattice
constants for the samples are summarized in TABLE 1.

The removal of $Li^+$ from $LiSbO_3$ resulted in a slight expansion of
unit cell volume i.e. from 215.28 $Å^3$ to 216.98 $Å^3$. The X-ray scattering
factor for antimony and oxygen were much larger than lithium and hydrogen,
thus the diffraction intensities of $HSbO_3$ are very similar to those of
$LiSbO_3$. This monoclinic cell of $HSbO_3$ was closely related to the orthorhombic
$LiSbO_3$ cell by the relations: $a_{mono} = b_{ortho}$, $b_{mono} = a_{ortho}$ and $c_{mono} = c_{ortho}$. It could be concluded that antimony and oxygen atoms in $HSbO_3$ had

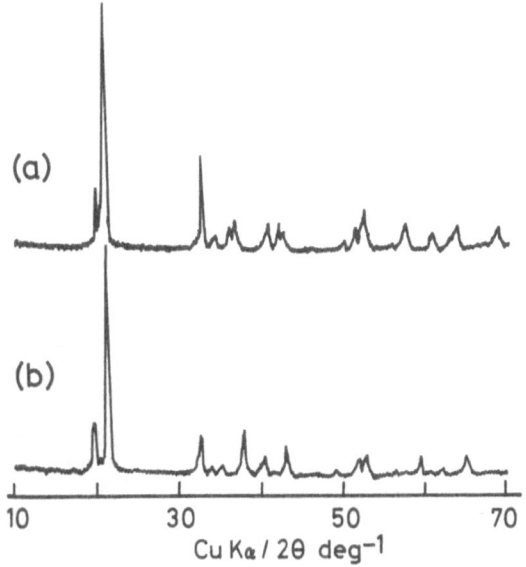

Figure 1. X-ray diffraction patterns of (a)LiSbO$_3$ and (b)HSbO$_3$.

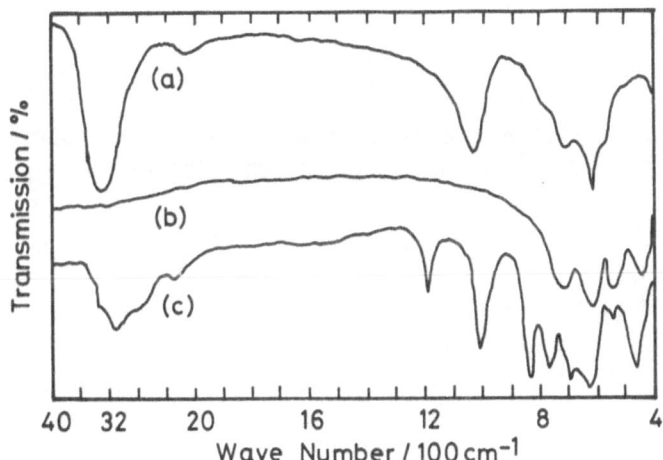

Figure 2. Infrared spectra of (a)LiSb(OH)$_6$, (b)LiSbO$_3$ and (c)HSbO$_3$.

TABLE 1

Comparison of Observed Unit Cell Dimensions for $LiSb(OH)_6$, $LiSbO_3$ and $HSbO_3$

| | |
|---|---|
| $LiSb(OH)_6$[18] | Hexagonal: a=5.351 Å, c=4.918 Å. |
| Observed | Hexagonal: a=5.350 Å, c=4.915 Å. |
| $LiSbO_3$[18] | Orthorhombic: a=4.893 Å, b=8.491 Å, c=5.182 Å. |
| Observed | Orthorhombic: a=4.899 Å, b=8.480 Å, c=5.182 Å |
| $HSbO_3$ | Monoclinic: a=8.676 Å, b=4.752 Å, c=5.263 Å, β=90.75°. |

essentially the same arrangement to $LiSbO_3$ in which the oxygen atoms form a distorted hexagonal close packing with antimony and lithium occupying some of the octahedral holes. Each oxygen octahedra around antimony shared two edges with two other antimony octahedra, thus forming a string of staggered octahedra along the c-axis in $LiSbO_3$ (13). The removal of most of the $Li^+$ in octahedral sites by exchange of $H^+$ with acid treatment resulted in formation of $HSbO_3$.

Infrared spectra. Infrared spectra of $LiSb(OH)_6$ obtained agreed well that reported for $LiSb(OH)_6$ [18]. A strong and broad band around 3300 $cm^{-1}$ was assigned to OH, SbOH and stretching water molecule with hydrogen bonding (FIG. 2). The band at 2100 $cm^{-1}$ may be due to an overtone of the SbOH deformation vibration [1]. The band at 1035 $cm^{-1}$ was due to SbOH deformation vibration and bands at 720, 620 $cm^{-1}$ were due to SbO stretching vibrations. When $LiSb(OH)_6$ was heated upto 900°C, all of the bands assigned to stretching and deformation vibrations of SbOH disappeared. Only the bands left were due to the SbO stretching vibrations. For $HSbO_3$, the bands around 2800–3120 $cm^{-1}$ were due to the OH, SbOH stretching vibration. A very weak band around 1600 $cm^{-1}$ was observed indicating the presence of free water. This result suggested that a much smaller amount of water was present in $HSbO_3$ sample. The bands at 1190 and 1000 $cm^{-1}$ were assigned to SbOH deformation and the remaining bands could be assigned to SbO stretching vibrations [1].

pH titration. The pH titration curves of M-SbA for alkali metal ions showed a large pH jump with MOH added indicating that M-SbA behaved as a strong monobasic acid type exchanger (FIG. 3, left). The amount of replaceable protons responsible for ion exchange reactions can be estimated by the difference between the blank run and each titration curve. The apparent capacity thus determined agreed well with the amount estimated from the pH titration curve (Fig. 3, right). This result indicates that the uptake of metal ions proceeded by an ion exchange mechanism. The uptake order was found to be $K^+ < Rb^+ < Cs^+ < Na^+ < Li^+$ throughout the pH range studied. A very low uptake was observed for $K^+$ and $Rb^+$ over the entire pH range studied. The capacities for $Na^+$ and $Cs^+$ at pH 1.6 were higher than those at pH 3 and, then decreased with increase in pH which was contrary to those observed on $Li^+$ exchange. The XRD patterns of the products exchanged with $Na^+$ and $Cs^+$ at lower pH showed the cubic antimonic acid (C-SbA) instead of monoclinic M-SbA. The high capacity for $Na^+$ and $Cs^+$ at low pH can be explained by transformation from monoclinic to cubic phase, but no phase transformations were observed at pH > 5. The phase transformation was not observed for other alkali metal ions over the entire pH range studied. A large ion exchange capacity of 4.3 mequiv/g was observed for $Li^+$ only at pH 11. The small uptake of $K^+$ and $Rb^+$ may be due to exchange only on the surface of M-SbA. These low uptake for $K^+$, $Rb^+$ $Cs^+$ and $Na^+$ at high pH might be due to steric or ion sieve effects for large unhydrated cations. The order of uptakes on M-SbA was different from those observed on C-SbA which showed $Li^+ < Cs^+ < Rb^+ < K^+ < Na^+$ from high loading of alkali metal ions in neutral solution[8]. The uptake order of the present study was similar to those observed on $\lambda - MnO_2$ at pH >7[12]. The $\lambda - MnO_2$ has high ion exchange capacity for $Li^+$ only while other alkali metal ions were adsorbed only on the surface.

Determination of equilibrium constant. In the present study, the equilibrium fraction of lithium in the exchanger was calculated on the basis of theoretical exchange capacity of 5.7 mequiv/g as done[8-10]. If the change in water content of the exchanger was negligible before and after ion exchange, then the titration data may be treated by the method of Argersinger et al.[19].

The reaction may be written as

$$\overline{H^+} + Li^+ \rightleftharpoons \overline{Li^+} + H^+ \qquad (1)$$

where bar refers to the exchanger phase. The equilibrium constant for this reaction was written as

$$K = \frac{\overline{a}_{Li} \, a_H}{a_H \, \overline{a}_{Li}}$$

where the barred represent activities of the ions in the solid phase and unbarred ones activities of the ions in the aqueous phase.

Let

$$K_C = \frac{\overline{X}_{Li} \, a_H}{\overline{X}_H \, a_{Li}}$$

where $K_C$ is the corrected rational selectivity coefficient for which

$$\log K = \int_0^1 \log K_c \, d\overline{X}_{Li} \qquad (4)$$

where X refers to equivalent ionic fractions in the exchanger phase. The contribution of the ionic activity coefficients in solution to the thermodynamic constant may be very small on uni-univalent ion exchange reactions. The activity coefficient of metal ions may be taken as unity at the temperature range studied[8,20]. Hydrogen ion activities were determined from the pH values at several loadings. A plot of log $K_c$ vs. $X_{Li}$ is shown in Fig. 4 and the area under the curve is the integral. The logarithm scale of overall thermodynamic equilibrium constant log K showed -2.04, -1.73 and -1.20 at 30, 45 and 60°C respectively. The plot of log K vs. 1/T showed a liner relationship. This permitted the calculation of approximate $\Delta S^o$.

As shown by Sherry and Walton[20], the ion exchange reaction can be separated into the following two reactions[21].

$$M^+_{aq} + H^+ \rightleftharpoons M^+_{gas} + H^+_{aq} \qquad -\Delta Y^o_{hyd} \qquad (5)$$

$$M^+_{gas} + \overline{H^+} \rightleftharpoons \overline{M^+} + H^+_{gas} \qquad \Delta Y^o_{ex} \qquad (6)$$

where Y represents thermodynamic functions such as G, H, and S. The numerical values of $\Delta Y^o$ contribute to the difference in the thermodynamic function of hydration ($\Delta Y^o$) of the ions in aqueous solution and that ($\Delta Y_{ex}$) of the exchanged cations. The values of $\Delta Y_{hyd}$ can be found from Rosseinsky table[22]. The calculated $\Delta Y_{ex}$ values are summarized in TABLE 2.

The equilibrium constant for Li$^+$/H$^+$ exchange at 25°C was $7.08 \times 10^{-3}$. Thus the overall exchange was not very favourable leading to $\Delta G_{298} = 2.93$ kcal/mol. However, the $\Delta G^o$ decreased with increase of temperature indicating that the reaction was favored by increase of temperature. Li$^+$ exchange might have occurred by shedding most of its water and diffusing

TABLE 2

THERMODYNAMIC DATA FOR Li$^+$/H$^+$ EXCHANGE ON M-SbA AT 298K

| $\Delta G^o_{298}$ kcal/mol | $\Delta H^o_{298}$ kcal/mol | $\Delta S^o_{298}$ eu/eq | $(\Delta H^o_{hyd})$ kcal/mol | $(\Delta H^o_{ex})$ kcal/mol | $(\Delta S^o_{hyd})$ eu/eq | $(\Delta S^o_{ex})$ eu/eq |
|---|---|---|---|---|---|---|
| 2.93 | 11.25 | 29.72 | 137.37 | 148.62 | -2.4 | 25.52 |

into the cavity in unhydrated state. Thus, a large portion of the energy was consumed in dehydrating Li$^+$ as well as breaking SbO-H bonds. These effects resulted in the large value of $\Delta H^o$. Again the release of water molecules by Li$^+$ resulted in an increase of entropy. The large $\Delta S^o_{ex}$

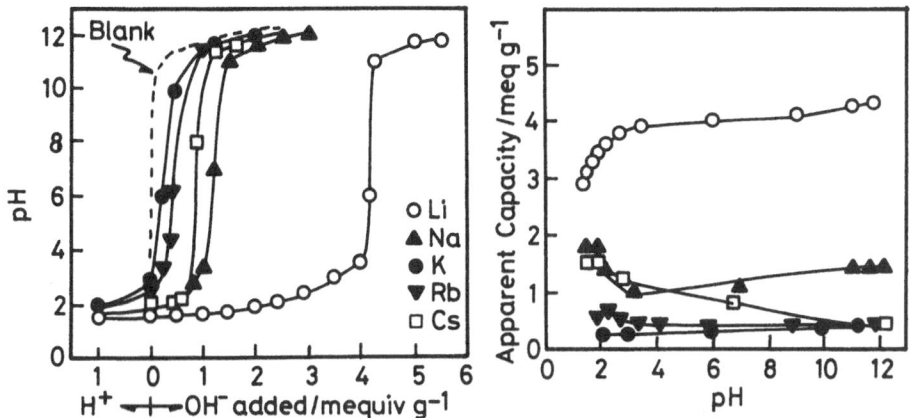

Figure 3. pH titration curves of alkali metal ions(left) and
pH dependence for uptakes of alkali metal ions(right)
on M-SbA.
Exchanger:0.10g, Soln.:0.1M (MNO$_3$ + HNO$_3$) or (MNO$_3$ +
MOH), M=Li$^+$, Na$^+$, K$^+$, Rb$^+$, Cs$^+$, Total vol.:10.0mL,
Temp.:30±0.5°C.

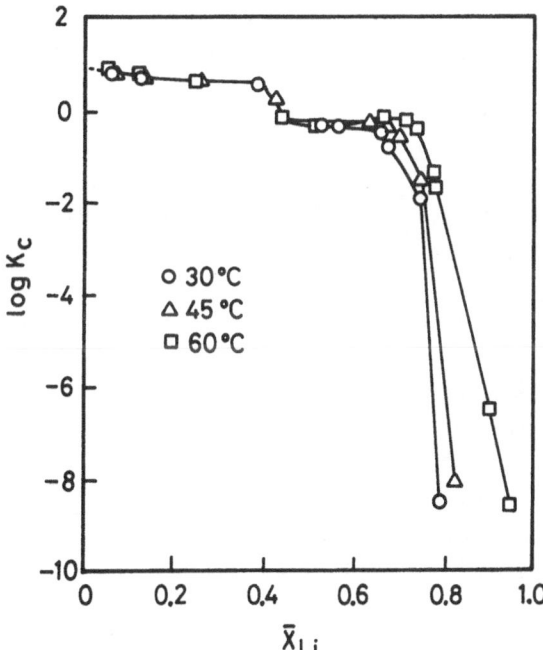

Figure 4. Plot of log K$_C$ vs. equivalent fraction of Li$^+$,
$\bar{X}_{Li}$, on M-SbA.

indicated that Li$^+$ was exchanged in the dehydrated state[22].

## Acknowledgments

This research was supported in a part of the Salt Science Research Foundation, Grant 8902. R. C. received fellowship support from Ministry of Education in Japan.

## REFERENCES

1. Abe, M., " Oxides and Hydrous Oxides of Multivalent Metals as Inorganic Ion Exchangers" in "Inorganic Ion Exchange Materials", A. Clearfield, Ed., CRC Press, Boca Raton, Fl., 1982. Chap. 6.
2. England, W. A., Cross, M. G., Hamnett, A., Wiseman P. J., and Goodennough, J. B., Solid State Ionics 1980, 1, 231-49.
3. U. Chowdhry, J. R. Barkley, A. D. English, and A. W. Sleight, Mat. Res. Bull., 1982, 17, 917-33.
4. Abe, M., and Ito, T., Bull. Chem. Soc. Jpn., 1986, 41, 333-42.
5. Abe, M., and Ito, T., Bull. Chem. Soc.Jpn., 1968, 41, 2366-71.
6. Abe, M., and Ito, T., Kogyo Kagaku Zasshi, 1967, 70, 2226-234.
7. Abe, M., Bull. Chem. Soc. Jpn., 1969, 42, 2683-89.
8. Abe, M., J. Inorg. Nucl. Chem., 1979, 41, 85-9.
9. Abe, M., and Sudoh, K., J. Inorg. Nucl. Chem., 1980, 42, 1051-5.
10. Abe, M., and Sudoh, K., J. Inorg. Nucl. Chem., 1981, 43, 2537-42.
11. Vol'khin, V. V., Leont'eva, G. V. and Onolin, S. A., Izu. Akad. Nauk. SSSR, Neog. Mater.,1972, 9, 1041-6.
12. Ooi K., Miyai, Y., Katoh, S., Maeda, H., and M. Abe, Langmuir, 1989, 5,150-157.
13. Tsuji, H.,and Abe, M., Solvent Extr. Ion Exch., 1984, 2, 253-74.
14. Abe, M., Chitrakar R., and Tsuji, M., kagaku to Kogyo, 1989, 42,1224-6.
15. Chitrakar, R., and Abe, M., Mat. Res. Bull., 1989, 23,1231-40.
16. Chitrakar, R., and Abe, M., Solvent Extr. Ion Exch., 1989, 7, 721-33.
17. Abe, M., and Hayashi, K., Solvent Extr. Ion Exch. 1983, 1, 97-112.
18. Franck, R., Thermochimica Acta 1970, 1, 261.
19. Argersinger, Jr. W. J.; Davidson, A. W.; Bonner, O. B. Trans. Kan. Acad. Sci. 1950, 53, 404-10.
20. Sherry, H. S., and Walton, H. F., J. Phys. Chem., 1967, 71, 1457-65.
21. Rosseinsky, D. R. Chem. Rev. 1965, 65, 467-88.
22. Abe, M.; Furuki, N., Solvent Extr. Ion Exch. 1986, 4, 547-65.

# UPTAKE OF SEVERAL TRACER CATIONS AND THE SEPARATION OF CARRIER FREE $^{140}$La FROM $^{140}$Ba AND ALSO $^{115M}$In FROM $^{115}$Cd USING A ZIRCONIUM(IV) OXIDE COLUMN

D.K.BHATTACHARYYA & N.C.DUTTA
Nuclear Chemistry Division, Saha Institute of Nuclear Physics, 1/AF,
Bidhannagar, CALCUTTA 700 064.

## ABSTRACT

Zirconium (IV) oxide of suitable column quality was prepared by precipitation with ammonia from a solution of $ZrOCl_2.8H_2O$ in 2M HCl, which was aged, washed, made chloride free and dried at $800^{0}$C for 12 hours. The composition of the material was ascertained by DTA and TGA to be $ZrO_2$, $1.7H_2O$. Uptake of tracer cations $^{24}$Na, $^{42}$K, $^{86}$Rb, $^{137}$Cs, $^{65}$Zn, $^{115}$Cd, $^{140}$Ba, $^{95}$Zr, $^{95}$Nb, $^{125}$Sb and $^{185}$W by zirconium oxide and the corresponding $K_d$ values were determined. It was observed that different tracer cations were adsorbed in the exchanger in different manners. The ion exchange capacity of the material was found to be 0.4 m moles gm. New methods of radiochemical separation of carrier free $^{140}$La and $^{115m}$In from $^{140}$Ba and $^{115}$Cd respectively over the columns of zirconium oxide are described. $^{140}$Ba in equilibrium with $^{140}$La was treated with 2% ammonium molybdate solution and fed in the zirconium oxide column when both isotopes were adsorbed. It was possible to remove $^{140}$La by eluting with 0.001N $HNO_3$ and later, $^{140}$Ba with 1:4 $HNO_3$. The $^{115}$Cd - $^{115m}$In pair in cold 0.002 M $HClO_4$ was poured in another column and both isotopes were found to be retained. $^{115m}$In was then removed by washing with 0.005M $HNO_3$, and $^{115}$Cd was later washed out by 1:2 $HNO_3$. The g-spectrum analysis showed that $^{140}$La and $^{115m}$In activities separated out were of high radionuclidic purity. The chemical procedures were simple but took less than 15 minutes and the yields are quantitative.

Zirconium (IV) oxide has long been known to behave both as a cation and anion exchanger [1,2]. Studies on thermodynamic equilibrium constants for alkali metal cations [3]; separation of cations using co-precipitations of Pu[4], Si, Ti, V[5], Mo[6], Zn[7], In[8] and separation of cation pairs such as U-U$_{x1}$[9], $^{115}$Cd- $^{115m}$In; $^{132}$Te- $^{132}$I[10]; $^{113}$Sn- $^{113m}$In; $^{44}$Ti- $^{44}$Sc, $^{68}$Ge - $^{68}$Ga[11]; Ag - Au(III); Cs - Eu[12]; $^{90}$Sr - $^{90}$Y; $^{233}$Ra $^{233}$Fr[13] have been described. The material was also utilised in the water throughout desalination; the Donnan electrolytic exclusion process and for preparing electron exchanging ion exchangers[14]. Zirconium oxide therefore might exhibit preferential adsorption of certain cations compared to others, and, with this in view, a study on the uptake characteristics of several cations has been undertaken to gather more information. The cations chosen were Na+, K+, Rb+, Cs+, Zn+2, Cd+2, Ba+2, Zr+4, Nb+5, Sb+5 and W+6 at tracer concentrations. In addition, separations of carrier-free $^{140}$La from $^{140}$Ba and $^{115m}$In from $^{115}$Cd have been carried out over the column of zirconium oxide by applying new and simple chemical procedures with the expectation of obtaining better yields in the shortest possible time.

It is well-known that the radionuclide generator [140]Ba - [140]La may have potential in the study of radiopharmaceuticals. Separation of carrier free [140]La from [140]Ba has so far achieved by methods like solvent extraction in toluene [15]; adsorption chromatography on $MnO_2$[16]; TLC on silica gel [17]; cation exchange over Dowex-50 [18]; $Al_2O_3$ [19]; stannic ferrocyanide [20]; anion exchanger Dowex-1 [21,22] and by reverse phase chromatography [23]. There is a need for a less complicated and quicker method for obtaining carrier free [140]La of higher yield and purity. An attempt was, therefore, made to utilise a zirconia column for this purpose. Since [115m]In-labelled compounds are being used in the field of nuclear medicine and, because the detection of single 390 Kev gamma energy of [115m]In is very advantageous, the separation of carrier free [115m]In from [115]Cd was also considered in the present study. This separation was earlier performed by several workers using the precipitation of indium as hydroxide [24]; sulphide [25]; chromate; oxalate; ferrocyanide; complex-carbonate [26]; basic acetate [27]; hydroxyquinolate [28] and reverse phase extraction chromatography [29]. Use of cation exchangers such as zirconium phosphate [30], tin (IV) phosphate [31], silica gel [32] and extraction in high molecular mass amines [33]; O-xylene [34] have also been described.

## MATERIALS

Zirconium (IV) oxychloride, ammonia solution and ammonium molybdate were of AnalaR grade. The radioisotopes [24]Na, [42]K, [86]Rb, [137]Cs, [65]Zn, [115]Cd in equilibrium with [115m]In, [140]Ba in equilibrium with [140]La, [95]Zr, [95]Nb, [125]Sb and [186]W were supplied by BARC, Bombay, India.

## METHODS AND RESULTS

Preparation of $ZrO_2$: A solution of $ZrOCl_2.8H_2O$ (15g) in water was treated with ammonia gas till precipitation was complete. The precipitate was then digested at 70⁰C for two hours, filtered and washed with water to remove Cl- ions, dried at 80⁰C for 48 hours. The hard solid substance thus obtained was found to be stable in water and dilute acids and was broken to sandy, granular pieces suitable for column operation.

Solutions of each tracer cation (containing ~7000 cpm) were separately and repeatedly evaporated to near dryness with a little quantity of water to remove as much acid as possible and then dissolved in 10ml. water. The uptake of each cation was then determined by shaking with 0.5g zirconia in 50ml water for 24 hours. The pH of each solution was approximately 6.5. β activity was then measured (with a Phillips β-type liquid counter) in the solution part and $K_d$ values were calculated in the usual way [30] (Table 1).

The water content of the exchanger was determined by TGA. The thermal stabilities and possible decompositions etc, were studied by DTA with a DT-40 thermal analyser, Shimadzu (Japan). The composition of the zirconia was ascertained to be $ZrO_2$, $1.7H_2O$ (Fig. 1).

The pH titration curve was obtained by equilibrating 0.5g samples in varying proportions of 0.1M NaCl + 0.1M NaOH/0.IM HCl for 24 hours at room

69

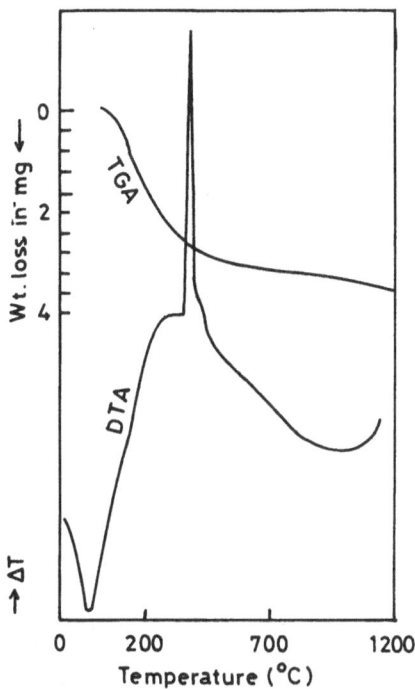

Fig.1. TGA and DTA curves of $ZrO_2$, 1.7 $H_2O$ exchanger. $\Delta T$ = Temp. difference between sample and reference ($\alpha$ - alumina).

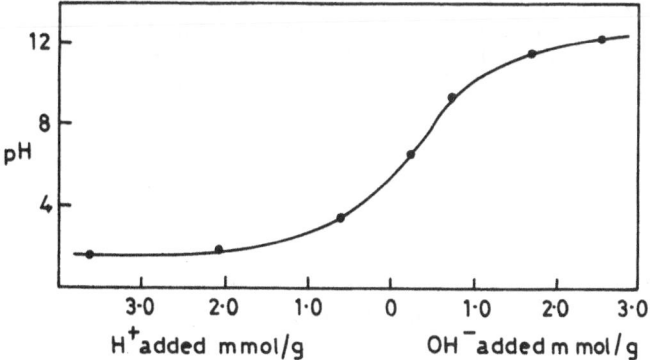

Fig.2. pH titration curve of Zirconia

temperature. The pH was measured in the supernatant liquid with a glass electrode and an ELICO digital pH meter (Fig.2).

The ion exchange capacity was determined by passing 0.IM NaCl through the exchanger column and titrating the acid thus liberated in the usual way. The obtained value was 0.4 m moles/gm.

TABLE 1

Distribution Co-efficients ($K_d$) of Several Tracer
Cations over Zirconium (IV) oxide (Measured by a
Radiochemical Method)

| Cations | $K_d$ ml/g | Cations | $K_d$ ml/g |
|---------|------------|---------|------------|
| $^{24}$Na | 10.75 | $^{140}$Ba | 1366.33 |
| $^{42}$K | 1.43 | $^{95}$Zr | 246.96 |
| $^{86}$Rb | 0.44 | $^{95}$Nb | 206.53 |
| $^{137}$Cs | 18.09 | $^{125}$Sb | 793.96 |
| $^{65}$Zn | 2254.40 | $^{99}$Mo | 9.54 |
| $^{115}$Cd | 10496.55 | $^{185}$W | 7932.01 |

## SEPARATION OF $^{140}$Ba AND $^{140}$La

The $^{140}$Ba - $^{140}$La solution cpm (ca.10,000) was repeatedly evaporated after successive addition of water to remove acid and dissolved in 10 ml. water into which 2% ammonium molybdate solution was added dropwise to reach the stage just before precipitation. The solution was cooled and fed into the exchanger column and this was followed by washing with a small quantity of water. When all the activity was adsorbed on the column, $^{140}$La was eluted by washing with only 25ml 0.00IM HNO$_3$. $^{140}$Ba was later eluted with 40 ml 1:4 dil HNO$_3$ (Fig.3a).

The eluents were analysed for g-radiation using a Princeton g-Tech Ge-Coaxial detector of 30Cm$^3$ active volume with a system resolution of 2.06 Kev FWHM for the 1.33 Mev $\gamma$-ray of $^{60}$Co coupled with a ND-8K multichannel analyser (Fig. 4).

## SEPARATION OF $^{115}$Cd from $^{115m}$ In

The solution of $^{115}$Cd in equilibrium with $^{115m}$In cpm (ca. 10,000) was treated as before; dissolved in 10ml. water (pH ~ 6.0) with 2 ml HClO$_4$ (0.002M). The solution was made neutral with a few drops of NaOH (2M) added dropwise until a very faint turbidity appeared. The solution was then fed over a zirconia column when all the activity was adsorbed. $^{115m}$In was then removed from the column by washing with 40ml HNO$_3$ (0.005 M); $^{115}$Cd

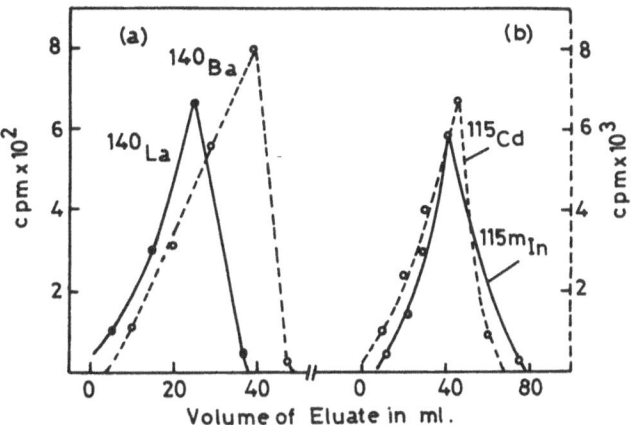

Fig.3. Elution curves of (a)$^{140}$Ba and$^{140}$La
and (b)$^{115}$Cd and$^{115m}$In.

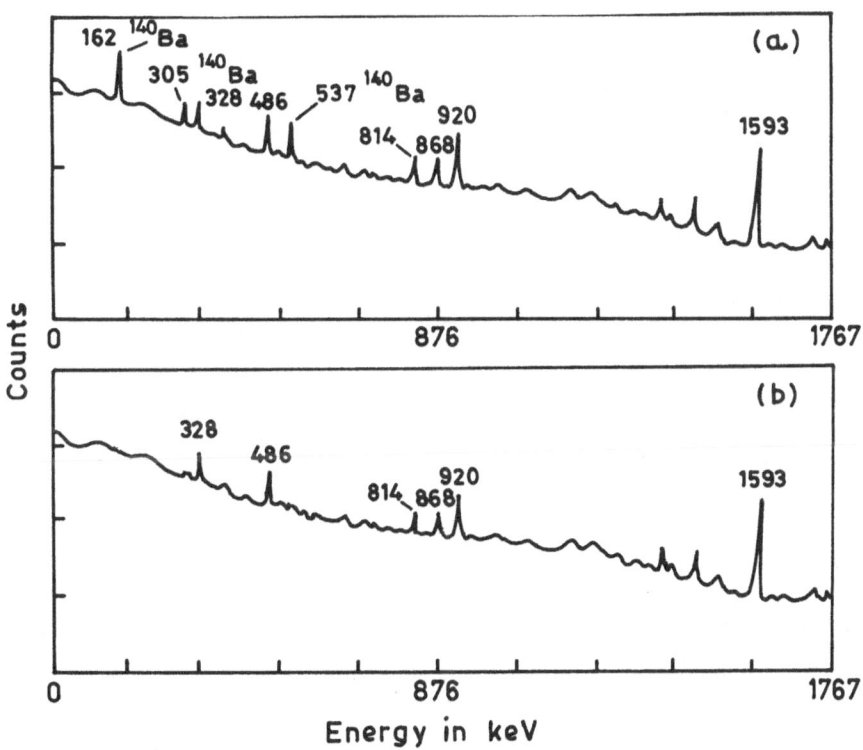

Fig.4. Gamma spectrum of (a) the equilibrium mixture
of $^{140}$Ba and $^{140}$La before separation,(b) $^{140}$La separated
from $^{140}$Ba.

was later completely removed by washing with 55ml $HNO_3$ (1:2) (Fig.3b). The eluted solution was analysed for γ-spectrum in the way described above (Fig.5).

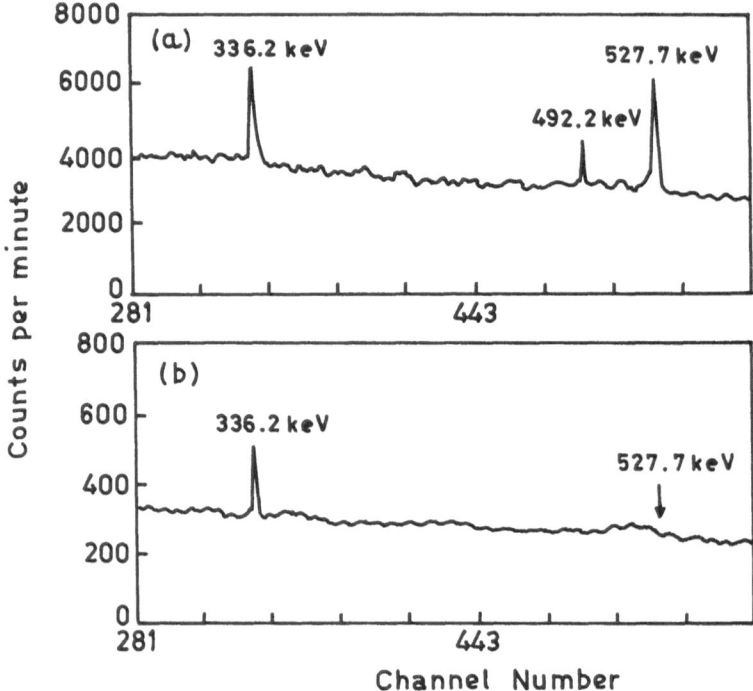

Fig.5. Gamma spectrum of (a) the equilibrium mixture of $^{115}Cd$ and $^{115m}In$ before separation, (b) $^{115m}In$ separated from $^{115}Cd$.

## CONCLUSION

From the study of the capacities, affinities and elution characteristics for different cations, it was considered that separation of the two title cation pairs should be possible at tracer scale in the pH range chosen. Table 1 shows that the uptakes of monovalent cations are insignificant whereas the corresponding figures for bivalent and $W^{+6}$ cations are very high. The capacities of other cations are different for each oxidation state. A glance at Figs.4 & 5 show that $^{140}La$ and $^{115m}In$ separated from $^{140}Ba$ and $^{115}Cd$ respectively were of high radionuclidic purity. The experimental procedures were very simple.

**REFERENCES**

1.  K.M.Pant, *J.Ind.Chem.Soc.,* 1969, 46, 541.
2.  J.Belloni-Cofler, D.Pavlov, *J.Chem.Phys.,* 1965, 62, 458.
3.  D.Britz and G.H.Nancollas, *J.Inorg.Nucl.Chem.,* 1969, 31, 3861.
4.  A.I.Novikov and I.A.Starovoit, *Radiokhimiya,* 1969, 11, 339.
5.  A.I.Novikov, V.I.Ruzankim & V.S.Dedova, *Ibid.,* 1969, 11, 347.
6.  V.I.Plotnikov & V.I.Kochetov, *IZV .Akad. Nauk. kaz. SSR. Ser, Fiz-Mat,* 1967, 5, 53.
7.  V.I.Plotnikov, V.P.Novikov & M.M.Novikov, *Ibid.,* 1967, 5, 90.
8.  V.I.Plotnikov, E.G.Gibova, *Ibid.,* 1967, 5, 58.
9.  D.K.Bhattacharya, S.Basu, *Int.J.Appl.Radiat.Isot.,* 1977, 28, 535.
10. D.K.Bhattacharya, S.Basu, *J.Radioanal.Chem.,* 1979, Vol. 52, No.2, 267-273.
11. V.E.Seidl, K.H.Lieser, Die Radionuklidgeneratomen $^{113}Sn/^{113}In$, $^{68}Ge/^{68}Ga$ and $^{44}Ti/^{44}Sc$, *Radiochim. Acta.,* 1973, 19, 196.
12  K.A.Kraus, H.O.Phillips, T.A.Carlson & J.S.Johnson, *Proc.2nd Int.Conf.Peaceful uses At.Energy,* Geneva, 1958, 28, 3.
13. M.Abe, B.A.Nasir & T.Yoshipa, *J.Chromatography,* 1978, 153, 295-301.
14. V.Vesely & V.Pekarek. *Talanta.,* 1972, 19, 219-262.
15. D.F.Peppard, G.W.Mason & S.W.Moline, *J. Inorg. Nucl. Chem.,* 1957, 5, 141.
16. C.Bigliocca, F.Girardi, J.Pauly & E.Sabbioni, *Anal.Chem.,* 1967, 39, 1634.
17. Y.Maki & Y.Murakami, *J.Radioanal.Chem.,* 1973, 14, 317.
18. P.Kruger & C.D.Coryll. *J.Chem.Educ.,* 1955, 32, 280.
19. D.K.Bhattacharyya & S.Basu, *Separation Science,* 1976, 11(1) 103.
20. J.S.Gill and S.N.Tandon, *J.Inorg.Nucl.Chem.,* 1972, 34 (12) 3885.
21. R.W.Perkins, *Anal.Chem.,* 1957, 29, 152.
22. S.Misumi & T.Taketasse, *J.Inorg.Nucl.Chem.,* 1961, 20, 129.
23. S.Sarkar, S.N.Bhattacharyya, *J.Radioanal.Chem.,* 1979, 54, 355.
24. E.Jacobi, *Helv.Phys.Acta.,* 1949, 22, 66.
25. D.Coryell & N.Sugarman, *Radiochemical Studies, The Fission Products,* McGraw Hill Book Co., Book 2, paper No.127, p.828, 1951.
26. I.P.Alimarin, E.P.Tsintsevich & V.P.Burlaka, *Zarodskaya Lab.,* 1959, 25, 1287.
27. G.Wilkinson & W.E.Grummit, *Nucleonics,* 1951, 9, 52.
28. A.A.Smales, J.Van R.Smit & H.Hirving, *Analyst,* 1957, 82, 539.
29. I.Stronski, *Radiochem.Radioanal.Letters,* 1970, 5, 113.
30. D.K.Bhattacharyya, S.Basu, *J.Radioanal.Chem.,* 1978, 47, 105.
31. D.K.Bhattacharyya, A.De, *J.Radioanal.Nucl.Chem.,* 1986, 2, 100.
32. K.R.Verma, K.B.Shah & R.S.Mani, *Radiochem.Radioanal. Lett.,* 1980, 43, 4, 255.
33. O.V.Singh & S.N.Tandon, *J.Radioanal.Chem.,* 1977, 36, 331.
34. W.Sonad & A.Haggag, *Ibid,* 1978, 43, 209.

# CATION AND ANION EXCHANGE PROPERTIES OF PILLARED CLAYS

A.Dyer and T.Gallardo
Department of Chemistry and Applied Chemistry, Cockcroft Building,
University of Salford, M54WT U.K.

## ABSTRACT
Cation and anion pH dependent exchange processes on an Aluminium Pillared Clay (Al–PILC) and Zirconium Pillared Clay (Zr–PILC) were studied by titrating the PILCS with four different solutions; 0.1N(NaCl+NaOH/HCl), 0.1N(KCl+KOH/HCl), 0.1N(CsCl+CsOH/HCl), and 0.1N(RbCl+RbOH/HCl). Results indicate that the ion exchange capacity of the original clay was increased due to the pillaring process.

## INTRODUCTION
A new kind of porous, high surface area material of potential interest as catalysts and adsorbents has been introduced by the preparation of pillared clays (PILCS).

Preparation of PILCS involves the exchange of the cations present in sites between the individuals layers of a clay with bulky, polymeric, inorganic cations such as $[Al_{13}O_4(OH)_{24}(H_2O)_{12}]^{7+}$ (ref.1) and $[Zr_4(OH)_{14}(H_2O)_{10}]^{2+}$ (ref.2). These precursor pillared clays are then calcined resulting in dehydration and dehydroxylation of the polymeric cation and producing hydroxy/oxide pillars between the silicate layers of the original clay. High surface area materials ($c.a$ 150–350 m$^2$/g) are formed in this way which exhibit basal spacings of about 18 A, mean pore diameters of 7–9 A, and have the ability to adsorb hydrocarbons and develop Bronsted and Lewis acidities (ref.3).

There has been considerable work aimed at characterising these materials including; pore–volume and surface area measurements (ref.4), neutron diffraction (ref.5), magic angle spinning N.M.R (ref.6) and I.R studies on the nature of acid sites (ref.7). The nature of the pillars and their relationship to the clay sheets, however, is still not very well understood.

Ancillary to this is the observation that, athough several publications refer to the exchange of catalytically active metal ions into PILCS, none clarify the ion exchange properties of the PILCS. Clearly a PILC may contain cation and anion exchange sites on the clay surface and/or on the pillars. A further complication arises due to the amphoteric nature of the pillaring materials *eg* "$Al_2O_3$" and "$ZrO_2$" as well as the presence of clay cation exchange capacity created by OH groups rather than at sites in the clay created by the presence of $Mg^{2+}$ and $Al^{3+}$ cations in interstitial positions in the clay layers.

This paper attempts to provide a clarification of these potential ion exchange sites in Zr and Al pillared smectites.

## MATERIALS AND METHODS

The clays used in this work were a natural Spanish bentonite, a Zr-PILC and a commercially available Al-PILC supplied by Laporte Inorganics, Widnes. The Zr pillared clay (Zr-PILC) was prepared using the Spanish natural bentonite treated with a fresh zirconyl chloride solution (0.1M) as a pillaring agent. The procedures used to prepare and characterize the Zr-PILC were described in a previous paper (ref.8). The Al-PILC came from the same natural Spanish bentonite.

Characterization of the Zr-PILC showed a basal spacing of 19.7 Å, a BET specific surface area of 134 m²/g and the presence of microporosity. The structural formula obtained from the chemical analysis of the three clays are shown in table 1.

TABLE 1
Structural formulae of the original Spanish bentonite, Zr-PILC and Al-PILC

| | |
|---|---|
| Clay | $[Si_{7.6}Al_{0.4}O_{20}(OH)_4Al_{3.02}Fe_{0.2}Mg_{0.96}]Ca_{0.33}Na_{0.07}K_{0.03}$ |
| Zr-PILC | $[Si_{7.6}Al_{0.4}O_{20}(OH)_4Al_{2.8}Fe_{0.2}Mg_{0.7}]Zr^*_{1.02}Ca_{0.01}Na_{0.06}K_{0.01}$ |
| Al-PILC | $[Si_{7.6}Al_{0.4}O_{20}(OH)_4Al_{3.02}Fe_{0.3}Mg_{0.94}]Al^*_{1.42}Ca_{0.02}Na_{0.06}K_{0.03}$ |

*Zr or Al as pillar

Titration curves are frequently used to study the cation and anion exchange properties of hydrous oxides. This technique was adopted to assess and compare the ion exchange properties of the original clay, Al–PILC and the Zr–PILC.

The pH titration curves were obtained by the method described by Amphlett *et al* (ref.9), using solutions containing varying proportions of NaOH and NaCl for measurements in alkaline solutions, and varying proportions of NaCl and HCl in acid solutions, all solutions were initially 0.1N with respect to their total Na$^+$ ion concentration.

Solutions (12.5 ml) containing 0.1N(NaCl+HCl) or (NaCl+NaOH) were added to a 0.25g sample of the PILC or clay. The mixtures were continuously stirred by means of a minerological roller for 3 days at room temperature. At equilibrium, the suspension pH of each sample was measured and the pH data of the entire batch plotted against meq of acid (or base) added. The process was repeated for three other salt–acid/base solutions (KCl, CsCl and RbCl). The pH of solution blanks were determined in the same way.

## RESULTS

The titration curves obtained are in Figure 1 and 2. Uptake of Na$^+$, K$^+$· Cs$^+$, Rb$^+$ and Cl$^-$ was calculated as appropriate from the differences in the concentrations of [OH–] or [H$^+$] between the blank and the sample suspension at the same amount of alkali or acid added to the sample. Graphs of the uptake of the different cations and chloride against pH then were constructed. An example of such a graph is shown in Fig.3.

Figure 4 shows the comparison of the maximum uptake of the different cations and anion (Na$^+$,K$^+$· Cs$^+$, Rb$^+$, Cl$^-$) for the clay and PILCS together with the pH at which these maxima were obtained.

The CEC's of the original clay and PILCS were calculated from their chemical analyses, based on the initial amount of exchangeable cations present (Na+, K+ and Ca++), giving a total CEC for the original clay of 96 meq/100g and a CEC of 4.81meq/100g for Zr–PILC and 10.1meq/100g for Al–PILC (*i.e* after pillaring).

The CEC's of the PILCS when expressed as a fraction of the total CEC in the original clay were 0.05 for Zr–PILC and 0.10 for Al–PILC.

The residual CEC in the PILCS for each cation were then calculated using the experimental maximum CEC of the clay for each cation (Fig.4) and the above mentioned fractions (0.05 and 0.10); the results are shown in the left hand side of Fig.5.

The CEC due to the pillars in the PILCS (shown in the right hand side of Fig.5.) were calculated by difference between the maximum CEC of the PILCS for each cation (Fig.4) and the corresponding residual CEC.

The percentage contribution to the maximum CEC due to pillars and residual from the clay are shown in Fig.6.

Table 2 shows the increase in cation (CEC) and anion (AEC) exchange capacities obtained by pillaring the clay. These percentages were based on the maximum cation and anion exchange capacities attained as shown in Fig.4.

Table 2
% Increase in cation and anion maximum exchange capacities due to
pillaring of the clay

| Sample | % Increase of CEC or AEC | | | | |
| --- | --- | --- | --- | --- | --- |
| | $Na+$ | $K+$ | $Cs+$ | $Rb+$ | $Cl-$ |
| Zr-PILC | 81 | 40 | 94 | 118 | 139 |
| Al-PILC | 37 | 22 | 43 | 51 | 161 |

## DISCUSSION

A large number of synthetic inorganic substances have been shown to exhibit ion-exchange properties and among these materials are the hydrous oxides (ref.10). Most hydrous oxides behave as 1) cation-exchangers in alkaline solutions and/or 2) anion-exchangers in acid solutions, and they exhibit characteristic titration curves (ref.11).

The titration curves of the clay and the two PILCS (Fig.1 and 2) resemble those of hydrous oxides indicating that cation and anion exchange occurs in the clay and PILCS.

By considering a surface hydroxyl group attached to the metal atom of the pillar (originated from the pillaring agent $[Al_{13}O_4(OH)_{24}(H_2O)_{12}]^{7+}$, $[Zr_4(OH)_{14}(H_2O)_{10}]^{2+}$) or on the clay surface, it is clear that ion exchange processes can take place (Fig.3), as in the case of hydrous oxides, in either of two ways, viz.

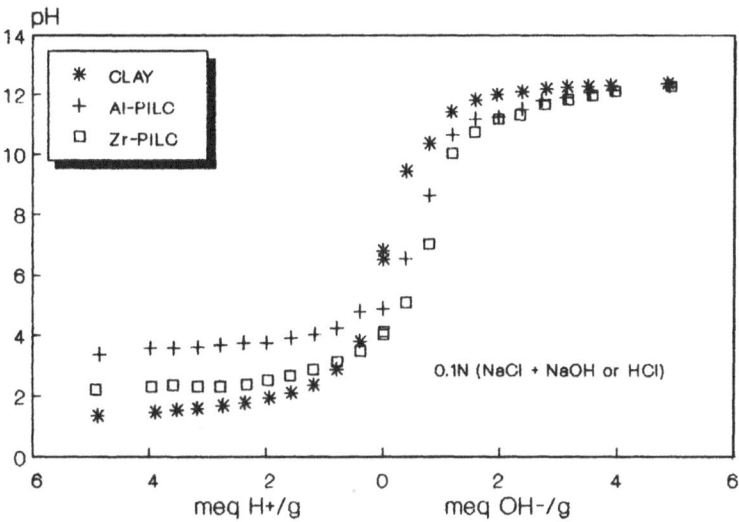

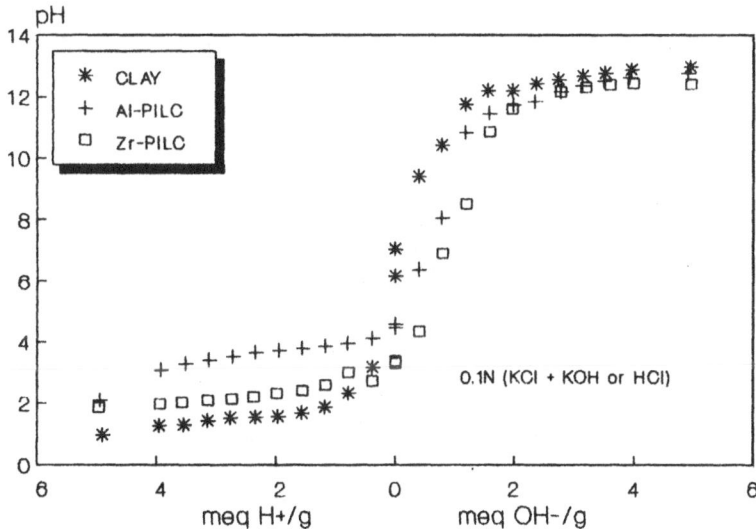

Figure 1. pH titration curves for the original clay, the Al–PILC and the Zr–PILC. Exchanger, 0.25g; total volume, 12.5 ml, ionic strength, 0.1N of Na and K solutions.

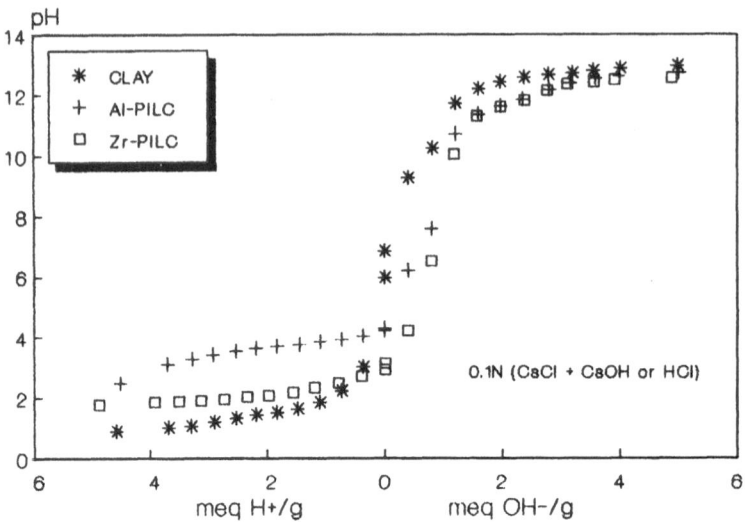

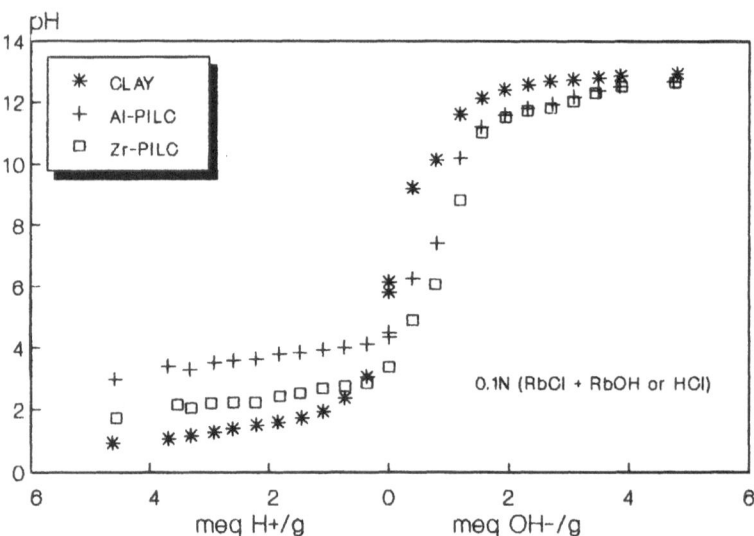

Figure 2. pH titration curves for original clay, Al–PILC and Zr–PILC. Exchanger, 0.25g; total volume, 12.5 ml, ionic strength, 0.1N of Cs and Rb solutions.

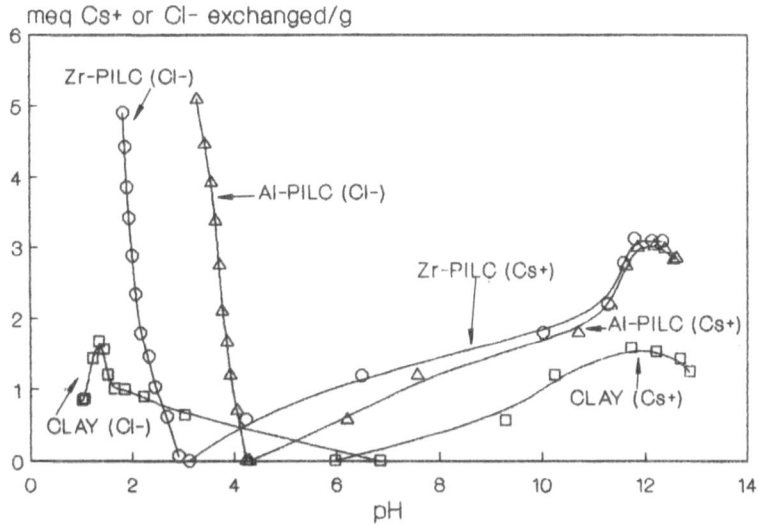

Figure 3. Uptakes of Cs+ and Cl− for the original clay, the Al–PILC and the Zr–PILC.

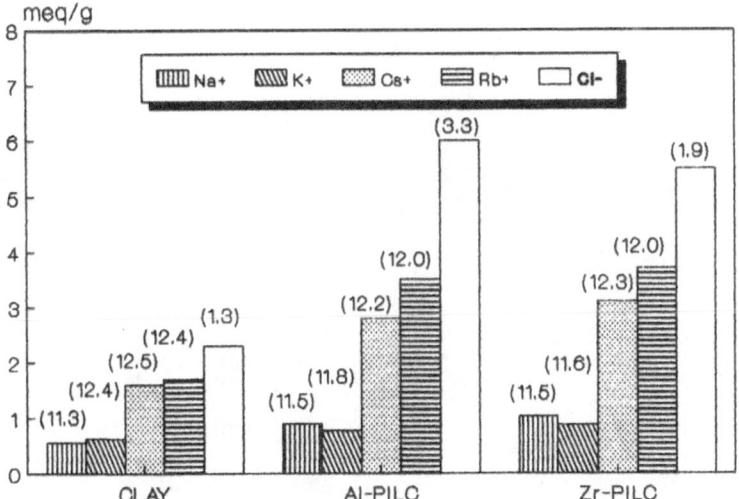

Figure 4. Comparison of the maximum cation and anion exchange capacity for the original clay, the Al–PILC and the Zr–PILC. Figures in brackets indicate the pH at which these maxima were obtained.

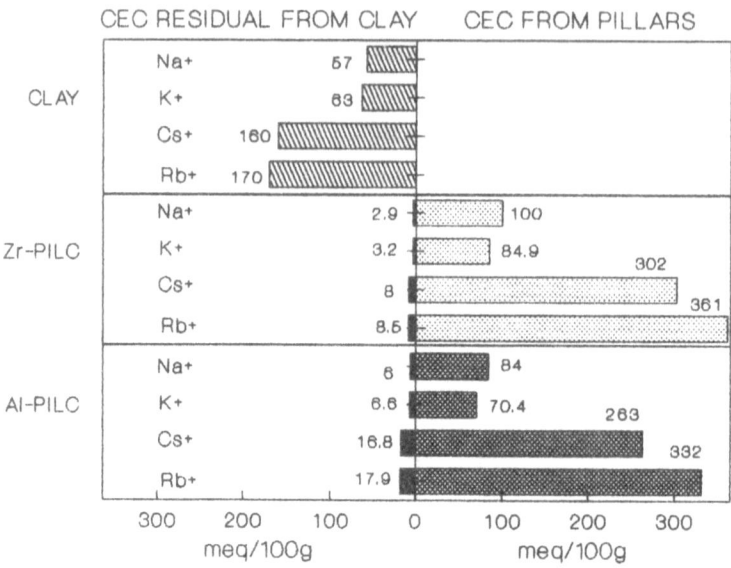

Figure 5. Cation exchange capacities from clay and pillars.

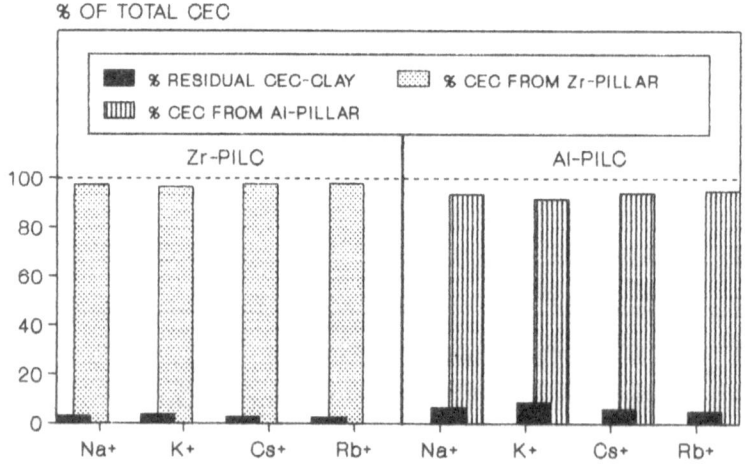

Figure 6. Percentage contribution to the total CEC of residual fraction from clay and pillars.

$$M - OH \leftrightarrow MO^- + H^+ \qquad (1)$$

$$M - OH \leftrightarrow M^+ + OH^- \qquad (2)$$

Case (1) favoured by low pH, would lead to anion exchange, while at high pH, case (2) would lead to cation exchange. Figure 1 and 2, also show that no sharp transition point from cation to anion exchange behaviour exists in the titration curves of both PILCS and clay, indicating that the exchange sites are likely to be heterogeneous, like those of $ZrO_2$ and hydrated $Al_2O_3$ (ref.11).

Comparing the maximum cation exchange capacity of the clay and PILCS (Fig.3 and 4) it is clear that both PILCS show higher exchange capacities than the original clay. Figure 5, shows that this increase in uptake of the cation/anion is due almost entirely to the introduction of OH groups originating from the intercalation of the Al-pillar or Zr-pillar between the silicate layer sheets of the starting clay. The residual CEC from the original clay, contributes only about 3% in the Zr-PILC and about 7% in the Al-PILC to their total exchange capacity (Fig.6).

According to Ruvarac (ref.12), the exchange capacity of hydrous zirconia is 1 to 2 meq/g of $ZrO_2$ which corresponds to 1 to 2 hydroxyl groups per eight zirconium atoms. If the maximum exchange capacity of Zr-PILC is taken as 6 meq/g of Zr-PILC (Fig.4), this value corresponds to 5 hydroxyl groups per zirconium atom in the PILC. These results shows an approx 20 to 40 fold increase in exchange capacity largely associated with Zr-pillar. A similar calculation shows that the Al-PILC has approx one hydroxyl group per Al atom but no analogous figure for hydrated alumina is available.

Behaviour as a cation or anion exchanger in metal hydrous oxides depends upon the basicity of the central metal atom (M-OH) and the strength of the metal-oxygen bond relative to that of the oxygen-hydrogen bond in the hydroxyl group (ref.13). In the case of PILCS, it seems that the same dependency occurs, since as it is shown in Fig.3 and Table 2., the Zr-PILC has a higher cation and lower anion exchange capacity than the Al-PILC at a given pH. This is due to the lower basicity of hydrous Zr oxide, as compared with the hydrated alumina.

From Fig.4 the ion selectivity for the clay is; Rb $\approx$ Cs > K > Na

which is parallel to the increase of the hydrated radii of the ions *i.e.*
$Cs^+$(3.29A), $Rb^+$(3.29 A), $K^+$(3.31A) and $Na^+$(3.58A) (ref.14).

In the case of both Al–PILC and Zr–PILC, the ion selectivity was:
Rb>Cs>>Na ≈ K

## CONCLUSIONS

Al–PILC and Zr–PILC exhibited both cation and anion exchange properties
depending on the pH of the exchanging solution, in a similar way as the
hydrous metal oxides. The ion exchange capacity of the clay is increased
due to the introduction of OH groups by the pillarization of the clay.
About 93–97% of the total exchange capacity was due to the pillars.
Pillaring of the clay increased the cation exchange capacity by 25–120%/g
depending on the exchanging cation.

## ACKNOWLEDGMENTS

One of the authors (T.Gallardo) would like to thank the National Council
of Science and Technology of Mexico (CONACYT) and the IPN (E.N.C.B) for
their support. Laporte Inorganics, Widnes are thanked for the gift of
materials.

## REFERENCES

1. D.E.W. Vaughan and R.J. Lussier. Proc. 5th. Int. Conf. Zeolites,
   L.V.C. Rees (Editor), Butterworths, London (1980) p.94.
2. S. Yamanaka and G.W. Brindley, *Clay and Clay Minerals*, 27 (1979) 119.
3. H. Ming-Yuan. L. Zhonghui and M. Enze, in Catalysis Today, vol.2. R.
   Burch (Editor), Elsevier, Amsterdam, (1988) p.321.
4. D.E.W. Vaughan, R.J. Lussier and J.S. Magee, U.S. Patent 4,176,090
   (1979).
5. T.J. Pinnavaia, V. Rainey, Ming-Shin Tzou and J.W. White. *J. Mol.
   Catal.*, 27 (1984) 213.
6. D. Plee, F. Borg, L. Gatineau and J.J. Fripiat, *J. Am. Chem. Soc.*,
   107 (1985) 2262.
7. M.L. Occelli and J.E. Lester, *Ind. Eng. Chem. Pro. Res. Dev.* 24
   (1985) 27.
8. A. Dyer, T. Gallardo and C.W. Roberts, in Zeolites: Facts, Figures,
   Future, Part 49-A, P.A. Jacobs and R.A van Santen (Editors),
   Elsevier, Amsterdam, (1989) p.389.
9. C.B. Amphlett, L.A. McDonald and M.J. Redman, *J. Inorg. Nucl. Chem.*,
   6 (1958) 236.
10. V. Vesely and V. Pekarek, *Talanta*, 19 (1972) 219.
11. M. Abe in Inorganic Ion Exchange Materials, A. Clearfield (Editor).
    C.R.C. Press Inc., Florida, (1981) p.161.
12. A. Ruvarac in Inorganic Ion Exchange Materials, A. Clearfield
    (Editor), C.R.C. Press Inc., Florida, (1981) p.145.
13. C.B. Amphlett. Inorganic Ion Exchangers. Elsevier, Amsterdam, (1964).
14. R.G. Gast, in Minerals in Soil Environments, J.B. Dixon (Editor)
    Soil Science Soc. Am., Wisconsin USA., (1977) p.44.

# ION EXCHANGE PROPERTIES OF α-ZIRCONIUM PHOSPHATE MODIFIED BY A PROCESS OF INTERCALATION/DE-INTERCALATION WITH MONOAMINES

G. Alberti, U. Costantino, F. Marmottini, R. Vivani and C. Valentini*
Department of Chemistry, University of Perugia, Via Elce di Sotto, 8,
06100 Perugia, Italy.*
* ENIRICERCHE S.p.A., Via Ramarini, 32, 00016 Monterotondo (Rome)
  Italy.

## ABSTRACT

Colloidal dispersions of delaminated zirconium phosphate, obtained as a result of intercalation of some amines, have been sonicated and then put under a spray-drying or freeze-drying treatment. Highly dispersed and hydrated powders of zirconium phosphate have been obtained. The water content of these new phases ranges from 1.2 to 3.5 mol/mol of Zr, as the relative humidity rises from 5 to 90%, correspondingly the interlayer distance changes from 8.0 to 10.5 A. The ion exchange properties of polyhydrated zirconium phosphates towards $Na^+$, $K^+$, $Cs^+$ and $Ba^{2+}$ cations have been investigated and compared with those of $\alpha\text{-Zr(HPO}_4)_2 \cdot H_2O$. Owing to the large interlayer distance, these polyhydrated phases exhibit good ion exchange and intercalation properties expecially towards large cationic or neutral species.

## INTRODUCTION

Alpha-zirconium phosphate, $\alpha\text{-Zr(HPO}_4)_2 \cdot H_2O$ (hereafter called ZrP) is a layered compound that behaves as a cation exchanger. The proton of the phosphate groups can be replaced by an equivalent amount of other counterions and the ion-exchange capacity (IEC) is 6.64 meq/g (1-3). Despite this high IEC and the good thermal and chemical stability, ZrP has been considered a poor exchanger since the free diffusion of counterions in the interlayer spaces is limited by the strong layer-layer interactions and the small passageways ( 2.6 Å) that interconnect the interlayer cavities. Accordingly $Li^+$, $Na^+$, $K^+$ are exchanged in acidic medium to form half exchanged phases of formula $ZrMH(PO_4)_2 \cdot nH_2O$, while $Rb^+$ and $Cs^+$ need a neutral or alkaline medium to be exchanged. It is thought that $OH^-$ tend to remove some protons from the exchanger and these large cations are forced to diffuse in the interlayer region with a concomitant increase of the interlayer distance. It has however been found that cations having a large size or being strongly hydrated are easily exchanged, even with high rates and good selectivity, if modified zirconium phosphates (i.e., $Zr(HPO_4)_2 \cdot 5H_2O$ or $ZrHNa(PO_4)_2 \cdot 5H_2O$) with an interlayer distance larger than that of the original ZrP, i.e. 7.6 Å, are used (4). Good precursors for the exchange and/or intercalation of large species have been found to be also intercalation compounds with ethanol (5) or butylamine (6).

Even higher rates of exchange and no steric hindrance to the diffusion of the counterions has to be expected if the ZrP microcrystals, used as exchangers, were completely exfoliated in single layers, so that all the ionogenic groups result in being surface groups. In previous work (7,8) performed in this laboratory, it has been shown that it is possible to exfoliate ZrP crystals by suitable intercalation of some amines in the interlayer region. Very stable colloidal dispersions containing single layers (or packets of few layers) of ZrP were obtained. It was thought that if the solvent were suddenly removed (i.e. by spray-drying or freeze drying), in order to maintain in the solid state the situation present in solution, materials with good ion exchange properties could be obtained.

This paper reports the preparation and characterization of zirconium phosphates obtained by a procedure of sonication-lyophilization of colloidal dispersions of ZrP and an account of their ion exchange and intercalation properties.

## MATERIALS AND METHODS

The following samples of ZrP having different degrees of crystallinity and average crystal size, were prepared:

(1) ZrP, obtained by refluxing amorphous zirconium phosphate in 12 M $H_3PO_4$ solution for 150 hrs (9). The microcrystals have an average size of 5-10 $\mu$m.

(2) ZrP prepared according to the HF procedure (10). The microcrystals have a crystal size 20-40 $\mu$m.

(3) ZrP obtained by decomposing the zirconium fluorocomplexes in a solution of phosphoric acid, by means of a slow and gradual increase of temperature (11). The dimensions of the crystal range between 0.1 - 3 mm. The colloidal dispersion was obtained by titrating the ZrP microcrystals, suspended in water, with n-propylamine up to 50% of intercalation (i.e. one propylamine every two phosphate groups) as described in (7). The mass/volume ratio at the end of titration was generally 1 g of ZrP per 200 ml of water. Sonication was performed at 7 MHz and 150 W. Fast removal of solvent was achieved by means of spray-drying at 120°C or lyophilization at -20°C. Titration curves were recorded with a Mettler DK automatic titrimeter operating by "equilibrium point" method. Alkaline and alkaline earth metal ions were determined by A.A.S., with a Perkin Elmer 305 apparatus. The X-ray diffraction (XRD) patterns were taken with a computer controlled Philips diffractometer using the Ni-filtered $CuK_\alpha$ radiation. The XRD patterns at high temperature were taken with a A. Paar HTK attachment. The TGA-DTA was performed with a Stanton STA781 Thermoanalyzer.

## RESULTS AND DISCUSSION

**Preparation and characterization of highly dispersed and hydrated zirconium phosphate powders.**

The strategy used to obtain exfoliated zirconium phosphates consists of the following steps:

1) Intercalation of n-propylamine in ZrP in order to obtain a colloidal dispersion of $Zr(HPO_4)_2 \cdot C_3H_7NH_2 \cdot nH_2O$ lamellae.

2) Treatment of the colloidal dispersion with ultra sounds in order to facilitate the breaking of aggregates of lamellae eventually present. The effect of sonication was followed by measuring the absorbance of the dispersion at 620 nm, until a constant value of 0.1-0.07 was obtained.

3) Regeneration of the lamellae with 0.1 M HCl solution in order to obtain the hydrogen form of the exchanger, followed by a washing with distilled water to eliminate the propylammonium ions and the excess of acidity.

4) Treatment of the dispersion of ZrP lamellae in hydrogen form with spray-drying at 120° C or with liquid nitrogen followed by a lyophilization process.

It is known that, under suitable conditions (room temperature and pH 4) only surface protons of ZrP are replaced by $Cs^+$ or other large cations such as methylene blue (BM) cation and these two species have been used as probes to determine the surface ion exchange capacity of the microcrystals (12). A preliminary characterization of the highly dispersed and voluminous powders obtained by the sonication-lyophilization (or spray drying) process has been made by determining the amount of $Cs^+$ and/or methylene blue cation taken up per g of product and Table 1 reports some of the data obtained together with the values referred to the original microcrystals. It may be seen that the treatment greatly modifies the ion-exchange properties of ZrP microcrystals since uptake values 300-500 times higher than those of the original microcrystals are obtained as if the steric hindrance to the diffusion of such large cations were for the most part, removed. Note that the ratio between $Cs^+$-uptake and BM-uptake also changes from 1.5-1.4 to 1.3-1.1.

Table 1

Methylene Blue and $Cs^+$ uptake of zirconium phosphates obtained as a result of sonication and spray-drying or freeze-drying treatment of colloidal dispersion. Data of the original ZrP are also reported.

| Preparative method of the microcrystals [a] | Treatment [b] | Uptake ($\mu$mol/g) | |
|---|---|---|---|
| | | $Cs^+$ | MB |
| HF | - | 5.1 | 3.5 |
| R | FD | 980 | 770 |
| R | SD | 1120 | 970 |
| HF | SD | 980 | 640 |
| HF | FD | 1250 | 1150 |
| HF | FD | 1250 | 1100 |
| HF | FD | 1050 | 900 |
| HF-LC | SD | 1260 | 1150 |
| HF-LC | FD | 1180 | 854 |

a) The marks mean: R = refluxing procedure; HF = hydrofluoric acid procedure; HF-LC = large crystals obtained from HF procedure.

b) FD = freeze-drying; SD = spray-drying.

The spray-drying treatment gives rise to products having a "surface" ion exchange capacity relatively lower than that obtained with those freeze-dried. Furthermore, the ZrP samples obtained with the HF procedures are more effective in obtaining good colloidal dispersion (7) and hence lyophilyzed products. Further characterization has thus been made with a product obtained from ZrP microcrystals prepared according the HF procedure (sample (2)) and having a crystal size of 20-40 $\mu$m, sonicated for 14 min (seven periods, each of 120 s , followed by a pause of 60 s to cool the dispersion in water) and lyophilyzed.

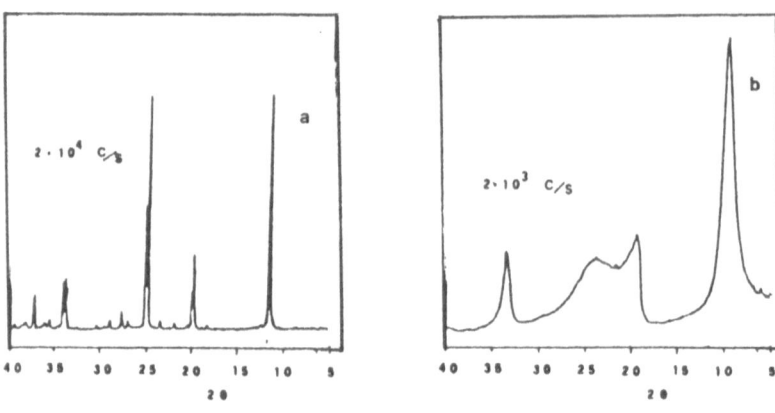

Figure 1 – X-ray diffraction patterns of α-Zr(HPO$_4$)$_2$ .H$_2$O (a)
and of the polyhydrated zirconium phosphate (b).

Fig. 1 reports the XRD patterns of the sample at room temperature in comparison with that of the original ZrP. The peaks are few and broad but the essential features of the α-phase are maintained.

The interlayer distance is much larger (10.4 A) and it has been found to depend on the water content that in turn depends on the relative humidity. As it can be seen from Table 2, for the relative humidity increasing from 5 to 90%, at r.t. the hydration water ranges from 1.2 to 3.5 mol/mol of Zr, correspondingly the interlayer distance increases from 8.0 to 10.5 A. It may be noted that the product, previously dried in the oven at 150°C, quickly takes up water from the air to form the polyhydrated phases. The interlayer distance of the anhydrous product has been thus determined with the aid of the high temperature X-ray attachment. It was found that, at 150°C, the interlayer distance is 7.7 A instead of 7.4 A, typical of the original ZrP microcrystals. This value remains essentially the same up to 320°C, and the reversible phase transition at 220°C, observed in the original microcrystal (interlayer distance changes from 7.4 to 6.8 A), is missed. At 650°C the d value is

Table 2

Hydration water and interlayer distance of polyhydrated zirconium phosphate at room temperature and different relative humidities.

| % relative humidity | Hydration water mol/ mol Zr | Interlayer distance d(A) |
|:---:|:---:|:---:|
| 92 | 3.5 | 10.5 |
| 75 | 3.0 | 10.0 |
| 59 | 2.7 | 9.8 |
| 25 | 1.5 | 8.2 |
| 5 | 1.2 | 8.0 |

typical of the layered $ZrP_2O_7$ while the transition to cubic pyrophosphate is observed at temperatures higher than 900°C.

**Ion exchange and intercalation properties of polyhydrated zirconium phosphates.**

The polyhydrated phases of ZrP, obtained from the colloidal dispersion, exhibit good ion exchange properties. The diffusion of the counterions is facilitated because of the large interlayer distance and the presence of weakly bonded water molecules; furthermore it was found that the products, dispersed in water, undergo a progressive ulterior swelling and the exchange should occur for the most part at the hydrated surfaces. Figures 2a and 2b report the potentiometric titration curves of the polyhydrated and original ZrP respectively, with the indicated alkaline and alkaline earth metal hydroxides. Several aspects are noteworthy of comments. In contrast to the behaviour of the ZrP microcrystals, where the pH remains constant with the ion uptake and

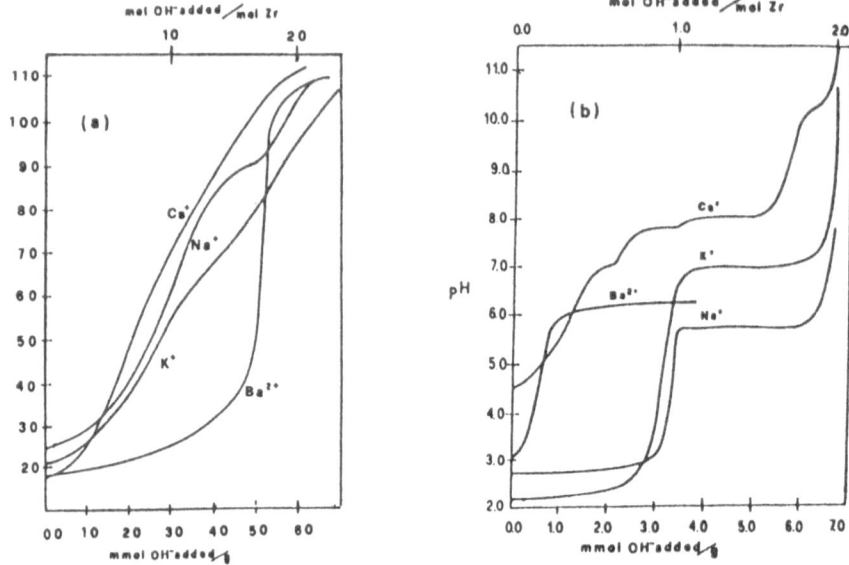

Figure 2 – Potentiometric titration curves of polyhydrated zirconium
phosphate (a) and of $\alpha-Zr(HPO_4)_2 \cdot H_2O$) (b) with the indicated
cations. Conditions: 1 g of product dispersed in 80 ml of
0.100 N metal chloride solutions and titrated with 0.100 N
solutions of the corresponding hydroxides, at room temperature.

formation of intermediate phases, in the case of polyhydrated ZrP, each
increment of metal ion uptake results in a increase in pH and no
definite plateaux as well as sharp end points are observed.

Large cations such as $Cs^+$ and $Ba^{2+}$, exchanged at relatively high pH
values or not exchanged in ZrP (see Fig. 2b), are taken up in acidic
medium and with good selectivity in the polyhydrated phase. At low
uptake the selectivity order is $Ba^{2+} \geq Cs^+ > K^+ > Na^+$; reversals in
selectivity ($Cs^+ - K^+$; $Na^+ - K^+$) are observed at higher uptake. The
calculated ion exchange capacity of polyhydrated zirconium phosphate
obviously depends on the drying conditions of the product; at 75% r.h.,
the IEC results 6.0 mmol/g. The experimental value, as derived from the
titration curves of Fig. 2a, is near to the calculated one, also taking

into account    the loss of some phosphate groups ( 10% of the IEC) as a result of the sonication treatment. The shape of the titration curves seems to indicate that the incoming cations are uniformly distributed within the interlayer space to form solid solutions (as it occurs in amorphous gels) rather than being regularly accommodated to give rise to new crystalline phases.

X-ray diffraction patterns of wet samples with increasing $Na^+$ content do not reveal phase transitions, the interlayer distance being 10.4 - 10.8 A in all the composition ranges.    Nevertheless if the half and fully exchanged samples are dehydrated at 180°C, the anhydrous phases $ZrHNa(PO_4)_2$ (d = 7.6 A) and $Zr(NaPO_4)_2$ (d = 8.4 A) are obtained.

Zirconium phosphate has received much attention for the separation of radioactive ions because of its resistance    to ionizing radiation, oxidizing media and strong acid solution (13). As a preliminary investigation of the possible use of the polyhydrated phase in the field of cation separation, the $Na^+/Cs^+$ isotherm performed at r.t., 0.100 M cations concentration and pH = 5 has been determined, using only half of the IEC (see Fig. 3).As expected the polyhydrated phase shows a marked

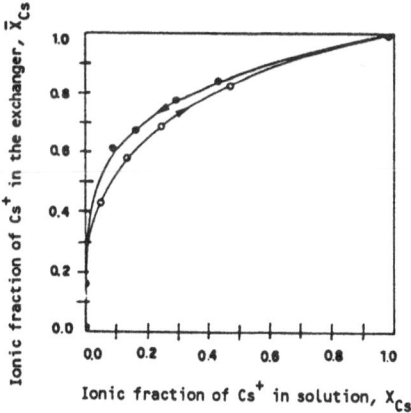

Figure 3 – Forwards and reverse $Na^+/Cs^+$ ion exchange
isotherms of polyhydrated zirconium phosphate.
Concentration: 0.100 M, temperature 20°C.

preference for the $Cs^+$ ion, expecially at low loading, and the hysteresis loop is very small.The polyhydrated phases are also good intercalating agents of polar molecules. Alkanols and glycols have been directly intercalated by dipping the host in the pure liquid, while alkylamines and larger species such as bornylamine and DABCO have been intercalated from diluted aqueous solutions.

## REFERENCES

1. A. Clearfield (Ed.), Inorganic Ion Exchange Materials, CRC Press, FL U.S.A., 1982 see Chapters 1-3.

2. G. Alberti, in Recent Developments in Ion Exchange (P.A. Williams and M.J. Hudson, Eds.) Elsevier Applied Science, London, 1987, pp. 233-48

3. A. Clearfield, Chem. Rev., 88, 125 (1988).

4. G. Alberti, Acc. Chem. Res., 11, 163 (1978).

5. U. Costantino, J. Chem. Soc. Dalton Trans., 402 (1979).

6. G.Z.-Peng and A. Clearfield, J. Incl. Phenom., 6, 49 (1988).

7. G. Alberti, M. Casciola and U. Costantino, J. Colloid Inter. Sci., 107, 256 (1985).

8. G. Alberti, M. Casciola, U. Costantino and F. Di Gregorio, Solid State Ionics, 32/33, 40 (1989).

9. A. Clearfield and J.A. Stynes, J. Inorg. Nucl. Chem., 26, 117 (1964).

10. G. Alberti and E. Torracca, J. Inorg. Nucl. Chem., 30, 317 (1968).

11. G. Alberti, U. Costantino and R. Giulietti, J. Inorg. Nucl. Chem., 42, 1062 (1980).

12. G. Alberti, M.G. Bernasconi, M. Casciola and U. Costantino, Ann. Chim. (Rome), 68, 265 (1978).

13. I.A.E.A. - Inorganic Ion Exchangers and Adsorbents for Chemical Processing in the Nuclear Fuel Cycle, Tecdoc-337, Vienna, 1985.

# NEW CROSS-LINKED LAYERED TIN PHOSPHATE EXCHANGERS.

Pedro Maireles-Torres, Pascual Olivera-Pastor,
Enrique Rodriguez-Castellon and Antonio Jimenez-Lopez
Departamento de Quimica Inorganica, Universidad de Malaga,
Apartado 59, 29080 Malaga, Spain.

and

Anthony A. G. Tomlinson,
I.T.S.E., Area della Ricerca di Roma, C.N.R.,
C.P. 10, Monterotondo Stazione, 00016 Roma, Italy.

## ABSTRACT

A new class of composite inorganic exchangers has been prepared by the intercalation and calcination of highly polymerised $Cr^{3+}$ polyhydroxide species into $\alpha$-tin phosphate. The materials have free heights of 6.7 and 7.6 Å and surface areas of 198 and 219 $m^2$ $g^{-1}$. There is some saturation of potential exchange sites by the chromia pillars. Nevertheless, cation exchange capacities for $Co^{2+}$, $Ni^{2+}$, $Cu^{2+}$, and $Pr^{3+}$ are still relatively high. They are promising as selective ion exchangers.

## INTRODUCTION.

Over the last decade there has been much interest in preparing solids based on layered materials in which the layers are propped apart, 'pillared', by oxide moieties so as to induce the formation of pores. To date they have been prepared by calcination of robust intercalated polyhydroxy cations in smectite clays [1]. Unlike rigid, zeolite molecular sieves, the pore structure can be controlled via :

i) changing the c.e.c. of the initial layered material;
ii) changing the shape and charge of the 'pillaring' precursor, and
iii) changing the amount of precursor utilised in initial intercalation[2].

The more ordered basal structures and higher c.e.c. of layered phosphates are expected to provide pillared porous materials in which the pore size is more strictly controlled than in clays. Recently, the stable colloid formed between

$Me_4N^+$ and $\alpha$-Sn$(HPO_4)_2H_2O$ has been found useful for inserting aluminium polyhydroxides, with subsequent formation of oxide pillars[3]. We now describe a strategy for preparing analogous chromia-pillared $\alpha$-tin phosphates the pore structure of which can also be manipulated to give desirable ion-exchange properties.

## EXPERIMENTAL.

A 1% aqueous suspension of $\alpha$-Sn$[NMe_4]_{0.9-1.0}H_{1.1-1.0}(PO_4)_2nH_2O$, prepared as reported previously[3], was contacted with $Cr(CH_3CO_2)_3$ solutions with ratios ranging from 5 to 180 meq/g starting $\alpha$-Sn$(HPO_4)_2H_2O$. The most crystalline products were obtained after consecutive contacting at 25°C (4-7 days), then 60°C (2-4 days) followed by refluxing (1-2 days) and with the ratios :

$$20 \text{ mmol } Cr^{3+}/g \text{ } \alpha\text{-Sn}(HPO_4)_2H_2O \quad (n = \text{ I})$$
$$\text{and } 33.2 \text{ mmol } Cr^{3+}/g \text{ } \alpha\text{-Sn}(HPO_4)_2H_2O \quad (n = \text{ II}).$$

After calcination, ion exchange of $Co^{2+}$, $Ni^{2+}$, $Cu^{2+}$ and $Pr^{3+}$ into the two series of cross-linked products obtained was followed using routine batch methods[3].

## Physical Methods.

Solids obtained during manipulations were monitored by XRD with a Siemens D501 diffractometer (graphite monochromator, Ni-filtered Cu-$K_\alpha$ radiation on powders and cast films. TG and DTA were measured on a Rigaku Thermoflex and $Cr^{3+}$ was analysed colorimetrically by the chromate method. Metal ion uptakes were followed spectro-photometrically and adsorption-desorption of $N_2$ at 77 K using a Carlo Erba sorptiometer.

## RESULTS AND DISCUSSION.

Unlike the situation with smectite clays, crystalline reproducibly preparable intercalated precursors are not obtained by simply contacting polyhydroxyCr$^{3+}$ (prepared as in ref 2) with $Me_4N$-SnP. By contrast, using the more forcing conditions described above always led to intercalated materials with expanded layers. From the uptake curve of Fig.1, it is clear that after initial sharp uptake of $Cr^{3+}$ species up to 30 mmol $Cr^{3+}$ added, there is a slow decrease in uptake. This behaviour is typical of association of a species in solution, following Giles et al's classification[4]. The solids obtained at the various points of the curve have high interlayer distances : d $_{002}$ = 31.3 Å (point 2), 25.6 Å (point 3) and 23.5 Å (point 4). Further, calcination of each showed that higher $[Cr(OAc)_3]$ : $[Me_4N$-SnP] ratios give materials with slightly lower interlayer distances and chromium contents.

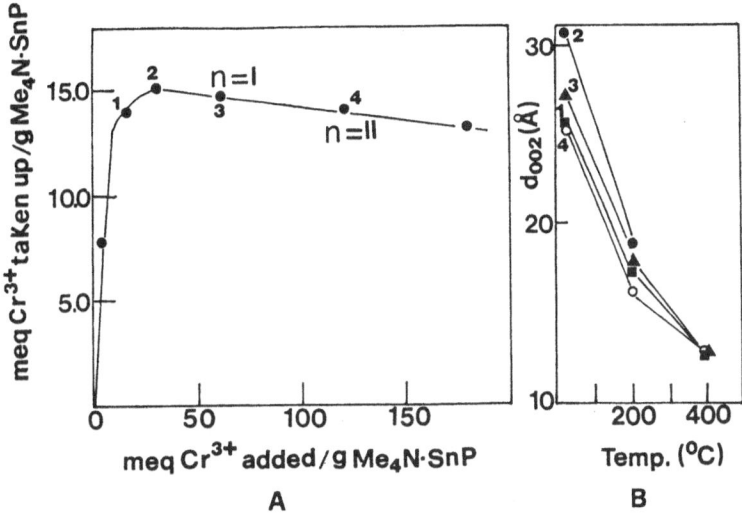

Fig. 1.- A. Cr$^{3+}$ uptake by "Me$_4$N-SnP" (r.t., 1 day, 60°C, 3 days/reflux, 1 day). B. Thermal behaviour of basal spacings in XRD of materials along curve A.

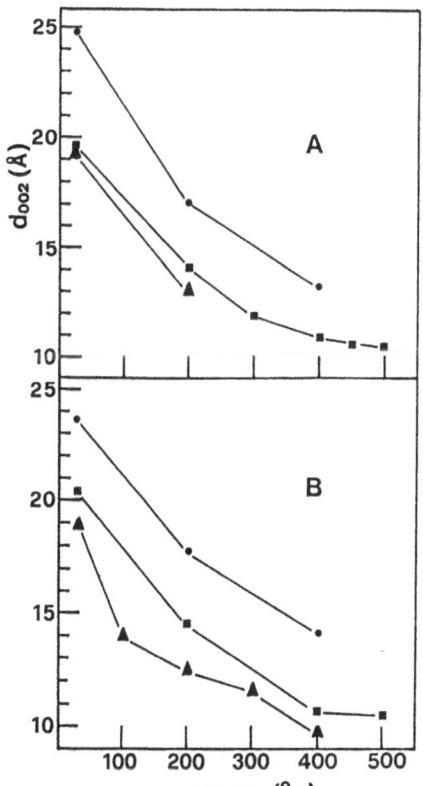

FIG.2 - Temperature dependence of basal spacings in XRD of polyhydroxo Cr$^{3+}$ intercalated into "Me$_4$N-SnP". A. Products prepared from hydrolysed precursor solutions with n = II; B. with n = I; at the contact temperature shown.

Figure 2 illustrates the temperature dependence of the XRD for materials prepared with n = I and n = II. Figure 3 gives an example of XRD showing the good crystallinity ($d_{001}$ progression to 3rd or 4th member). In both cases the refluxing method gives the highest free heights of the pillared final oxide at 400°C:

$$n = I : 13.9 - 6.3 \text{ Å} = 7.6 \text{ Å}$$
$$n = II : 13.0 - 6.3 \text{ Å} = 6.7 \text{ Å}$$

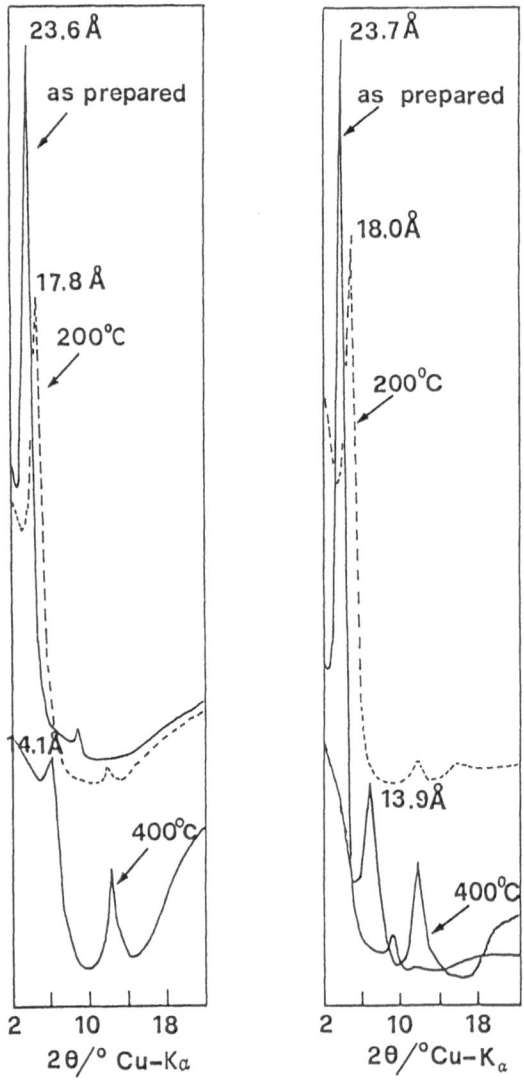

Fig. 3.

XRD of $Cr(OAc)_3$ + $Me_4N-SnP$ (n = I) film at r.t. and after calcination. A. Calcined in $N_2$ atmosphere; B. Calcined in air.

Important as regards ion exchange properties is the fact that there is no characteristic weight loss and exothermic peak at 400°C expected for the $2PO_4^{3-} \longrightarrow P_2O_7^{2-}$ condensation. This presumably reflects the fact that formation of the oxide pillar leaves interlayer -P-OH groups too far from one another, preventing condensation. Apart from intercalation, and interlayer distance expansion and its retention after calcination, evidence for pillaring also comes from adsorption-desorption measurements. Both chromia-pillared materials show relatively high specific surface areas: I 198, II 219 $m^2g^{-1}$. These compare with values of 300 - 400 $m^2g^{-1}$ found in chromia-pillared montmorillonite[5].

The polymerisation of $Cr^{3+}$ species in solution is believed to give rise to higher oligomers, although their nature is still controversial[6]. However, the high interlayer distances and the i.r. spectra (presence of OAc ions) show that the intercalated precursors do not contain any of the known simple polyhydroxy species (e.g. $(Cr_4(OH)_7^{5+})$[7]. The electronic spectrum of polyhydroxo-$CrMe_4NSnP$ (preparation with n = I) is shown in Figure 4. The d-d band region is very similar to that in $Cr(OAc)_3$. On calcination (and thorough washing), however, the electronic spectrum is more complex and suggests the presence of both $Cr^{3+}$ and $Cr^{5+}$. That the combustion of the polyhydroxyacetato$Cr^{3+}$ to chromia$^{3+,5+}$ pillar goes through complex redox steps is also demonstrated by the fact that calcination leads to production of $Cr_2O_7^{2-}$.

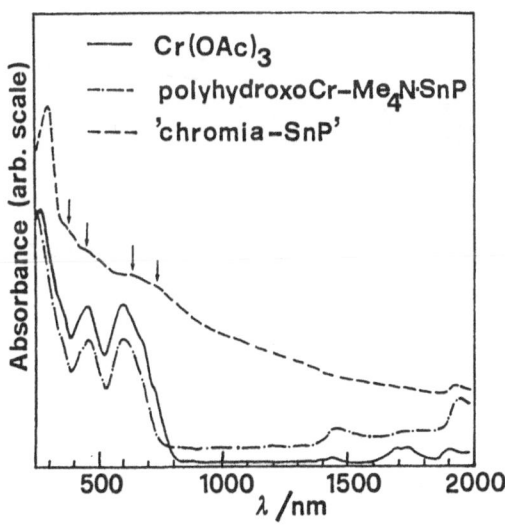

Fig. 4.- Diffuse reflectance spectra of $Cr(OAc)_3$, PolyhydroxoCr-$Me_4N$-SnP and pillared "Chromia-SnP".

Figure 5 shows the ion-exchange curves for $Co^{2+}$, $Ni^{2+}$, $Cu^{2+}$, and $Pr^{3+}$, from acetate solutions, for the chromia-pillared n = II material. The first point of note is that oxide pillaring has not entirely destroyed the c.e.c. of $\alpha$-$Sn(HPO_4)H_2O$, whereas in alumina-pillared clays (e.g. $Na^+$-exchanged gelwite and hectorite[8]) the migration of protons into the intralayer region makes the materials inaccessible to ion exchange. However, the c.e.cs are much lower than those for starting $\alpha$-$Sn(HPO_4)H_2O$ (6.08 meq $g^{-1}$, for $Co^{2+}$, $Ni^{2+}$, and $Cu^{2+}$)[9], a result of the saturation of potential exchange sites by the chromia pillars. Further, the retention ratios are in the order : $Cu^{2+}$ > $Co^{2+}$ > $Ni^{2+}$, following the Irving-Williams order. This is as expected for diffusion of the hydrates. Finally, the material also exchanges $Pr^{3+}$.

Further work is under way on the separation of metal ions by both the n = I and n = II materials.

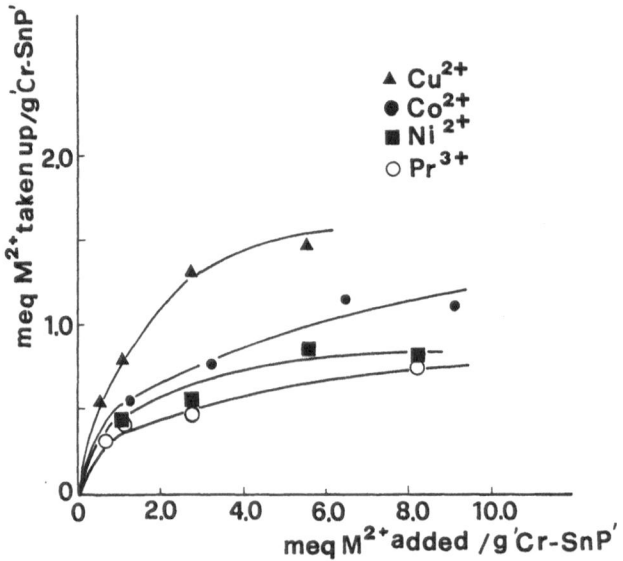

Fig. 5. Cation uptake by pillared "chromia-SnP" at r.t.

## CONCLUSIONS.

Thermally stable chromia-pillared $\alpha$-$Sn(HPO_4)_2$ materials has been prepared, using $Cr(OAc)_3$ as the precursor, colloidal $\alpha$-$Sn[NMe_4]_{0.9-1.0}H_{1.1-1.0}(PO_4)_2 \cdot nH_2O$ as substrate and refluxing methods. The materials are porous and have free heights similar to the industrially important zeolites ( 8-,10-and 12 membered rings, with effective diameters of ca 5,6 and 7 Å respectively). They retain part of the exchange capacities for transition metal ions and rare earth ions.

## ACKNOWLEDGEMENTS

We thank the E.E.C. Twinning Programme (contract No. ST2J-0383-C(A)) and C.I.C.Y.T. (Contract No. PB86/244) for financial support.

## REFERENCES.

1. D.E.W. Vaughan, in 'Catalysis Today', Ed. R. Burch, Elsevier, 1988, p.187.

2. A. Clearfield, in 'Surface Organometallic Chemistry: Molecular Approaches to Surface Catalysis', Kluwer, 1988, p.271.

3. P. Mairele-Torres, P. Olivera-Pastor, E. Rodriguez-Castellon, A.Jimenez-Lopez, L.Alagna and A.A.G. Tomlinson, Chemistry of Materials, submitted for publication. Ibid; J. Chem. Soc., Chem Commun., 1989, 751.

4. C.H. Giles, T.H. MacEwan, S.N. Nakhwa and D. Smith, J. Chem. Soc., 1960, 3973.

5. M.S. Tzou and T.J. Pinnavaia in 'Catalysis Today' R. Burch. Elsevier, 1988, p. 243.

6. L. Spiccia, Inorg. Chem., 1988,27, 432.

7. L. Monsted, O. Monsted and J. Springborg, Inorg. Chem., 1985, 24, 3496.

8. D.T.B. Tennakoon, W. Jones and J.M. Thomas, J. Chem. Soc., Faraday Trans., 1986, 308.

9. E. Rodriguez-Castellon, A. Rodriguez-Garcia and S. Bruque, Mat. Res. Bull., 1985, 20, 115.

# FORMATION OF DIAMINE-COMPLEX PILLARS IN $\tau$-Ti(HPO$_4$)$_2$2H$_2$O

By

C. Ferragina[1], A.La Ginestra[2], M.A.Massucci[3*],
P.Cafarelli, P.Patrono[1], and A.A.G.Tomlinson[4*].

[1]. IMAI /[4]ITSE, Rome Research Area, C.N.R., C.P.10
Monterotondo Staz., 00016 Rome, Italy
[2] Dipartimento di Chimica, 'La Sapienza' University, Rome
[3] Dipartimento di Chimica Inorganica, University of Cagliari,
Cagliari. Italy

## INTRODUCTION.

Our interest in the properties of layered phosphates in which the layers are pillared by in-situ formed complexes has recently extended to the $\tau$-series[1]. $\alpha$-Zr(HPO$_4$)$_2$· H$_2$O was found to give well-ordered composites with 2,2'-bipyridyl, 1,10-phenanthroline and 2,9-dimethyl-1,10-phenanthroline. In turn, these exchange Cu$^{2+}$, although only the first two provide evidence for subsequent coordination to form complex-pillars. This behaviour, which is different from that in the $\alpha$-series[2], induced us to investigate $\tau$-Ti(HPO$_4$)$_2$2H$_2$O as a host. It also provides ordered composites with these ligands, showing interesting ion-exchange properties.

## EXPERIMENTAL.

Crystalline $\tau$-Ti(HPO$_4$)$_2$2H$_2$O (d$_{002}$ = 11.62 A) was prepared by treating the amorphous product under hydrothermal conditions at 300° C for 24h. The amines were commercial products, and were used as received.

Analyses and physical measurements (atomic absorption spectrophotometry, XRD, electronic absorption)[2] were obtained as before.

### Intercalation of Diamine.

$\tau$-Titanium phosphate was first pre-swelled with EtOH (1g $\tau$-Ti(HPO$_4$)$_2$ 2H$_2$O + 100 ml anhydrous EtOH contacted with agitation for 24h). The suspension was centrifuged, and immediately contacted whilst still wet with 100 ml of a 0.1 mol dm$^{-3}$ H$_2$O/EtOH solution of the ligand. After varying contact times, the solids were filtered off and air-dried. Table 1 lists stoichiometries, preparation conditions and interlayer distances of the materials obtained.

Table 1. Diamine-Intercalated τ-Titanium Phosphates.

| Material | Contact time (days) | temp. (°C) | $d_{002}$ (Å) |
|---|---|---|---|
| τ-Ti(HPO$_4$)$_2$[2,2'-bipy]$_{0.45}$0.4H$_2$O | 15<br>9 | 25<br>43 | 14.47 |
| τ-Ti(HPO$_4$)$_2$[1,10-phen]$_{0.48}$ 1.5H$_2$O | 15<br>8 | 25<br>43 | 18.0 |
| τ-Ti(HPO$_4$)$_2$[2,9-dmp]$_{0.33}$2.3H$_2$O | 15<br>9 | 25<br>43 | 17.4 |

**Exchange of Metal Ions.**

Batch methods were used. In a typical preparation, a solution of 0.01 mol dm$^{-3}$ metal acetate was contacted with agitation with 1g of diamine-intercalated material at 43° C for 10 days. The supernatant was analysed (atomic absorption spectrophotometry) after filtration and the solid then dried in air. Ag$^+$ was exchanged from a AgNO$_3$ solution (pH 4.3-4.4, as for the other metal acetates).

Figure 1 shows an example of the thermal behaviour - for the case of bipy. The other two materials are similar and the behaviour is analogous to that in α- and τ-zirconium phosphate analogues[1,2]. Water of crystallisation (which is present in slightly varying amounts in the preparations) is lost between 25 and 250° C, followed by loss of amine + H$_2$O of condensation of phosphate to pyrophosphate between 250 and 600° C (note the clean exotherm of the phase change to cubic α-TiP$_2$O$_7$ at 810 °C). The final, small, losses at 600 - 1100° C are ascribed to carbonisation of remaining traces of diamine.

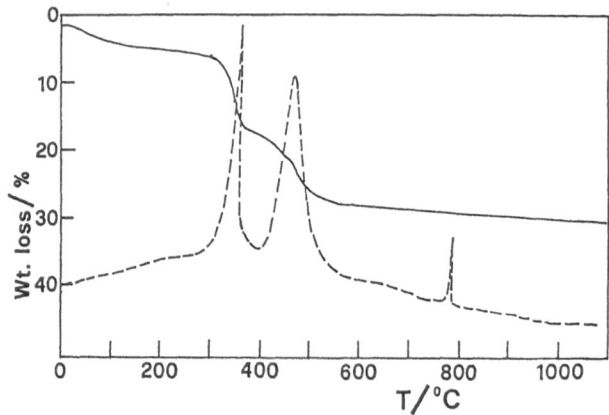

Figure 1. TG/DTA of τ-Ti(HPO$_4$)$_2$[2,2'-bipy]$_{0.45}$·0.4H$_2$O.

These bulky diamines intercalate into τ-titanium phosphate to give products different from those given by the τ-zirconium analogue[1]. τ-Titanium phosphate gives only one form with 2,9-dmp, whereas the zirconium analogue gives two composites, one with -[dmp]$_{0.25}$- and the other with -[dmp]$_{0.50}$[1]. The interlayer distance for the 2,2'-bipy material is similar to that found for the τ-zirconium analogue (14.7 Å). A similar, almost flat orientation of the diamine in the interlayer may be suggested. In the 1,10-phenanthroline case, the interlayer is rather smaller than that in the τ-zirconium analogue, 17.4 versus 18.6 Å. This presumably means that the 1,10-phenanthroline is oriented more closely vertical to the phosphate sheets than in the τ-zirconium phosphate case. Finally, the interlayer distance in the 2,9-dmp is not very different to that - 17.7 A - in the τ-zirconium analogue, again suggesting that the orientation of the diamine within the sheets is very similar, i.e almost vertical.

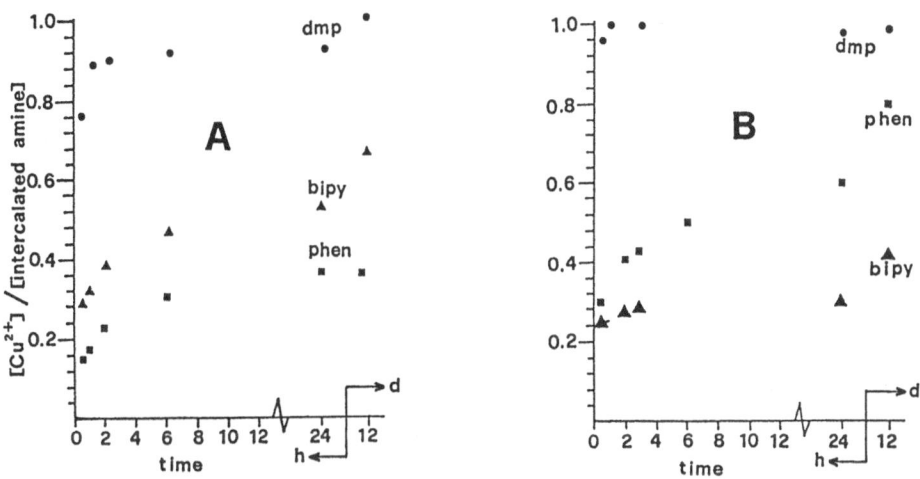

Figure 2. Exchange of $Cu^{2+}$ by Diamine-intercalated τ-titanium phosphate (A) and τ-zirconium phosphate (B)

Figure 2 shows the uptake of $Cu^{2+}$ by the three materials, compared with that for the τ-zirconium analogue from ref.1. In both cases, the 2,9-dmp intercalated material exchanges $Cu^{2+}$ the fastest, although slightly slower into τ-titanium phosphate than into the τ-zirconium analogue (as might be expected for a higher density of intercalated diamine). Instead, the uptake order for the 2,2'-bipy and 1,10-phen composites are inverted for the two phosphates, despite the fact that the interlayer distances are very similar. Clearly, the interlayers in the two matrices are different.

The exchange of $Co^{2+}$ and $Ni^{2+}$ into the three intercalates has also been followed. For all three metal ions there is no change in interlayer distance (i.e.the final materials cannot be considered "pillared"). For the as-prepared $Ni^{2+}$ and $Co^{2+}$ cases, the visible absorption spectra are as expected for hydrated species (i.e. exchange occurs in the cavities formed). Conversely, $Cu^{2+}$ appears to be coordinated to the diamine in all three cases. This is particularly clear for the dmp case. The visible absorption spectrum of 1:1 $[Cu^{2+}]$ : $[\tau\text{-}Ti(HPO_4)_2(dmp)_{0.33} \ 2.3H_2O]$ is shown in Fig.3. After 4h contact the material is orange-red and after 24 days it turns green. In both cases, $Cu^{2+}$ and $Cu^+$ are simultaneously present (differently from the case of the $\alpha$-zirconium phosphate analogue, where the $Cu^+$ grows in[3]). The $Cu^{2+}$ d-d band positions and $Cu^+$ MLCT are characteristic of N-coordination[4].

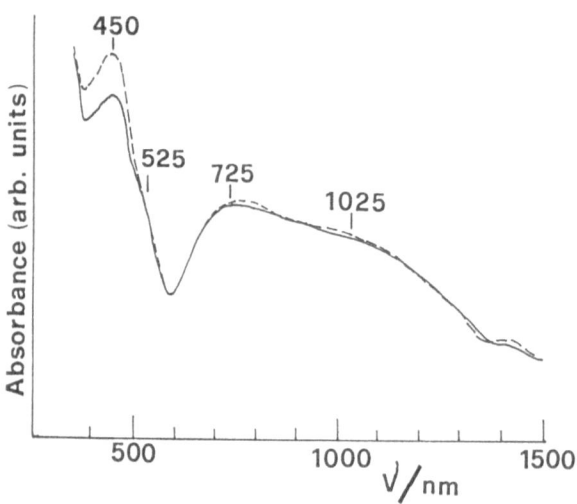

Figure 3. Visible absorption spectra of 1:1 $[Cu^{2+}]$ : $[\tau\text{-}Ti(HPO_4)_2 \ [2,9\text{-}dmp]_{0.48} \ 2.3H_2O]$. (——) after 4h contact; (--) after 28 days contact.

$Ag^+$ exchanges into the 1,10-phen and 2,9-dmp intercalates to give 1:1 materials with large interlayer distances : 1,10-phen 20.5 Å; 2,9-dmp 21.3 Å, i.e. "pillaring" occurs. These are very large free heights (interlayer distance - layer thickness ca 6.3 Å) :14.2 Å (phen) and 15.0 Å (dmp). The as-prepared dmp material is colourless,but on dehydration becomes a pale yellow, and has two bands at 375 and 460 nm in the visible spectrum (Fig.4).

This suggests the presence of juxtaposed exchanged $Ag^+$ ions in the former and $Ag_n^+$ clusters in the latter . The XRD of the as-

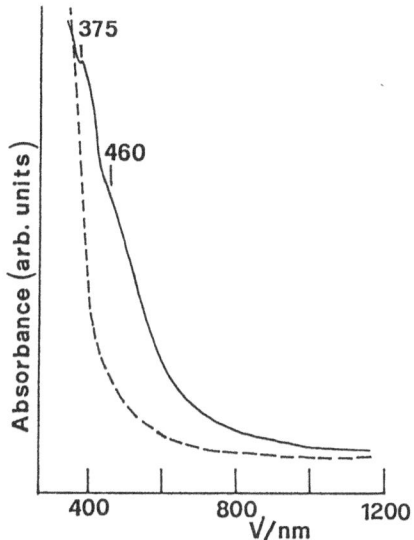

Figure 4. Visible absorption spectra of 1:1 $[Ag^+]$: $[\tau$-Ti$(HPO_4)_2[1,10$-phen$]_{0.48} \cdot 1.5H_2O]$ (---) as prepared; (——) after dehydration at 200°C.

prepared material, Fig.5, shows an enhanced intensity $d_{004}$ reflection, an indication that the silver ions have exchanged throughout alternate layers, i.e. that the material is a stage 2 layered material.

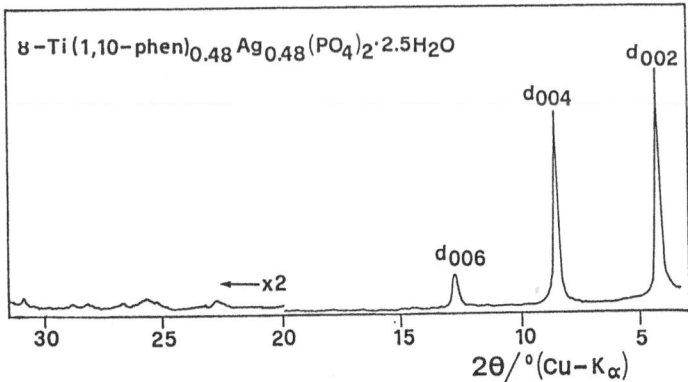

Figure 5. XRD of 1:1 $Ag^+$ exchanged $\tau$-Ti$(HPO_4)_2[1,10$-phen$]_{0.48} \cdot 1.5$ $H_2O$.

## CONCLUSIONS

$\tau$-Ti$(HPO_4)_2 \cdot 2H_2O$ forms layered composites with 2,2'-bipyridine, 1,10-phenanthroline and 2,9-dimethyl-1,10-phenanthroline. Their ion-exchange behaviour with $Cu^{2+}$ is different to that of the $\tau$-Zr$(HPO_4)_2$ $2H_2O$ analogue. Stoichiometries, XRD and spectroscopic evidence indicates that exchangeable P-OH groups in $\tau$-Ti$(HPO_4)_2 2H_2O$ may be arranged in alternate layers.

## REFERENCES

1. C. Ferragina, M.A. Massucci and A.A.G. Tomlinson, J. Chem. Soc. Dalton Trans., 1990, in press

2. C. Ferragina, M.A. Massucci, A. La Ginestra, P. Patrono and A.A.G. Tomlinson, J. Chem. Soc. Dalton Trans., 1988, 851 & refs therein.

3. C.Ferragina, M.A.Massucci, P.Patrono, A.A.G.Tomlinson, and A.La Ginestra, Mat. Res. Bull., 1987, **22**, 29.

4. A.T.Jacob and A.B.Ellis, Inorg. Chem., 1989, **28**, 3896.

5. R.A.Schoonheydt, J.Phys.Chem.Solids, 1989, **50**, 523.

ION EXCHANGE BEHAVIOUR OF SOME AMINE INTERCALATES OF α-TIN(IV)
HYDROGEN PHOSPHATE MONOHYDRATE

M.J. HUDSON[a], E. RODRIGUEZ-CASTELLON[b], and P. SYLVESTER[a]

[a]Department of Chemistry, University of Reading, P.O. Box 224,
Whiteknights, Reading, Berks RG6 2AD, U.K.

[b]Departmento di Quimica Inorganica, Facultad de Ciencias,
Universidad de Malaga, 29080 Malaga, Spain.

ABSTRACT

The ion exchange behaviour of some amine intercalation compounds of
α-tin(IV) hydrogen phosphate (α-SnP) has been investigated in the
presence of transition metal ions. It has been found that the behaviour
of these intercalates depends upon whether the guest molecule is a mono-
or a polyamine. For each of the intercalation samples studied the rate
of uptake of the metal ion is rapid ($t_{1/2}$ less than five minutes). When
the guest molecule is a primary monoamine the uptake of the metal ion is
accompanied by the loss of the amine in the form of a soluble alkyl
ammonium ion. The resultant solid product of this reaction is essentially
amorphous due to the delamination of the lamellar α-SnP. The surface
areas, however, are approximately the same (α-SnP 51.6 $m^2 g^{-1}$, the
delamination product 55.9 $m^2 g^{-1}$). The retention capacities of the
α-SnP allylamine intercalate for $Ni^{2+}$, $Co^{2+}$ and $Cu^{2+}$ were 6.3, 5.9 and
3.5 meg/g respectively. This intercalate appeared to be selective for
$Cu^{2+}$ especially at low copper concentrations in mixed solutions.

INTRODUCTION

There is a continuing need to find new ion exchangers which are capable
of treating radioactive nuclear effluent [1] and to remove toxic substances
from aqueous effluents. Hydrogen phosphates of tetravalent metals have
been the object of considerable study in recent years because of their
ion-exchange and intercalation properties [2]. These materials possess
ion-exchange properties several times higher than those of clays like
montmorillonite or vermiculite. Tin(IV) hydrogen phosphate, $\alpha\text{-Sn(HPO}_4)_2 \cdot$
$H_2O$ (α-SnP), belongs to this group of exchangers and possesses a layered

structure similar to that of α-zirconium(IV) hydrogen phosphate [3].
Although α-SnP can take up transition ions from aqueous solutions by
replacing protons of =POH groups of the interlayer surfaces, high
temperatures are required to completely saturate the phosphate with such
cations [4]. It has been shown [5-11] that mono- and polyamines can be
rapidly intercalated into α-SnP leading to an increase of the interlayer
distance; the monoamines forming bilayers and bisamines normally monolayers
in which amine groups are bonded to opposite faces of the layers. The
exchange reaction is appreciably facilitated when a n-butylamine intercalate of α-SnP is used as the exchanger since fully exchanged phases
with transition ions can be obtained at room·temperature by employing
solutions of the corresponding acetates [8].


## EXPERIMENTAL

α-SnP was prepared using the method of Costantino and Gasperoni [11].

The amine intercalates were prepared by shaking a solution of the amine,

in an appropriate solvent such as water or toluene, (see Table 1).


## RESULTS AND DISCUSSION

### INTERCALATION COMPOUNDS WITH AMINES

The intercalation compounds with alkylmonoamines of layered tetravalent

metal phosphates have a great interlayer distance (Table 1) [5,12] in

comparison to that of the parent α-SnP (0.78nm). In these intercalation

compounds an easy exfoliation of the microcrystal for the co-intercalation

of a convenient amount of solvent could be expected.


### ADSORPTION ISOTHERMS OF $Ni^{2+}$, $Co^{2+}$ and $Cu^{2+}$

The extraction process of metal ions from solutions by amine intercalates

of α-SnP was rapid, $t_{1/2}$ was under five minutes. Figure 1 plots the

adsorption isotherms of $Ni^{2+}$, $Co^{2+}$ and $Cu^{2+}$ by the α-SnP/allylamine

intercalate. At low concentrations of metal ions in solution, the

phosphate showed a strong affinity for $Ni^{2+}$ and $Cu^{2+}$, the slope being

practically 1, whereas $Co^{2+}$ was adsorbed to a lesser extent. Figure 2

shows the evolution of the equilibrium solution pH as a function of

the amount of metal ion added. In all cases, pH decreases initially with

the addition of the metal ion and the curves become asymptotic, abrupt pH

decreases being apparent between metal additions of 0-25 and 70-125% of

the C.E.C. The different final pH values are a consequence of the

different acidities of the metal ions; $pk_a$ values are 10.5, 10.2 and 8.0

TABLE 1

Interlayer distance and composition of some $\alpha$-SnP/amine intercalation compounds.

| AMINE | SOLVENT | INTERLAYER SPACING (nm) | COMPOSITION MOL AMINE PER MOL $\alpha$-SnP |
|---|---|---|---|
| Allylamine | water | 1.56 | 2.0 |
| Allylamine | toluene | 1.67 | 2.0 |
| Propylamine | water | 1.75 | 2.0 |
| Propylamine | toluene | 1.86 | 2.0 |
| DAE[1] | toluene | 1.08 | 1.2 |
| Diethylamine | toluene | 1.36 | 1.2 |
| Triethylamine | toluene | 1.34 | 0.7 |
| TEPA[2] | water | 1.89 | 0.4 |

[1]DAE = 1,2-diaminoethane

[2]TEPA = Tetraethylenepentamine

for $Ni^{2+}$, $Co^{2+}$ and $Cu^{2+}$ respectively [13]. Adsorption maxima of $Co^{2+}$ is only about 60% of the theoretical C.E.C. when the hydrogen phosphate is used as the exchanger (figure 3). Final equilibrium pH values were, in this case, about 2.5 (figure 2).

Figure 4 shows the adsorption isotherm for $Co^{2+}$ by the $\alpha$-SnP/allylamine intercalate as a function of the equilibrium solution pH. From this isotherm, it can be seen that the uptake of the metal ion is constant at pH values higher than 3.5. Below this value, the uptake decreases as a consequence of the proton competition. The use of acetate solutions minimizes the proton competition and therefore, permits a full utilisation of the exchange sites of the phosphate [8]. Although the $\alpha$-SnP/allylamine intercalation compound gave a clear and well defined X-ray diffraction

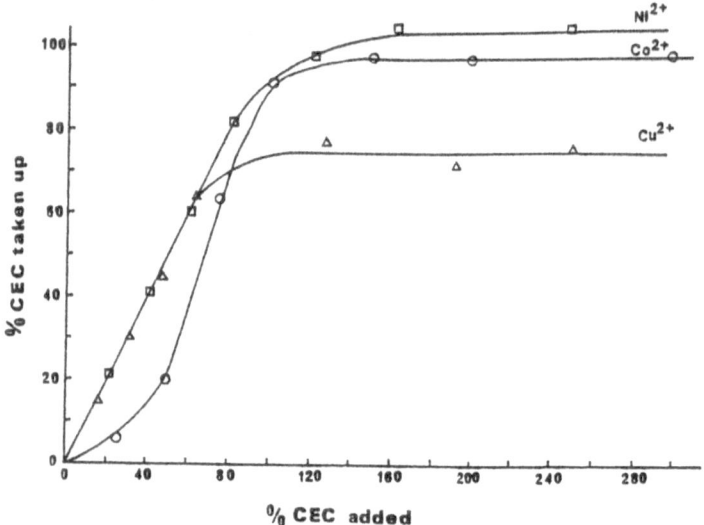

Figure 1.   Adsorption isotherms of $Ni^{2+}$, $Co^{2+}$ and $Cu^{2+}$ by the
α-SnP/allylamine intercalate.

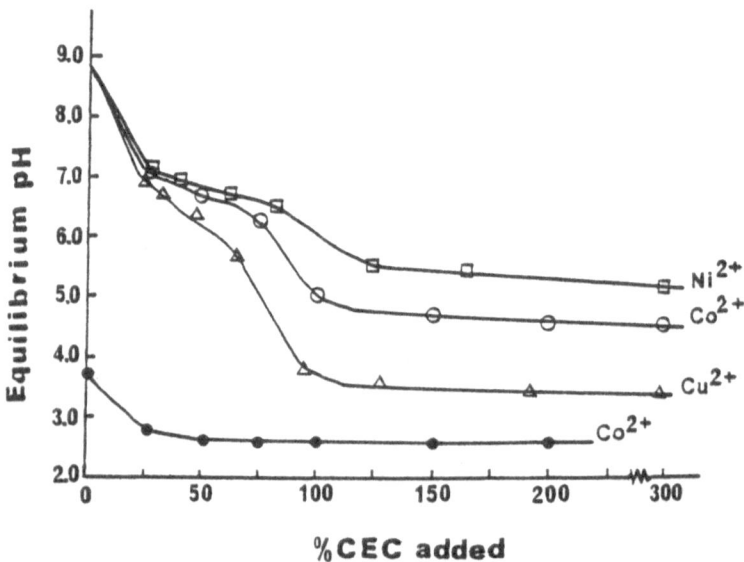

Figure 2.   Variation of the equilibrium pH as a function of $M^{2+}$ added
to the α-SnP/allylamine intercalate (□,○,△,)  or pure α-SnP (●).

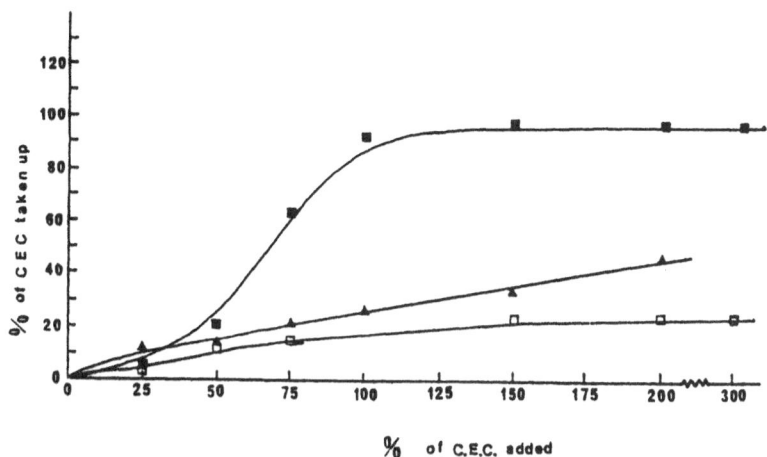

<u>Figure 3.</u>  Adsorption isotherms of Co$^{2+}$ by ■ α-SnP/allylamine; ▲ pure α-SnP; ▢ α-SnP/TEPA.

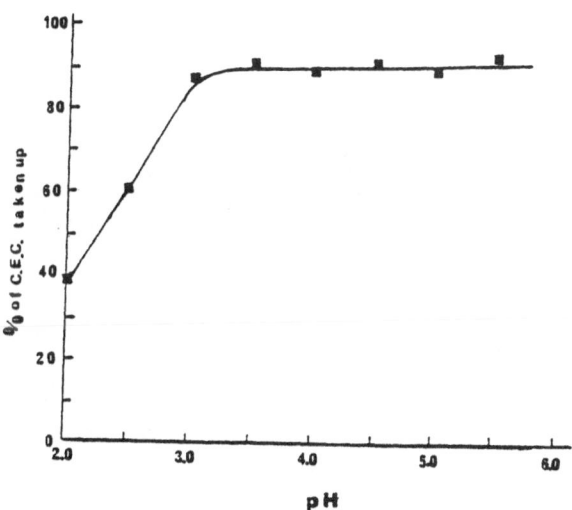

<u>Figure 4.</u>  Adsorption of Co$^{2+}$ by α-SnP/allylamine as a function of the pH.  (Co$^{2+}$ added = 200% C.E.C.)

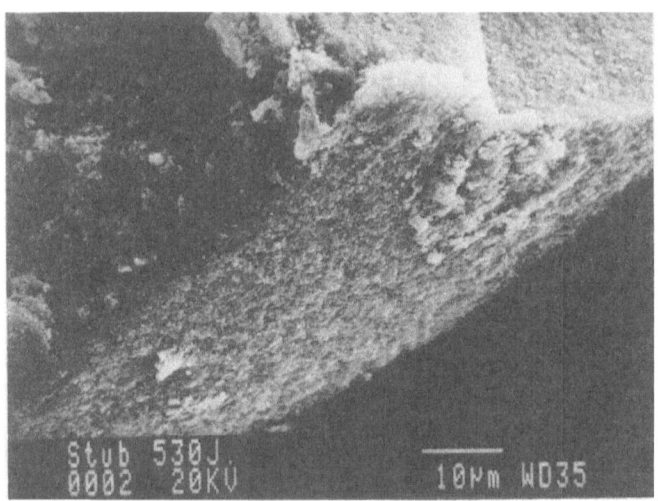

Figure 5A.  Electronmicrograph of the α-SnP.

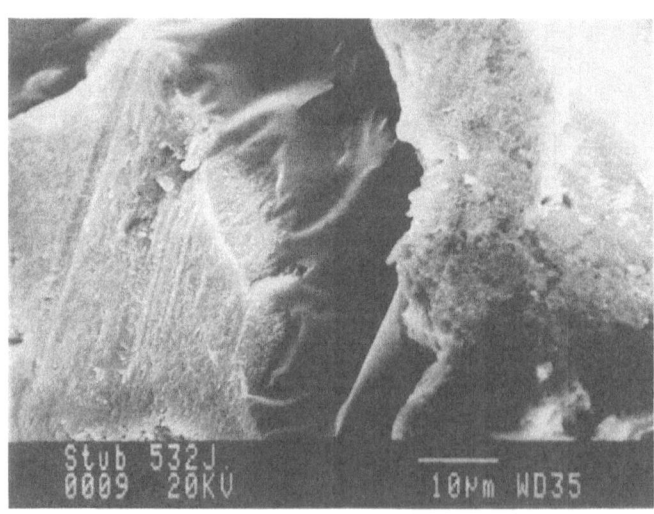

Figure 5B.  Electronmicrograph of the α-SnP/allylamine intercalation
compound.

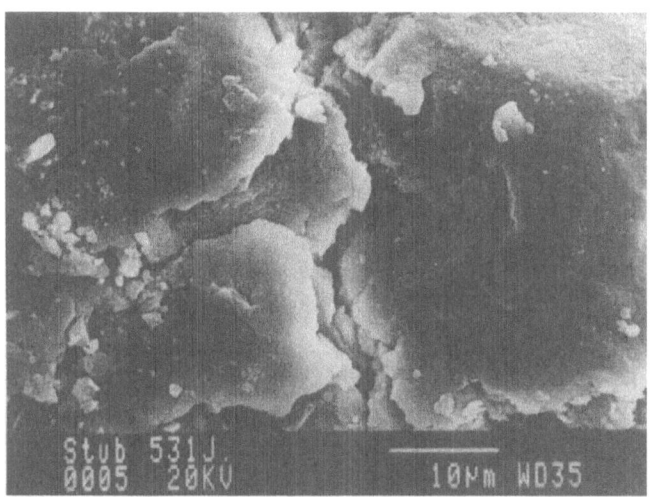

**Figure 5C.** Electronmicrograph of the α-SnP/allylamine intercalation compound after reaction with $Co^{2+}$.

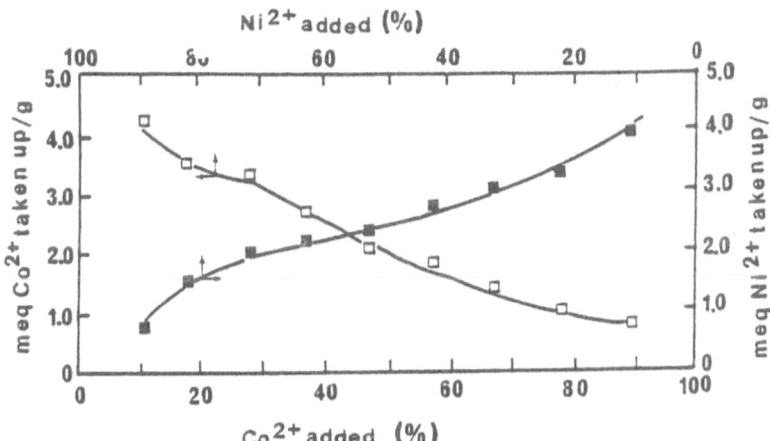

**Figure 6.** Adsorption of $Co^{2+}$ and $Ni^{2+}$ by α-SnP/allylamine intercalate from mixed solution. (Total amount of metal ions = 200% C.E.C.)

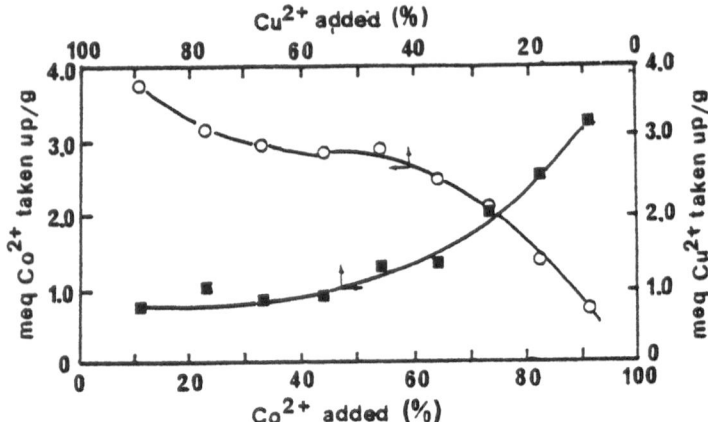

Figure 7. Adsorption of $Co^{2+}$ and $Cu^{2+}$ by α-SnP/allylamine intercalate from mixed solution. (Total amount of metal ions = 200% C.E.C.)

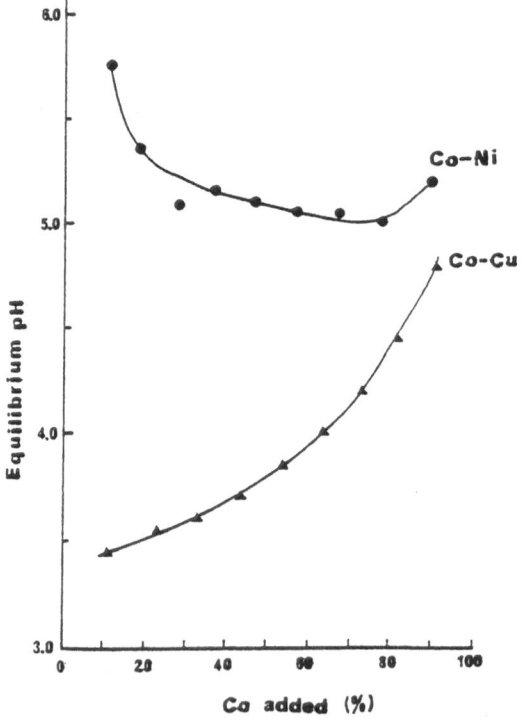

Figure 8. Variation of equilibrium pH of mixed solutions $(Co^{2+}-Ni^{2+})$ and $(Co^{2+}-Cu^{2+})$ as a function of the percentage of $Co^{2+}$ added.

pattern, the metallic phosphates obtained after the reaction with aqueous transition metal ions were virtually amorphous. It appears that the reaction of metal ions with the colloidal suspension of the allylamine intercalation compound provokes an irreversible delamination and to some extent this is confirmed by electron microscopy as shown in figure 5, where it can be seen that there is a marked change in the morphology of the three samples shown; only a-SnP and the allylamine intercalation compound show defined areas of microcrystallinity.

When metal ions were equilibrated with the polyamine intercalation compounds, the extraction capacities were less than 20% of the theoretical C.E.C. (figure 3). This low capacity, in the case of the $\alpha$-SnP/TEPA intercalate, was accompanied by the formation of a complex in solution similar to that formed when an aqueous solution of the free ligand was contacted with such ions. Data concerning the selective extraction of metal ions from mixed solutions by the $\alpha$-SnP/allylamine intercalation compound are shown in figures 6 and 7. In the adsorption of mixtures of $Co^{2+}$ and $Ni^{2+}$, the intercalation compound shows only slight selectivity for $Co^{2+}$ at low concentrations. Variations of the equilibrium pH of the $Co^{2+}/Ni^{2+}$ mixed solutions was not significantly important, which is consistent with a similar acidity of these metal ions (figure 8).


## CONCLUSIONS

The intercalation of monoamines such as allylamine or n-propylamine into $\alpha$-SnP doubles the extraction capacity relative to the parent $\alpha$-SnP. Ion-exchange reactions of the intercalates are rapid with $t_{1/2}$ typically being less than five minutes.

REFERENCES

(1) Cross, J.E. and Hooper, E.W., in "Ion Exchange for Industry", Streat M. (Editor), Conference Proceedings, Cambridge, Ellis Horwood 1988.

(2) Alberti, G., in "Recent Developments in Ion Exchange", P.A. Williams and M.J. Hudson (Eds.), Elsevier Applied Science, London, 1987.

(3) Chernorukow, N.G., Mochalova, I.R., Moscuichev E.P. and Sibrina, G.B., Zh. Prikl. Khim., 1977, 50, 1618.

(4)   Rodriguez-Castellon, E., Bruque S. and Rodriguez-Garcia, A., An.
      Quim., 1985, 81, 316.

(5)   Rodriguez-Castellon, E., Bruque, S. and Rodriguez-Garcia, A.,
      J. Chem. Soc. Dalton Trans., 1985, 213.

(6)   Fotheringham, I., MSc Thesis, University of Reading, 1986.

(7)   Fotheringham, I., Giwa, C.O. and Hudson, M.J., J. Chem. Soc.:
      Chem Commun., 1986, 1554.

(8)   Rodriguez-Castellon, E., Rodriguez-Garcia, A. and Bruque, S.,
      Mat. Res. Bull., 1985, 20, 115.

(9)   Rodriguez-Castellon, E., Rodriguez-Garcia, A. and Bruque, S.,
      Inorg. Chem., 1985, 24, 1187.

(10)  Whittingham, M.S. and Jacobson, A.J., (Editors), "Intercalation
      Chemistry", Academic Press, New York, 1982.

(11)  Costantino, U. and Gasperoni, A., J. Chromatogr., 1970, 51, 289.

(12)  Clearfield, A. and Tindwa, R.M., J. Inorg. Nucl. Chem., 1979,
      41, 871.

(13)  Baes, C. and Mesmer, F., "The Hydrolysis of Cations", Wiley
      Interscience Ed., New YOrk, 1976.

(14)  Killman, E., Korn, M. and Bergmann, M., in "Adsorption from
      Solution", R.H. Ottewill, C.H. Rochester and A.L. Smith (Editors),
      Academic Press, London, 1983, 259.

ACKNOWLEDGEMENTS

This work has been supported by the U.K. Department of the Environment
as part of their Radioactive Waste Management Programme and by CICYT of
Spain (Project PB-86-244).  The results may be used in the formation
of government policy but at this stage do not necessarily represent
government policy.  We also thank Acciones Integradas Hispano-Britanicas
(Ministerio de Education y Ciencia, British Council).

# Part 3

# NUCLEAR INDUSTRY

# ION EXCHANGE IN THE NUCLEAR INDUSTRY

JANE P. BIBLER
Westinghouse Savannah River Company
Aiken, South Carolina  29808  USA

## ABSTRACT

Ion exchange is used in nearly every part of the nuclear fuel cycle - from the purification of uranium from its ore to the final recovery of uranium and transmutation products.  Ion exchange also plays a valuable role in the management of nuclear wastes generated in the fuel cycle.

## INTRODUCTION

The nuclear industry encompasses a variety of operations and ion exchange processes play an important role in nearly every operation.  Uranium, recovered from its ore, must be purified and concentrated.  Enriched uranium is fabricated into fuel elements and placed in nuclear reactors where it produces energy, fission products, and transmutation products.  If the fuel cycle is completed, uranium and useable transmutation products are separated from the fission products and from each other, to be recycled or used elsewhere.  Finally, new waste disposal requirements have prompted studies for the use of ion exchange in remediating wastes generated in each of the  above operations.

## URANIUM PURIFICATION AND CONCENTRATION

Triggered by the U.S. nuclear defense efforts in the late 1940's and early 1950's, extensive pioneering research and development efforts [1-8] laid down the relevant process chemistry of anion exchange purification of uranium.  The first actual plant-scale recovery, concentration, and purification of uranium from $H_2SO_4$ ore leach liquors by anion exchange was demonstrated in South Africa in 1952. [9]  This demonstration was quickly followed in many countries by construction and operation of similar ion exchange processes for clarified and unclarified (resin-in-pulp) ore leach liquors, both acidic and alkaline carbonate.  Subsequent investigations of improvements to this earliest technology have been concerned with the combination of ion exchange with solvent extraction and with development and testing of new resins and new equipment for countercurrent applications.  Recently, ion exchange technology has been directed at the recovery of uranium from mine waters [10] and mill waste streams.[11]

### Important Ion Exchange Resins:Types and Properties

Strong base Type I resins were the first ion exchange resins used for uranium recovery [6] and still remain as resins of choice in the industry.[12] Typically, the exchange site on a Type I resin is a symmetrical quaternary ammonium ion. Strong base Type II resins use an exchange site similar to Type I resins except that the quaternary ammonium ion is no longer symmetrical, one methyl group having been substituted by an ethanolic group. The advantage in sorption offered by Type II resins is due to their lower affinity for $Cl^-$ or $NO_3^-$ compared to Type I resins.[12] Vinyl pyridine strong base resins are effective in removing uranium species from solutions but allow much easier stripping of uranium by $Cl^-$ or $NO_3^-$ than either Type I or Type II resins. Weak base resins used for uranium recovery have tertiary amine functional groups. New waste disposal regulations have precipitated studies of the use of ion exchange for removal of very small concentrations (2 to 10 ppm) of uranium from mine waters and waste streams. The resins that have been used in these studies are conventional strong base anion exchangers, weak acid (carboxylic) cation exchangers, and those containing chelating functional groups (e.g. iminodiacetate, amidoxime, picotylamine, aminophosphonic).

The earliest ion exchange uranium recovery operations used gel-type 16-50 mesh anion resin beads in conventional fixed-bed columns. Modern continuous or countercurrent ion exchange (CIX) processes employ macroporous beads of 12-20 mesh because of their greater resistance to attrition.

**Ion Exchange Chemistry**

Dilute $H_2SO_4$ is generally used to leach uranium from ground ores. Acidic sulfate leach liquors contain $UO_2^{2+}$, $UO_2SO_4$, $[UO_2(SO_4)_2]^{2-}$, and $[UO_2(SO_4)_3]^{4-}$ in equilibrium.

$$UO_2^{2+} + nSO_4^{2-} \leftrightarrow UO_2(SO_4)_n^{2-2n} \tag{1}$$
$$n = 1, 2, \text{ or } 3$$

Ion exchange of uranium from these solutions is represented on either strong or weak base anion resins by the following general reaction:

$$4\,R\text{-}X + UO_2(SO_4)_3^{4-} \leftrightarrow (R)_4(SO_4)_3 + 4\,X^- \tag{2}$$

where $X = Cl$ or $NO_3$ and $R$ = the resin support network.

For mined or ground ores that are high-lime ores (~15% $CaCO_3$), leaching with $H_2SO_4$ would consume very large quantities of the acid and carbonate solutions are, instead, used to leach uranium. (Some *in situ* leach operations have used ammonium carbonate/bicarbonate mixtures.)[13] Carbonate leach liquors contain the $UO_2(CO_3)_3^{4-}$ ion as the predominating species. This ion is sorbed on the weak and strong base ion exchange resins according to equation (3). The symbols R and X have the same meaning as in equation (2).

$$4\,R\text{-}X + UO_2(CO_3)_3^{4-} \leftrightarrow (R)_4UO_2(CO_3)_3 + 4\,X^- \tag{3}$$

Process variables such as pH, concentrations of sulfate or carbonate, competing anions, temperature, and resin poisons have been extensively documented for the reactions shown in equations (2) and (3).[5]

When a strong base anion exchange resin has been loaded with uranyl sulfate, it can be efficiently eluted using acidic chloride (0.9 M $NH_4Cl$ or NaCl and 0.1 M HCl or $H_2SO_4$) or acidic nitrate (1.0 M $NaNO_3$ and 0.1 M $H^+$) solutions.[2] Elution can also be effected using dilute (~1.0 M) $H_2SO_4$ although with less efficiency than with chloride or nitrate. Sulfuric acid elution is useful in conjunction with a subsequent solvent extraction process because the eluate can be directly used as feed for that process and the step of converting the resin from the nitrate or chloride form to the sulfate form before the next adsorption cycle can be eliminated.[2]

Acid elution of sorbed $UO_2(CO_3)_3^{4-}$ from anion resins would lead to gas generation and consequent disruption of the resin bed. Thus, elution of this species is accomplished with NaCl or $NH_4NO_3$ solutions.

After elution of uranium, common resin poisons such as thiocyanate ion, $S_2O_6^-$, molybdenum, and iron are removed by treatment with NaOH or $NH_4OH$ solutions. [2,5]

**Important Ion Exchange Equipment and Methods**

The traditional ion exchange systems designed for collecting and concentrating uranium used fixed-bed columns arranged in a "merry-go-round" array. These systems require clarified feeds for smooth operation. Clarification of feeds after leaching is very expensive, accounting for about half of the operating costs of a uranium milling plant.[14] Costs of the clarification process are driven up by the presence of very small particles (ore slimes) of the ore. One way to lower costs is to first separate the ore slimes from the easily washed and clarified sands fraction of the leachate. Sands fraction wash water and ore slimes are then combined to produce a slurry of up to 30% suspended solids. Such slurries are used as feedstock for resin-in-pulp (RIP) or countercurrent ion exchange processes which have been successful processes for treating unclarified feedstock.

In all CIX systems, the resin bed is fluidized by an upflow stream of feedstock. Continuous ion exchange processes have been described as a major development in ion exchange technology. The success of CIX processes is directly related to the careful experimental and theoretical work of a variety of investigators.[15-22] Three particular fluidized column systems have been used commercially with great success:

> NIMCIX System-This system was developed from Cloete-Streat concepts by the National Institute for Metallurgy in South Africa.[23,24]

> USBM-MCIX System-This system was developed by the U.S. Bureau of Mines. It is used for *in situ* uranium mining processes.[17-19]

> HIMSLEY COLUMN-The most recently proven fluidized bed contacting system, the Himsley column, has seen use in South Africa, Canada, and the U.S.[21,22]

Porter has developed an upflow, multi-tank fluidized bed ion exchange system for Namibia.[25,26] A downflow pulsed-bed ion exchange method (Higgins Loop CIX) has been used to recover uranium from copper dump leach liquors at Bingham Canyon, Utah.[11]

Continuous RIP processes mix resin beads with desanded pulp, separate the uranium-rich beads by a screening step, and then elute uranium from the beads in a separate system, recycling the regenerated beads for reuse in uranium sorption. One or more of these operations may be used in any given process. An upflow RIP process has been used to recover uranium in France.[27] Several RIP processes have been used successfully over the past 25 years in uranium mills.[27] All use screen-mix systems, which are the only technology capable of handling medium or high density slurries. No basket RIP systems, developed in the 1980's, are in operation. Development of high specific gravity ($\geq 1.25$) ion exchange resins would significantly extend the scope of RIP applications.

A combination of ion exchange and solvent extraction, based on pioneering work done at Oak

Ridge National Laboratory [28], represents the most widely used method for uranium recovery from sulfate ore leach liquors. [22,29-31] The uranium is eluted from ion exchange resins using $H_2SO_4$ and the eluate is used as feed for an extraction by a tertiary amine.

## SEPARATION PROCESSES

Uranium, concentrated in fuel assemblies, undergoes fission, after capturing a neutron in nuclear reactors, to generate fission products, energy, and more neutrons. In power reactors, energy is the desired product; in production reactors, the additional neutrons are the desired product because they react to produce new isotopes. The main isotope of the neutron irradiation of natural uranium is [239]Pu, produced when a neutron is captured by [238]U (accompanied by two beta decay reactions). The [237]Np isotope is also produced in much lower yields.

After the irradiated fuel elements are removed from the reactor they are stored in cooling basins to allow short lived fission product isotopes to decay. The last step in the nuclear fuel cycle is the separation of uranium, plutonium, and useful byproducts from each other and from unwanted fission products. Ion exchange cannot compete with solvent extraction in the primary process of uranium recovery, but it is used in several important secondary recovery steps.

### Nuclear Fuel Reprocessing

Throughout the world, reprocessing of nuclear fuels employs some form of the Purex or Thorex processes, solvent extraction processes using tri-n-butylphosphate (TBP) as the extracting agent. Although no reprocessing of power reactor fuels is planned in the U.S., the Savannah River, Hanford, and Idaho Falls defense sites use TBP-based solvent extraction. European, Japanese, and Indian reprocessing facilities use Purex or modified Purex processes. The Thorex process was developed to process thorium-based fuels.

In Purex reprocessing, such as at the Savannah River Site, irradiated uranium is first dissolved in nitric acid and then separated from plutonium and fission products during solvent extraction with TBP. The high activity waste concentrate from the first Purex cycle contains most of the [237]Np and some of the [239]Pu. It also contains aluminum and most of the fission products. As shown in Figure 1 [32], the neptunium and plutonium are separated from other ions, except for thorium, by an agitated bed anion exchange step using a strong base resin. This is possible because Pu(IV) and Np(IV) both form stable anionic hexanitrato species in the concentrated (7-9M) nitric acid solution.

The Pu and Np are then eluted with dilute (0.35M) nitric acid as the Pu(IV) and Np(IV) ions and reconverted after elution to the hexanitrato anions with 8M nitric acid. A second anion column is used to resorb $Pu(NO_3)_6^{2-}$ and $Np(NO_3)_6^{2-}$. Plutonium is separated from neptunium by elution as Pu(III) with 5.5M nitric acid, ferrous sulfamate, and hydrazine. Neptunium ion is not reduced under these conditions and remains sorbed on the resin.

A strong acid cation resin is used next to separate [234]Th that is often found with the [237]Np. The neptunium is oxidized to Np(V) in 1-2M nitric acid and, in that oxidation state, is not strongly sorbed by the cation exchanger. Thorium, $Th^{4+}$, is retained on the resin and is periodically eluted with $NaHSO_4$. Any [239]Pu is then recycled back to the rest of the process and the [237]Np is used in targets for irradiation to make [238]Pu.

The second cycle of the Purex process produces a dilute (1-3 g/L) plutonium stream. Plutonium is concentrated (Figure 1) on a strong acid cation resin after reduction to Pu(III) with hydroxylamine. Sorbed plutonium is eluted with 5M nitric acid to yield a solution (40-60 g/L) that is concentrated enough for subsequent Pu recovery by precipitation.

**FIGURE 1**

**ION EXCHANGE USED IN REPROCESSING URANIUM**

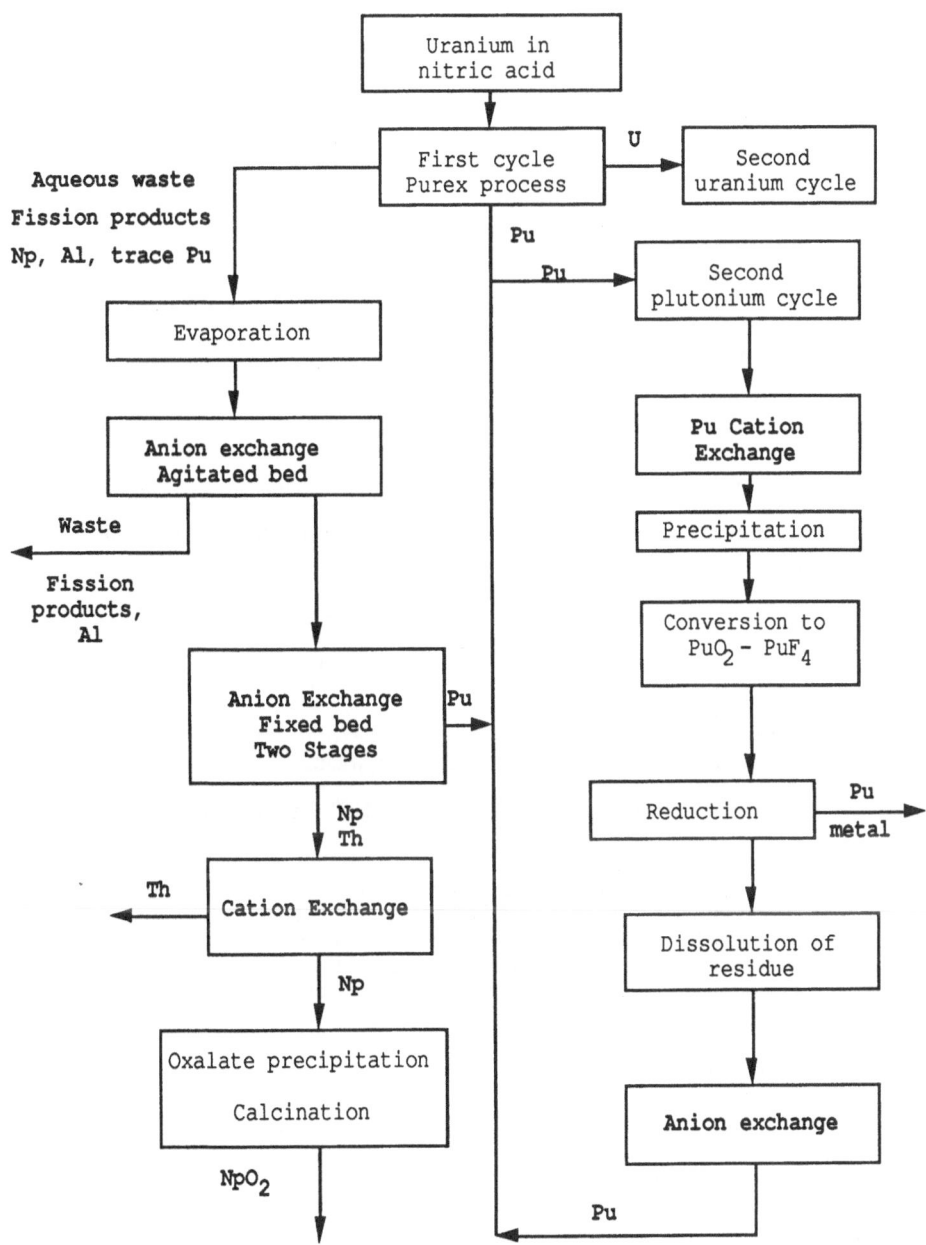

When $^{239}$Pu metal is produced by reduction using Ca metal in the recovery process, unreacted plutonium compounds and metal fragments remain in the reaction residue. The residue is treated with concentrated nitric acid to dissolve the plutonium, filtered, and the Pu is sorbed as Pu(IV) on a strong base anion resin. Plutonium collected in this way is recycled to the main process after elution with dilute $HNO_3$.

## ION EXCHANGE IN NUCLEAR POWER PLANTS

A wide variety of reactor designs have been used throughout the world to produce power. Regardless of the design, the basic operation employs the use of a controlled nuclear reaction to supply energy to produce steam to drive turbine generators. Ion exchange is used extensively to purify water entering and leaving the reactor.[33,34] These applications are briefly noted below:

- Plant makeup water demineralizers usually are strong base or strong acid nuclear grade resins. Nuclear grade resins are distinguished from regular grade resins by their uniform particle size, their purity relative to leachable ions, and their high degree of conversion to the desired ionic form. These very pure resins generate effluent water of the high quality needed to minimize neutron activation product formation and corrosion.

- Reactor coolant water is purified by mixed bed cation and anion exchange systems, normally found in the $H^+$ and $OH^-$ forms, respectively. In pressurized water reactors (PWR) the anion resin also serves to remove borate ion, a nuclear poison.

- Steam condensates are polished using mixed bed, powdered resins that act as filters as well as ion exchangers. Often, deep bed strong acid/strong base systems are used to treat condensate. Evaporator condensates are usually discharged or recycled as makeup water after treatment.

- Contaminated water from fuel pools and other dilute miscellaneous wastes can be treated by ion exchange to yield high purity water.

## DECONTAMINATION OF NUCLEAR WASTES

Waste streams from the reprocessing of nuclear fuels contain essentially all of the fission products found in irradiated fuel. The number of variations of the Purex or Thorex reprocessing processes used throughout the world leads to an equal number of nuclear wastes, each having its own characteristic properties. It is not possible to consider them all here. Instead, the uses of ion exchange in treatment of some general catagories of nuclear wastes are described. In addition, specific waste treatments of interest involving ion exchange technology are noted.

### High Activity Waste Streams

Waste streams from fuel reprocessing are acidic in nature. At Savannah River and Hanford the pH of the waste is adjusted to 12 or greater with NaOH and the waste is stored in carbon steel tanks. During aging, insoluble materials precipitate and the primary radionuclide left in the supernate, as the soluble portion of the waste is called, is $^{137}$Cs.

At Hanford [35], cesium ion was recovered and purified by cation exchange. The first ion exchanger used was a zeolite but this was later changed to a phenolic resin, Duolite® ARC-359.[36] Feed for the process consisted of supernate blended with the product from a phosphotungstate precipitation performed on acid waste. These solutions contained as much as

250 Ci of $^{137}$Cs per gallon and were about 5M in sodium ion. After sorption, the sodium ion was preferentially eluted from the resin using 0.25M ammonium carbonate. Cesium was eluted with a 3M $(NH_4)_2CO_3$/2M NaOH mixture at 60°C. Ammonium carbonate in the eluate was thermally decomposed and the resulting solution was concentrated by evaporation and sent to storage.

At the Savannah River Site [37], a resorcinol/formaldehyde condensation resin has been developed that is capable of selectively sorbing 2.5 E-4M $Cs^+$ from supernate containing up to 6M $Na^+$. The cesium can be recovered by elution with 1M HCOOH or 0.5M $HNO_3$. Two modes of final disposal of the waste cesium have been proposed:

- Incorporate the eluate containing cesium ion with feed to the glass melter for final encapsulation in glass and then reuse the resin.

- Incorporate spent resin with the feed to the glass melter for final vitrification.

At West Valley Nuclear Services Co., a subsidiary of Westinghouse Electric Corp., cleanup has begun on about 550,000 gallons of neutralized Purex waste. The initial treatment step involves passing waste supernate through zeolite ion exchange columns to remove $Cs^+$ and $Sr^{2+}$. The loaded zeolite will be combined with washed hydroxide sludge and about 8,000 gallons of acidic Thorex waste, generated during processing of thorium based fuel, for incorporation into glass.[38]

## Low Activity Waste

High activity wastes are usually concentrated by evaporation to minimize the storage volume required. Condensate from waste evaporators usually contains radioactive cesium due to entrainment. Removal of the cesium can be effected using a zeolite or an organic ion exchange resin, depending on the pH of the waste stream and the fate of the decontaminated waste stream. For example, evaporator condensate from nuclear power plants is decontaminated using mixed bed ion exchange in order to generate very pure effluent that can be recycled as makeup water.

At Savannah River Site, condensate is decontaminated first by passage through a zeolite bed and then treatment in an effluent treatment facility. A final polishing step in the effluent treatment uses strong base cation exchange to remove the small quantity of cesium ion that remains in the waste after a reverse osmosis step that preceeds the ion exchange. The cation resin is eluted with 2M $NaNO_3$ to remove $Cs^+$ and the eluate is incorporated in concrete for storage in vaults.[39]

At Oak Ridge National Laboratory, cesium ion is separated from dilute waste by passage through a column of a phenolic/carboxylic acid resin. It is subsequently eluted with 0.5M $HNO_3$.[40]

After the accident at Three Mile Island Unit 2 Nuclear Power Station (TMI), large volumes of contaminated water remained in the facility that required decontamination. The Containment Building sump held approximately 600,000 gallons of contaminated water; about 500,000 gallons were stored in the Auxiliary and Fuel Handling building and another 90,000 gallons were in the Reactor Coolant system. The primary radionuclides of concern were $^{137}$Cs and $^{90}$Sr in each case but the activity level in the Auxiliary and Fuel Handling building was less by a factor of ten than that in the other two facilities. A slightly different ion exchange treatment was used for the two streams.

Waste water from the Auxiliary and Fuel Handling building was decontaminated by the Epicor II system [41] that was composed of a primary series of carbon steel ion exchange liners (columns) followed by two polishing steps. The liners contained organic ion exchange resins, although some also contained zeolite. The Epicor II system removed approximately 55,000 Ci of gamma emittors, generating 50 highly loaded liners that were disposed of at the commercial waste burial ground at Richland, Washington.

A submerged demineralizer system was used to decontaminate the waste held in the Containment Building sump and the Reactor Core system. Two parallel trains of four ion exchange columns in series were operated in carousel mode while submerged in the reactor fuel pool, which provided shielding. The sorbing medium was a 60/40 mixture of Linde IONSIV™ IE-96 and A-51 zeolites, designed to remove $Cs^+$ and $Sr^{2+}$, respectively, with nearly simultaneous breakthrough of the two radionuclides. The effluent from the zeolite beds was further polished with the Epicor system and stored in tanks for reuse at the TMI facility.

Ion exchange technology has been proposed as a method of removing other fission product radionuclides from nuclear wastes. The pertechnetate ion, $TcO_4^-$, is readily sorbed on a strong base resin. Large quantities of 4M nitric acid are required to elute the $^{99}Tc$ ion from the resin and the process by itself has not been cost effective. Recent work has shown that the pertechnetate ion can be removed from the eluate by sorption onto a weak base resin, allowing recycle of the nitric acid as a cost saving measure. Elution from the weak base resin with a small volume of 1M NaOH completes the isolation of this long-lived isotope from the waste stream.[42]

Chelating resins can be used to complex and retain many cationic radionuclides such as $^{90}Sr$, $^{60}Co$, $^{51}Cr$, and $^{95}Zr$ that may be found in trace quantities in low activity wastes. The $^{106}Ru$ (nitrosyl) species found in waste streams have historically resisted efficient isolation by ion exchange. Recent work with somewhat flexible sulfur-containing copolymers [43] gives hope that specially designed functional groups may be able to effect removal of nitrosylruthenium ions from nuclear waste.

On a much smaller scale, except for the case of zeolite ion exchangers, several inorganic exchangers have been used for the sorption and separation of radionuclides. These include the zeolites, synthetic zeolites, hexacyanoferrates, oxides and hydrous oxides, and acidic salts of multivalent metals.[44]

### Non-Radioactive Waste-Hg

Fuel elements containing alloys of aluminum with enriched uranium are reprocessed by dissolution in nitric acid using mercuric nitrate as a catalyst.[45] The mercury introduced at this point ultimately finds its way into both high- and low-activity wastes. Strict regulation of dispensing and storing waste streams containing toxic materials such as mercury has led to studies of ways to isolate and immobilize dissolved mercury from the waste to protect the environment. Duolite® GT-73 cation exchange resin is used entensively at the Savannah River Site [39] to remove $Hg^{2+}$, $Hg_2^{2+}$, and dissolved $Hg^0$ from dilute waste streams.

## OTHER ACTINIDE SEPARATIONS OF INTEREST

Ion exchange has been used in several applications to separate actinide isotopes produced in special irradiation campaigns.

### Transplutonium Elements

A decade and a half ago, kilogram quantities of $^{243}Am$ and $^{244}Cm$ and milligram amounts of $^{252}Cf$ were produced at Savannah River.[46] At about the same time, Oak Ridge produced those radionuclides as well as $^{249}Bk$, $^{253}Es$, and $^{257}Fm$ in their High Flux Isotope Reactor.[47] The heavy transplutonium elements are made by neutron irradiation of a mixture of Pu isotopes [46,48] or a mixture of Pu, Am, and Cm.[47] The ion exchange separations of actinides at both Oak Ridge and Savannah River were accomplished using high pressure ion exchange.[46,48-51] There are two major advantages in operating high pressure ion exchange for this particular application. First, radiation damage to the resin is minimized because very small resin beads are used, allowing for rapid flow rates. Second, high pressure keeps radiolytic gases, that otherwise would disrupt the bed, in solution.

At Oak Ridge, the lanthanides and actinides were separated from each other on a strong base ion exchanger. Lanthanides were eluted with 10M LiCl and most of the Am and Cm were then eluted with 9M LiCl. The rest of the americium and curium and the heavier actinides were eluted with alpha-hydroxyisobutyrate (AHIB).[50] The heavier nuclides elute first because they form more stable complexes with AHIB.

The Savannah River process used displacement development, similar to that used at Hanford [52] to purify Am, on cation exchange resin to effect separation and purification. The lanthanides and actinides were first sorbed on the resin column behind a barrier band of zinc ion. Development of the band with diethylenetriaminepentaacetic acid (DTPA) moved the more stable actinide complexes to the front of the band, leaving the lanthanides behind. The actinides were resorbed on the resin when the band reached the zinc because zinc forms an even more stable complex with DTPA. The net result was that the actinides separated into discrete bands on the resin. They were eluted, resorbed and developed on a series of smaller columns, and finally eluted in very pure form.

## Concentration of $^{233}$U from $^{232}$Th Irradiation Products

Neutron irradiation of $^{232}$Th produces $^{233}$U.[53,54] The two isotopes are separated from each other by the Thorex solvent extraction process using TBP. The uranium stream is concentrated by sorption of $^{233}$U on a strong acid cation resin followed by elution with a $NH_4NO_3/HNO_3$ mixture.

## $^{241}$Am from Aged $^{241}$Pu

A product of the beta decay of $^{241}$Pu is $^{241}$Am, which is used in commercial applications such as smoke detectors, thickness gauging, and oil well logging.[55] The separation of plutonium and americium is possible by solvent extraction [56,57], molten salt extraction [55], or precipitation.[58] Each of these separation processes, however, must be coupled with at least one ion exchange process step. Cation exchange allows a preliminary separation of Pu(III) and Am(III) from ionic impurities of +1 or +2 charge. Also, anion exchange of $Pu(NO_3)_6^{2-}$ from concentrated nitric acid solution effects its separation from Am(III).

## REFERENCES

1.    Uranium in South Africa, 1946-1956," Associated Scientific and Technical Societies of South Africa, Johannesburg, South Africa (1957).

2.    Merritt, R. C., "The Extractive Metallurgy of Uranium," Colorado School of Mines Research Institute, Golden, CO (1970).

3.    Jacobs, S. A., S. J. Schnexailder, C. Urban, Jr., T. C. Smith, and D. E. Gonzales, "Recovery of Low-Level, Dissolved Uranium Values from Treated Mine Waters" in Proc. Int. Conf. on Uranium Mine Waste Disposal (1980).

4.    Brooke, J. N., "Uranium Recovery from Copper Leaching Operations," Min. Congress J., 63(8) 38-41, August 1977.

5.    Gittus, J.H.,"Uranium", Butterworths, Washington, DC, p. 51 (1963).

6.   Kunin, R., "Ion Exchange Resins," John Wiley & Sons, Inc., New York, NY (1958).

7.   Browning, S. J., "Processing of Uranium Ores," Australian Min., 48 (1972).

8.   Kennedy, R. H., "Status Report from the United States of America," in Processing of Low-Grade Uranium Ores, Vienna, 1966, IAEA, Vienna (1967).

9.   Phillips, C. R. and Y. C. Poon, "Status and Future Possibilities for the Recovery of Uranium, Thorium, and Rare Earths from Canadian Ores, with Emphasis on the Problem of Radium: Pt. 2, Solvent Extraction and Ion Exchange," Min. Sci. Eng., 12 (1980).

10.  Himsley, A., "Recovery of Uranium by Ion Exchange Process," in Extractive Metallurgy of Uranium, a Short Course, Dept. of Met. and Mat. Sci., University of Toronto and Met. Soc. of Canad. Inst. of Min. and Met., Hydromet. Section (1978).

11.  "Uranium Extraction Technology," OECD Nuclear Energy Agency and IAEA, Paris, France (1983).

12.  McGarvey. F. X. and J. Ungar, "The Influence of Resin Functional Group on the Ion-Exchange Recovery of Uranium., J.South African Inst. Min. Met., 93 (1981).

13.  Naden, D. ard M. R. Bandy, "Choice and Design of Solid Ion Exchange Plants for Recovery of Uranium," J. Chem. Biotechnol., 29 (1979).

14.  Cloete, F.L.D., "The Relix Process for the Resin-in-Pulp Recovery of Uranium," J. South African Inst. Min. Met., 81, 66 (1981).

15.  Cloete, F.L.D. and Streat,M, U.S. Patent 3,551,118 (1970).

16.  Streat, M. and R. Y. Qassim, "Recovery of Uranium from Unclarified Liquors by Ion Exchange," in Proc. Int. Symp. on Hydromet (1973).

17.  George, D. R. and J. B. Rosenbaum, "New Developments and Applications of Ion Exchange Techniques for the Mineral Industry," in Proc. Conf. on Ion Exchange in the Process Industries. London, July 1976, Soc. Chem. Ind., London (1976).

18.  Traut, D. E., I. L. Nichols, and D. C. Seidel, "Design Requirements for Uranium Ion Exchange from Ammonium Biocarbonate Solutions in a Fluidized System," U.S. Bureau of Mines Report RI 8280, Washington, DC (1978).

19.  Traut, D. E., I. L. Nichols, and D. C. Seidel, "Design Requirements for Uranium Ion Exchange from Acidic Solutions in a Fluidized System," U.S. Bureau of Mines Report RI 8282, Washington, DC (1978).

20.  Himsley, A., "Performance of Himsley Continuous Ion Exchange System," in Proc. Symp. on Hydromet., Manchester, England, July 1981, Soc. Chem. Ind., London (1981).

21.  Himsley, A. and E. J. Farkas, "Operation and Design Details of a Truly Continuous Ion Exchange System," in Proc. Conf. on the Theory and Practice of Ion Exchange, Cambridge, England, July 1976, Soc. Chem. Ind., London (1976).

22.  Higgins, I. R., "Update of Chem-Seps Continuous Ion Exchange Operation Relations

to Uranium Recovery," in Proc. South Texas Uranium Sem. 1979, AIME, New York (1979).

23. Haines, A. K., "The South African Programme on the Development of Continuous Fluidized Bed Ion Exchange with Specific Reference to its Application to the Recovery of Uranium," in S. Africa Inst. Min. and Met. Orange Free State Colloquium, (November 1977}.

24. Nicol, D. I., "Design snd Operating Characteristics of the NIMCIX Contactor," in Proc. Annual Conf. Metallurgists, Can. Inst. Min., Montreal, Canada (1978).

25. Porter, R.R. and Rössing Uranium Ltd., U.K. Patent 1,382,450.

26. Wylie, R. J. M., " Rössing Uranium Reaches Rated Production," World Mining, 52 (1979).

27. Simonsen, H.A., D.W. Boydell, and H.E. James, "The Impact of New Technlogy on the Economics of Uranium Production from Low-Grade Ores," in 5th Int. Symp. of Uranium Soc., London, September 1980, Westbury House, London (1981).

28. Crouse,D.J. and K.B. Brown, "Amine Extract Process for Uranium Recovery from Sulfate Liquor, Vol. I., ORNL-1959 (1955).

29. Fisher,J. W. and A. J. Vivjurka, "Combined Ion-Exchange - Solvent Extraction Process (ELUEX) for Ammonium Diuranate Production," in Proc. Conf. on Ion Exchange in the Process Industries, London, England, July 1969, Soc. Chem. Ind., London,, 163 (1970).

30. Gilmore, A. J., V. M. McNamara, H. W. Parsons, and R. Semard, "Production of High Purity Ammonium Diuranate by the Sulphuric Acid Elution - Amine Extraction Process," Mines Branch Investigation Report IR-60-95, Canada Dept. of Mines and Technical Surveys, Ottawa, Canada (1960).

31. Lloyd, P. J., "Solvent Extraction in the South African Uranium Industry," J. South African Inst. Min. Metall., 465 (1962).

32. Shulz, W.W., Wheelwright, E.J., Godbee, H., Mallory, C.W., Burney, G.A., and Wallace, R.M., "Ion Exchange and Adsorption in Nuclear Chemical Engineering", Proc. Conf. on Progress and Prospects in Adsorption and Ion Exchange, 75th Annual Meeting of the American Institute of Chemical Engineers, Washington, D.C., 1983.

33. Radwaste Management Workshop, New Orleans, LA, January 12-14, 1977.

34. Short Course: "Radioactive Waste Management for Nuclear Power Reactors and Other Facilities," Alexandria, VA, May 2-6, 1983.

35. Larson, D.E. and P.W. Smith, "Treatment of Hanford Nuclear Processing Wastes to Permit Lohg-Term Storage," ARH-SA-51 (1969).

36. Schulz, W. W., "Distribution Data for Various Cs, Sr, and Tc Sorbents," Atlantic Richfield Hanford Company Quarterly Report, ARH-ST-110D, pp. 34-37 (1975).

37. Bibler, J.P., Wallace, R.M., and Bray, L.A., "Testing a New Cesium-Specific Ion Exchange Resin for Decontamination of Alkaline High-Activity Waste", to be presented at the 1990 Waste Management Meeting, Tucson, AZ, February 25-March 1, 1990.

38. Barnes, S.M., Pope, J.M., and Chapman, C.C., "Three Year's Progress of the West Valley Demonstration Project Vitrification System", in Waste Management Eighty Eight, R.G.Post, Ed., vol.2, pp 195-202, 1988.

39. Bibler, J.P., and Wallace, R.M., "Ion Exchange Processes for Clean-up of Dilute Waste Streams by the F/H Effluent Treatment Facility at the Savannah River Plant", Recent Developments in Ion Exchange-Proceedings IONEX'87, P.A.Williams and M.J.Hudson, Eds., pp 173-180, 1987.

40. King, L. J. and Ichikawa, M., "Pilot Plant Demonstration of the Decontamination of Low-Level Process Wastes by a Recycle Scavenging-Precipitation Ion-Exchange Process," ORNL-3863 (1965).

41. D'Amrosia, J. T., "Three Mile Island Waste Management, a DOE Perspective," presented at the American Institute of Chemical Engineers 1981 Summer National Meeting, August 19, 1981.

42. Walker, D.D., Bibler, J.P., Wallace, R.M., Ebra, M.A., and Ryan, J.P.,Jr., "Tecnetium Removal Processes for Soluble Defense High-Level Waste", Symposium Proceedings, Materials Research Society, C.M.Jantzen, J.A.Stone, R.C.Ewing Eds., vol. 44, pp 810-809, 1986.

43. Hudson, M.J. and Dyer, A., "The Use of Coordinating Copolymers to Extract [106]Ru from Simulated Nuclear Wastes", Solvent Extraction and Ion Exchange in the Nuclear Fuel Cycle, D.H.Logsdail and A.L.Mills Eds., pp 157-168, 1985.

44. Hooper, E.W., "The Application of Inorganic Ion Exchangers to the Treatment of Medium Active Effluents", Solvent Extraction and Ion Exchange in the Nuclear Fuel Cycle, D.H.Logsdail and A/L/Mills Eds., pp 157-168, 1985.

45. Hyder,M.L.,et al,"Processing of Irradiated, Enriched Uranium Fuel at the Savannah River Plant", DP-1500, 1979.

46. Burney, G. A., "Separation of Maro-Quantities of Actinide Elements at Savannah River by High Pressure Ion Exchange," Sep. Sci. Technol. 15, 763 (1980).

47. King, L. J., J. E. Bigelow, and E. D. Collins, "Experience in the Separation and Purification of Transplutonium Elements in the Transplutonium Processing Plant at Oak Ridge National Laboratory" in Transplutonium Elements - Production and Recovery, J. D. Navatril and W. W. Schulz, Eds., American Chemical Society, Washington, DC, pp. 134-145 (1981).

48. Lowe, J. T.. W. H. Hale, Jr., and D. F. Hallman, "Development of a Pressurized Cation Exchanger Chromatographic Process for Separation of Transplutonium Actinides," Ind. Eng. Chem. Process Design Develop. 10, 131 (1971).

49. Collins, E. D., D. E. Benker, F. R. Chattin, P. B. Orr, and G. R. Ross, "Multigram Group Separation of Actinide and Lanthanide Elements by LiCl-Based Anion Exchange," in Transplutonium Elements - Production and Recovery, J. D. Navatril and W. W. Schulz, Eds., American Chemical Society, Washington, DC, pp. 148-160 (1981).

50. Benker, D. E., F. R. Chattin, D. E. Colins, J. B. Knauer, P. B. Orr, R. G. Ross, and J. T. Wiggins, "Chromatographic Cation Exchange Separation of Decigram Quantities of Californium and Other Transplutonium Elements" in

Transplutonium Elements - Production and Recovery, J. D. Navatril and W. W. Schulz, Eds., American Chemical Society, Washington, DC, pp. 161-171 (1981).

51.  Campbell, D. O., "The Application of Pressurized Ion Exchange to the Separations of Transplutonium Elements," in Transplutonium Elements - Production and Recovery, J. D. Navatril and W. W. Schulz, Eds., American Chemical Society, Washington, DC, pp. 190-201 (1981).

52.  Wheelwright, E. J., "Kilogram Scale Purification of Americium by Ion Exchange," Sep. Sci. Technol. 15, 783 (1980).

53.  Prout, W. E. and A. E. Symonds, "Recovery of $^{233}U$ from Irradiated Th Oxide and Metal," DP-1036 (1967).

54.  Burney, G. A., "Cation Exchange Concentration of Aqueous $^{233}U$ Nitrate and Conversion to $^{233}UO_3$," DP-1047 (1966).

55.  Knighton, J. B., P. G. Hagan, J. D. Navatril, and G. H. Thompson, "Status of Americium-242 Recovery at Rocky Flats Plant" in Transplutonium Elements - Production and Recovery, J. D. Navatril and W. W. Schulz, Eds., American Chemical Society, Washington, DC, pp. 53-74 (1981).

56.  Gray, L. W., G. A. Burney, T. A. Reilly, T. W. Wilson, and J. M. McKibben, "Recovery of Am-241 from Aged Plutonium Metal," in Transplutonium Elements - Production and Recovery, J. D. Navatril and W. W. Schulz, Eds., American Chemical Society, Washington, DC, pp. 93-108 (1981).

57.  Doto, P. C., L. E. Bruns, and W. W. Schulz, "Solvent Extraction Process for Recovery of Americium-241 at Hanford" in Transplutonium Elements - Production and Recovery, J. D. Navatril and W. W. Schulz, Eds., American Chemical Society, Washington, DC, pp. 109-128 (1981).

58.  Ramsey, H. D., D. G. Clifton, S. W. Hayter, R. A. Pennman, and E. L. Christensen, "Status of Americium-241 at Los Alamos National Laboratory," in Transplutonium Elements - Production and Recovery, J. D. Navatril and W. W. Schulz, Eds., American Chemical Society, Washington, DC, pp. 75-91 (1981).

# ACTIVITY REMOVAL FROM LIQUID WASTE STREAMS BY SEEDED ULTRAFILTRATION

E W HOOPER* AND R M SELLERS+

*AEA Technology, Harwell Laboratory, Harwell, Oxfordshire
+ Nuclear Electric plc, Technology Division,
Berkeley Nuclear Laboratories, Berkeley, Gloucestershire

## ABSTRACT

Modern cross-flow filtration techniques allow the removal of very fine
particles of solid material from aqueous streams. When the streams contain
radionuclides, some small additional activity removal may be achieved by
cross-flow filtration even after the stream has been treated by
conventional filtration and ion exchange, and despite the concentration of
radionuclides being exceedingly small. Much greater activity removal may
be achieved, however, if the stream is dosed with additives which absorb or
coprecipitate the radionuclides. This paper describes the development and
characteristion of several such seed materials (finely divided inorganic
ion exchange materials) and an investigation of factors which influence
their effectiveness. Individual seeds are, in general, targetted on
specific radionuclides (nickel hexacyanoferrate(II) is specific for caesium
for instance) and are unlikely, therefore to offer an across-the-board
approach to activity removal. Combinations or cocktails of seeds may,
however, allow such an approach and this possibility has also been
investigated. It is found that under certain circumstances seeds can
interfere with one another and actually give worse results than when used
individually. With careful selection of materials and conditions, however,
it is possible to achieve substantial reductions in a variety of
radionuclides simultaneously.

## INTRODUCTION

Ultrafiltration is a relatively new development in filtering science that

promises to give much better separations of small particles from aqueous

streams than has hitherto been possible. Large scale applications to date

have been mainly in the dairy and food industry; the process is, however,

largely untested in the nuclear industry, though it forms a key part of the Enhanced Actinide Recovery Plant (EARP) being built at British Nuclear Fuels plc Sellafield site. The aqueous waste streams arising at nuclear power plants are typically contaminated with very low levels of radioactive nuclides, and when necessary these are treated by conventional filtration and ion exchange prior to discharge to the environment. Many of these radionuclides are insoluble and so potentially can be readily removed by ultrafiltration. Soluble species cannot be filtered directly but if additives or "seeds" which absorb the species are added then even soluble radionuclides can be dealt with. Seeds specific to a number of radionuclides have been identified (eg nickel or copper hexacyanoferrate (II) for $Cs^{137}$ [1]), and optimum conditions of, for instance, pH and concentration identified. The use of a mixture or cocktail of seeds to remove a variety of radionuclides simultaneously is an obvious extension of the seeded ultrafiltration principle. This paper describes a study of single and multiple seeding for the removal of a range of radionuclides from an aqueous stream. Studies in the absence of seeds are also reported for comparative purposes.

The seed materials can enhance radionuclide removal by one of three basic mechanisms: precipitation, co-precipitation or absorption. Precipition and co-precipitation imply the generation of insoluble species in the aqueous waste either by pH change or addition of suitable chemicals in solution. With absorption, however, pre-formed solid material (or more usefully slurries of solid material) can be added to the aqueous waste to be treated. The absorption process is usually one of ion exchange (eg with surface hydroxyls) and this is generally the most versatile and useful of the three basic mechanisms. Seeded ultrafiltration is thus a novel way of utilising ion exchange materials.

A variety of ultrafilter membrane materials is available. This study has been conducted with membranes consisting of $ZrO_2$ on a carbon support, but similar results are to be expected with other materials. The pore size of the membrane (often quoted as a molecular weight cut off) is an important parameter characterising membranes but by and large seed materials produced by precipitation in aqueous solution (as here) have particle sizes much larger than the membrane pore size and hence virtually none of the seed can pass through the filter. The selection of more porous membranes can give higher permeation rates but can seriously compromise the effectiveness of the seed.

**EXPERIMENTAL**

### Apparatus

The ultrafiltration tests were carried out in a small rig consisting of a solution reservoir, a pump (Micropump type 120-445-10A) and a 140 mm long x 5 mm internal diameter (inner surface area 22 cm$^2$) vertically mounted ultrafiltration membrane as shown diagrammatically in Fig 1. The unit was constructed of stainless steel, apart from the inlet and outlet tubes which were polyethylene. For most of the work a "Carbosep" M4 membrane with a nominal pore size of 2 nm (molecular weight cut off 20,000) was used. The operating pressure was 20 psi (1.3 bar), well below that likely to produce a rapid build up of a fouling layer. Dead space in the system was 30 ml (~10% of the feed volume) and permeation rates were 100 ml h$^{-1}$ (~ 1.0 m$^3$/m$^2$/d). After each experiment the membrane was washed clean by pumping first 0.5 M nitric acid for one hour, followed by 0.5 M sodium hydroxide solution for one hour and then by distilled water for one hour. At the start of each new experiment the membrane was conditioned by pumping through water adjusted to the pH of the experiment by the addition of sodium hydroxide or nitric acid. This procedure has been found to clean the membrane very effectively, removing > 95% of any adherent material.

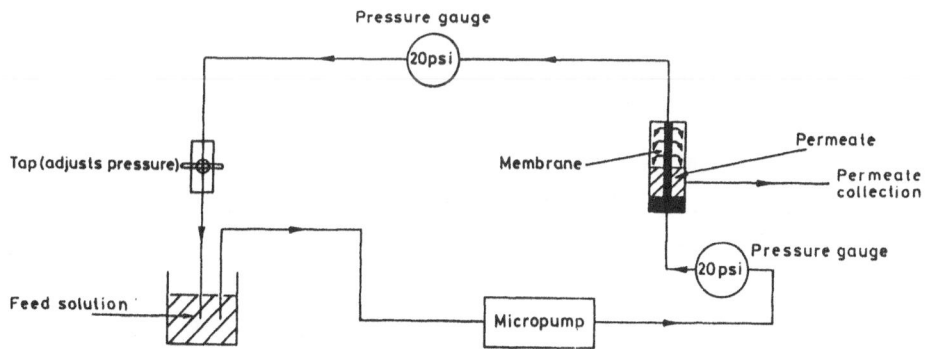

FIG. 1  SCHEMATIC DIAGRAM OF SMALL SCALE ULTRAFILTRATION UNIT.

## Materials

The four seed materials considered here were prepared as follows:

Sodium Nickel Hexacyanoferrate(II) (NHCF):[*] Equal volumes of 0.094 M $Na_4Fe(CN)_6$) and 0.113 M $Ni(NO_3)_2.6H_2O$ solutions were mixed well and a portion of the resulting slurry was added to the waste simulant to provide the required $Fe(CN)_6^{4-}$ concentration. The nominal composition of the nickel hexacyanoferrate as prepared was $Na_{1.6}Ni_{1.2}Fe(CN)_6$. (In practice the composition of the precipitated material does not follow exactly that of the solutions as made up - but for the present purposes this is irrelevant). (*Tetra sodium hexacyanoferrate (II)).

Manganese Dioxide ($MnO_2$):[+] 20 ml of an aqueous solution of potassium permanganate (14.4 g $KMnO_4$ per litre) were adjusted to pH 11 using sodium hydroxide solution. 400 mg of solid sodium dithionite were then slowly added over 15-20 minutes whilst maintaining the pH at 11 ± 0.5. The slurry was centrifuged with water and then washed with water adjusted to the pH of the solution to be treated. Addition of the precipitate to 1 litre of solution gave a manganese concentration of 100 ppm. Lower concentrations were obtained by slurrying the precipitate of $MnO_2$ in 20 ml of distilled water and taking an appropriate aliquot.[+] ([+] Manganese (IV) oxide).

A number of experiments were carried out with a sample of $MnO_2$ prepared by alternative route involving the oxidation of manganous ion by permanganate.

$$3Mn^{2+} + 2MnO_4^- + 2H_2O \rightarrow 5MnO_2 + 4H^+$$

1 ml of a manganese nitrate solution (93.8 g $Mn(NO_3)_2.6H_2O$ per litre) was added to 10 ml of a potassium permanganate solution (3.45 g $KMnO_4$ per litre). The slurry was centrifuged and washed with water adjusted to the pH of the solution to be treated. Addition of the precipitate to 300 ml of solution gave a manganese concentration of 100 ppm.

Hydrous Titanium Oxide (HTiO):[X] 10 ml of a 15% w/v titanium sulphate solution were treated with 1 M sodium hydroxide solution until no further precipitate formed. The precipitate was centrifuged, washed with water to remove excess alkali and then washed with water adjusted to the pH of the

([X]Titanium (IV) oxide)

solution to be treated. The slurry was then made up to 10 ml with distilled water. 1 ml of this slurry when added to 300 ml of solution gave a titanium concentration of 100 ppm.

Iron(III)Hydroxide (FH): For most experiments this was prepared in situ by addition of a solution of (Iron III)nitrate to the waste solution followed by adjustment of the pH to 7. In other tests the iron was added as Iron(III) sulphate and the Iron(III)hydroxide precipitated by bubbling with oxygen.

All absorbers except $MnO_2$ were prepared within an hour of addition. $MnO_2$ was prepared in ca 20 ml portions which was used over a two week period.

For the multiple seeding trials the seeds were added in the following order : ferric hydroxide, nickel hexacyanoferrate(II), hydrous titanium oxide, manganese dioxide.

In this work the term "1 ppm of absorber" means:

|  |  |
|---|---|
| for cyanoferrate (NHFC) | 4.7 µmolar |
| for $MnO_2$ | 18.2 µmolar |
| for HTiO | 20.8 µmolar |
| for ferric hydroxide | 17.0 µmolar |

Feed solution: All experiments were carried out with a simulated waste having the composition shown in Table 1. The isotopes were as supplied by Amersham and were present as either their chloride or nitrate salts.

TABLE 1
Composition of waste simulant solution (pH7)

| Nuclide | Activity (Bq l$^{-1}$) | Nuclide | Activity (Bq l$^{-1}$) |
|---|---|---|---|
| $Cr^{51}$ | 41100 | $Zr^{95}$ | 8000 |
| $Mn^{54}$ | 25200 | $Ru^{106}$ | 3200 |
| $Fe^{59}$ | 6700 | $Ag^{110m}$ | 22500 |
| $Co^{60}$ | 24200 | $Sb^{125}$ | 98400 |
| $Ni^{63}$ | 2400 | $Cs^{137}$ | 84000 |
| $Zn^{65}$ | 6100 | $Ce^{139}$ | 3900 |
| $Sr^{90}$ | 80 | $Pu^{239}$ | 0.59 |

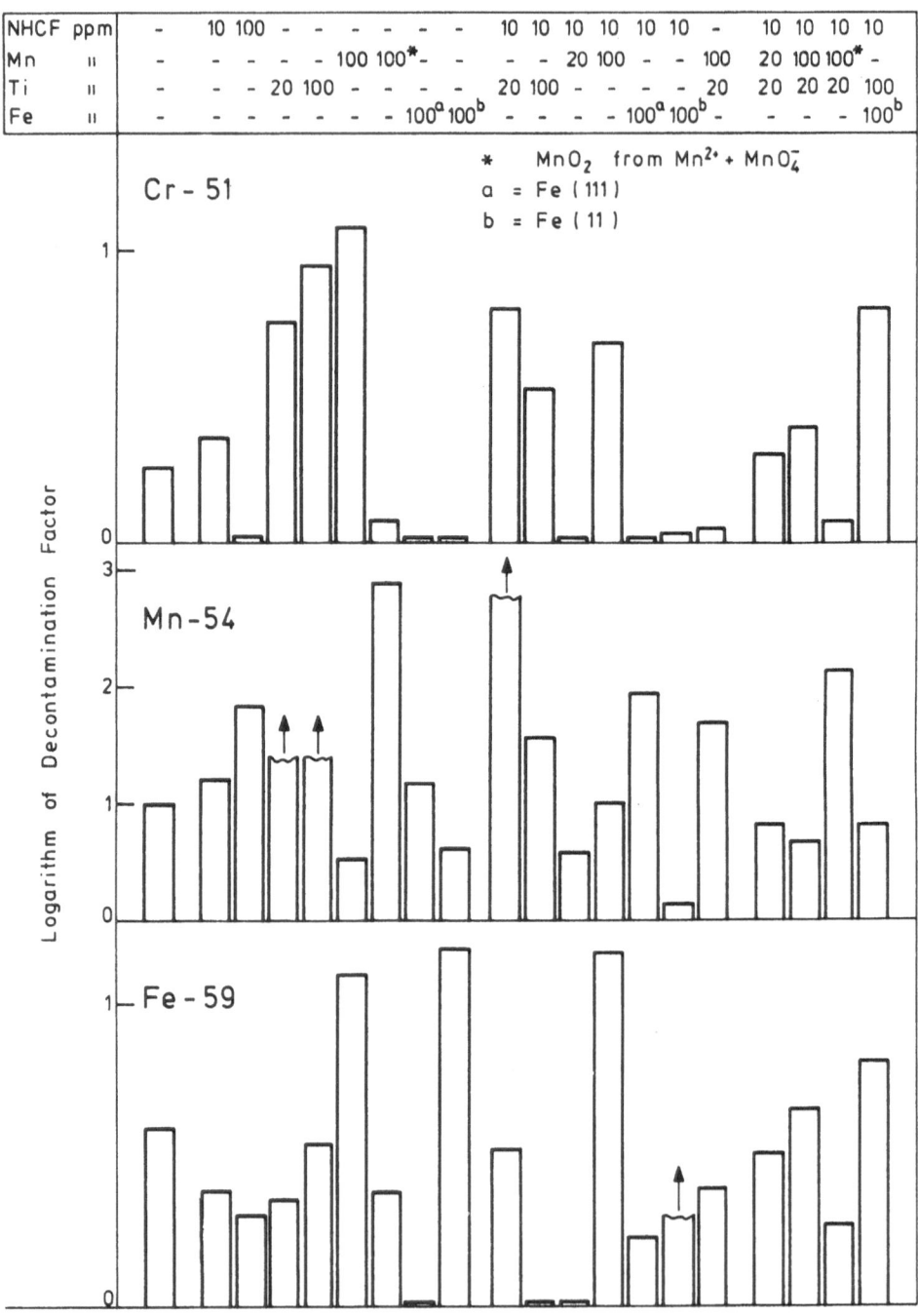

FIG. 2.

## Procedure

Concentrations of gamma-emitting nuclides were determined by high
resolution gamma spectrometry using an EG and G Ortec Gamma x HP Ge coaxial
detector coupled via a Model 918 Adcam Multichannel buffer (MCB) to a
Digital PC 350 computer. Samples for actinide analysis by high resolution
alpha spectrometry were prepared by electrodeposition onto stainless steel
planchets after anion exchange separation as described in Cross and Hooper
[2]. These were then counted under vacuum using silicon detectors via a
model 918 MCB into a Zenith HX-12E computer. Strontium analysis was
achieved by counting $SrCO_3(s)$ samples after radiochemical separation from
the test solutions using a carrier. The samples were then counted in a
Sugarman detector under a methane/argon atmosphere.

Radionuclide concentrations were determined in the feed and permeate
for all runs and these values used to determine the
decontamination factor (DF) defined as follows:

$$DF = \frac{\text{feed activity (Bq ml}^{-1})}{\text{permeate activity (Bq ml}^{-1})}$$

### RESULTS

### Activity Removal in the Absence of Seeds

Even in the absence of seeds, appreciable removal of several radionuclides
was found. For $Mn^{54}$, $Fe^{59}$, $Co^{60}$, $Ce^{139}$ and $Pu^{239}$ DFs were typically 3-5
indicating that some, but not all, of these species were present in some
microcolloidal form too large to pass through the ultrafiltration membrane
(but not removed by more conventional filtration methods). For $Zn^{65}$ and
$Ag^{110m}$ DFs were very high (> 30) and practically all of these nuclides are
removed. $Cr^{51}$, $Sr^{90}$, $Zr^{95}$, $Ru^{106}$, $Sb^{125}$ and $Cs^{137}$ were practically
unaffected by ultrafiltration and these species can be assumed to exist
primarily in solution.

### Activity Removal in the Presence of Individual Seeds

Nickel Hexacyanoferrate (II)DFs for caesium in excess of 50 were found
with this seed at concentrations of 10 ppm; only slightly higher DFs were
found at 20 ppm, and only a little smaller at 5 ppm indicating only a very

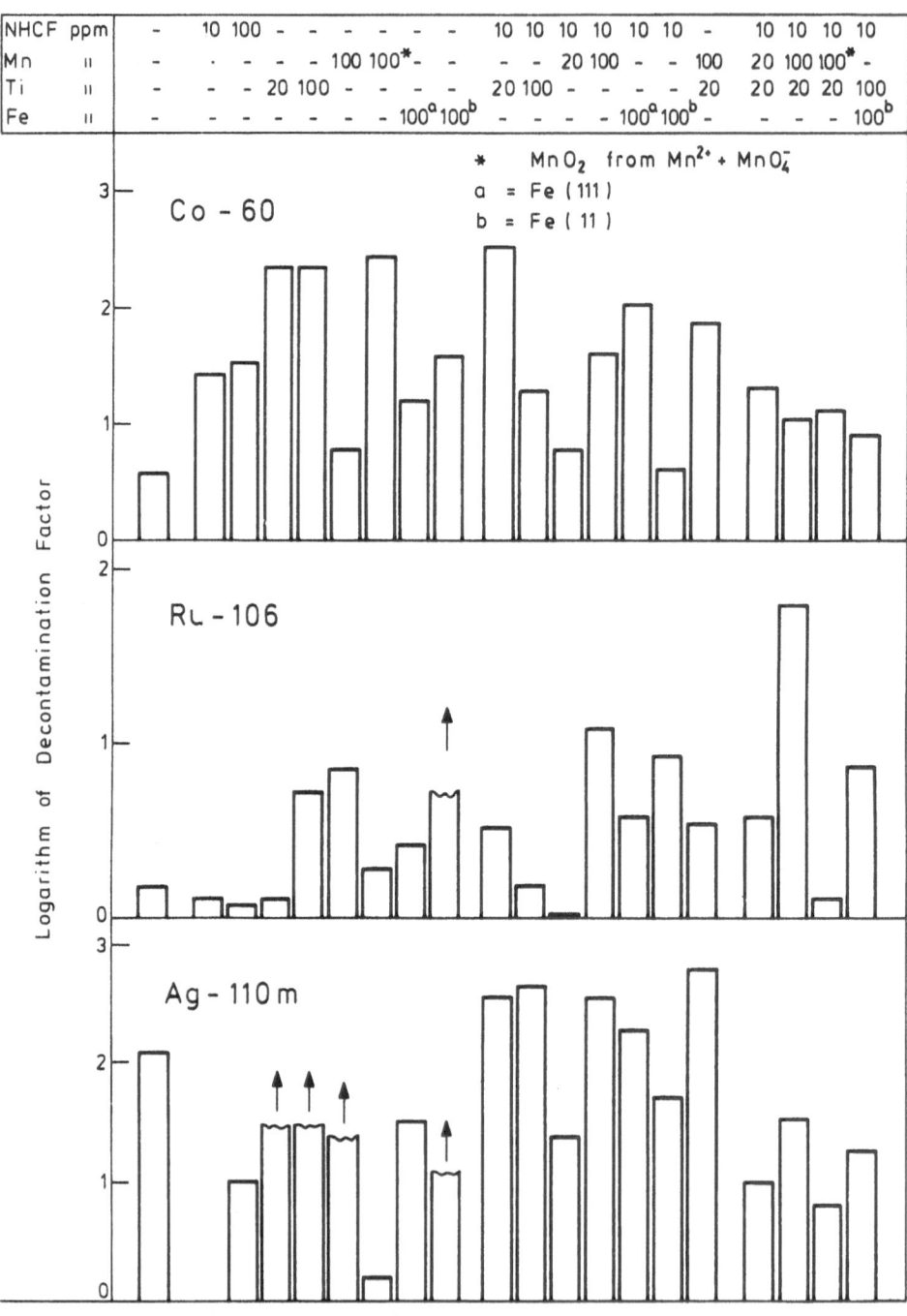

FIG. 3.

weak dependence on concentration. Removal of $Mn^{54}$ and $Co^{60}$ was slightly enhanced in the presence of 10 ppm of this seed; the remaining nuclides were unaffected.

When a solution of sodium hexacyanoferrate(II) was added to the waste to give a final concentration of 10 ppm hardly any caesium was removed, but removal of $Mn^{54}$, $Co^{60}$ and $Zn^{65}$ was again higher than in absence of seeds. We interpret these observations as indicating that caesium is removed by an ion exchange mechanism on nickel hexacyanoferrate(II), whereas manganese, cobalt and zinc are made filterable through the formation of insoluble hexacyanoferrates(II) (note that the nickel hexacyanoferrate(II) contains some free $Fe(CN)_6^{4-}$).

Hydrous Titanium Oxide: This seed was investigated at both 20 and 100 ppm concentrations and brought about a marked improvement in the removal of $Cr^{51}$, $Mn^{54}$, $Co^{60}$, $Sr^{90}$ and lesser increases in the removal of $Fe^{59}$, $Zr^{95}$, $Ru^{106}$ and $Ce^{139}$. Removal was generally slightly higher at the higher concentration but for none of these nuclides was the dependence on concentration pronounced.

Manganese Dioxide: This additive was investigated only at concentrations of 100 ppm, and gave DFs of about 10 for $Cr^{51}$, $Fe^{59}$ and $Ce^{139}$ and somewhat lesser values for $Co^{60}$, $Sr^{90}$ and $Ru^{106}$. The removal of the other radionuclides, including (perhaps surprisingly) $Mn^{54}$, was unaffected by the presence of this seed.

Iron(II)Hydroxide: Precipitation of this seed to give a concentration of 100 ppm in the waste stream had practically no effect on the removal of any of the radionuclides except $Ru^{106}$ for which the efficiency of removal was enhanced by about a factor of two. When the iron(II) hydroxide was precipitated in the waste solution by addition of a solution of iron(II) sulphate and bubbling with oxygen, $Ru^{106}$ removal was again enhanced by about a factor of two, and removal of $Fe^{59}$ increased by about 8-fold. In the latter case it seems likely that a co-precipitation mechanism was operating.

## Activity Removal With Seeds Used in Pairs

Of the four seed materials tested on their own, nickel hexacyanoferrate(II) was the only one capable of removing a significant fraction of the $Cs^{137}$ and so these paired tests were done with this absorber and each of the other three in turn. In all cases 10 ppm nickel hexacyanoferrate(II) was used together with either 20 or 100 ppm of the second seed. The results obtained are shown in Figure 2-4. Within the experimental errors involved in determining very low levels of activity, the behaviour of the paired seeds was roughly as expected by summing that found with the seeds when used individually, i.e. no evidence for any interference or synergism between the seeds was found. For $Mn^{54}$, $Cs^{137}$, $Ag^{110m}$, $Co^{60}$ and total βγ activity the best results were obtained with a combination of nickel hexacyanoferrate(II) and hydrous titanium(IV)oxide and for $Fe^{59}$ and $Ru^{106}$ with nickel hexacyanoferrate(II) and manganese (IV)oxide. Only quite small DFs were obtained for the other nuclides.

## Activity Removal With Seeds Used in Multiple

The results obtained with seeds used in pairs suggest that a combination of nickel hexacyanoferrate(II) (for $Cs^{137}$), hydrous titanium(IV)oxide (for $Mn^{54}$, $Ag^{110m}$, and $Co^{60}$) and manganese(IV)oxide (for $Fe^{59}$ and $Ru^{106}$) might give the best overall reduction in activity and this combination was therefore selected for initial investigation. The mixtures tested contained 10 ppm nickel hexacyanoferrate(II) plus either 20 or 100 ppm of each of the other two seeds. For five of the six main nuclides for which high DFs had been obtained in the paired-seeds tests removal was less than expected. The exception was $Cs^{137}$. The experiments were repeated using manganese (IV)oxide prepared by oxidation of $Mn^{2+}$ ion by permanganate and much more satisfactory results were obtained.

A mixture of 10 ppm nickel hexacyanoferrate plus 100 ppm hydrous titanium (IV)oxide and 100 ppm iron(II) hydroxide (generated _in situ_ from the aerial oxidation of iron(II) ion) gave results similar to those expected if the DFs obtained with the seeds individually were combined.

In terms of overall βγ activity removal, the optimum combination was 10ppm NHCF + 100 ppm HTiO + 100 ppm FH.

## Ni⁶³ Removal

Ni[63] is a ß-emitting radionuclide and is difficult to assay at the concentrations of interest in the presence of other active species. The behaviour of this isotope was, therefore, investigated in some separate experiments using a feed solution containing just this isotope at a concentration of 2400 Bq l⁻¹ at pH 7. In the absence of any seeds a DF of 14 was achieved and this increased to 100-200 with 100 ppm NHCF, 100 ppm HTiO or 100 ppm $MnO_2$. Using only 10 ppm of these seeds gave DFs in the range 30-50. A combination of 10 ppm NHCF + 100 ppm $MnO_2$ (prepared from $Mn^{2+}$ + $MnO_4^-$) gave a DF of only ~80 and a combination of 10 ppm NHCF + 20 ppm HTiO one of only ~30, whilst with a mixture of all three (10 ppm NHCF + 100 ppm $MnO_2$ + 100 ppm HTiO) the figure was ~25, even though the improved $MnO_2$ preparation had been used.

## DISCUSSION

All of the four absorber materials tested in this study enhanced the removal of some of the fourteen main radionuclides considered when used individually and different absorbers enhanced the removal of different radionuclides. A summary of the main enhancements in activity removal is given in Table 2. Only very small enhancements in removal were achieved with $Cr^{51}$, $Sb^{125}$ etc hence the comparatively modest overall DFs for total ßγ activity. $Pu^{239}$ is insoluble at the pH investigated and, although the seeds did not enhance removal significantly, reasonable DFs were achieved by ultrafiltration alone.

Mixtures of nickel hexacyanoferrate(II) with the $MnO_2$, $HTiO_2$ and $Fe(OH)_3$ gave results equivalent to the sum of those obtained when the seeds were used singly. This was not true however when mixtures of NHCF, $MnO_2$ and HTiO were used; similar behaviour was found with mixtures of $MnO_2$ and HTiO. The most probable explanation for this is that the two seeds interfere with one another, probably through precipitation of material from one seed onto particles of the other. The use of $MnO_2$ prepared by oxidation of $Mn^{2+}$ by $MnO_4^-$ gave a seed which did not cause nearly such a large decrease in DFs when used with NHCF and HTiO, implying that reprecipitation of $MnO_2$ or adsorption of other components in the mixture (such as dithionite or its oxidation products) may be responsible. If the former is the case it is difficult to understand why material from the two

TABLE 2

Summary of most significant enhancements in

activity removal using seed materials

| Radionuclide | Most suitable absorber(s) |
|---|---|
| $Cr^{51}$ | none |
| $Mn^{54}$ | $MnO_2$, HTiO |
| $Fe^{59}$ | $Fe(OH)_3$#, $MnO_2$, |
| $Co^{60}$ | HTiO, $MnO_2$* |
| $Ni^{63}$ | NHCF, $MnO_2$, HTiO |
| $Zn^{65}$ | high DF by UF alone |
| $Sr^{90}$ | generally low DF, but some removal with $MnO_2$* |
| $Zr^{95}$ | $Fe(OH)_3$# |
| $Rn^{106}$ | $MnO_2$, HTiO, $Fe(OH)_3$# |
| $Ag^{110m}$ | high DF by UF alone |
| $Sb^{125}$ | none |
| $Cs^{137}$ | NHCF |
| $Ce^{139}$ | small DF in absence of seeds, some enhancement with $Fe(OH)_3$# |
| $Pu^{239}$ | small DF in absence of seeds, some enhancement with $Fe(OH)_3$# |

*Prepared by $Mn^{2+}$ + $MnO_4$

#Prepared by aerial oxidation of $Fe^{2+}$

different sources behaved so differently. Multiple seeding is thus feasible, but careful attention needs to be given to the mutual compatability of the materials selected. Loss of effectiveness due to interaction between seeds can be overcome to an extent by increasing seed concentrations, though in none of the cases investigated, was there a

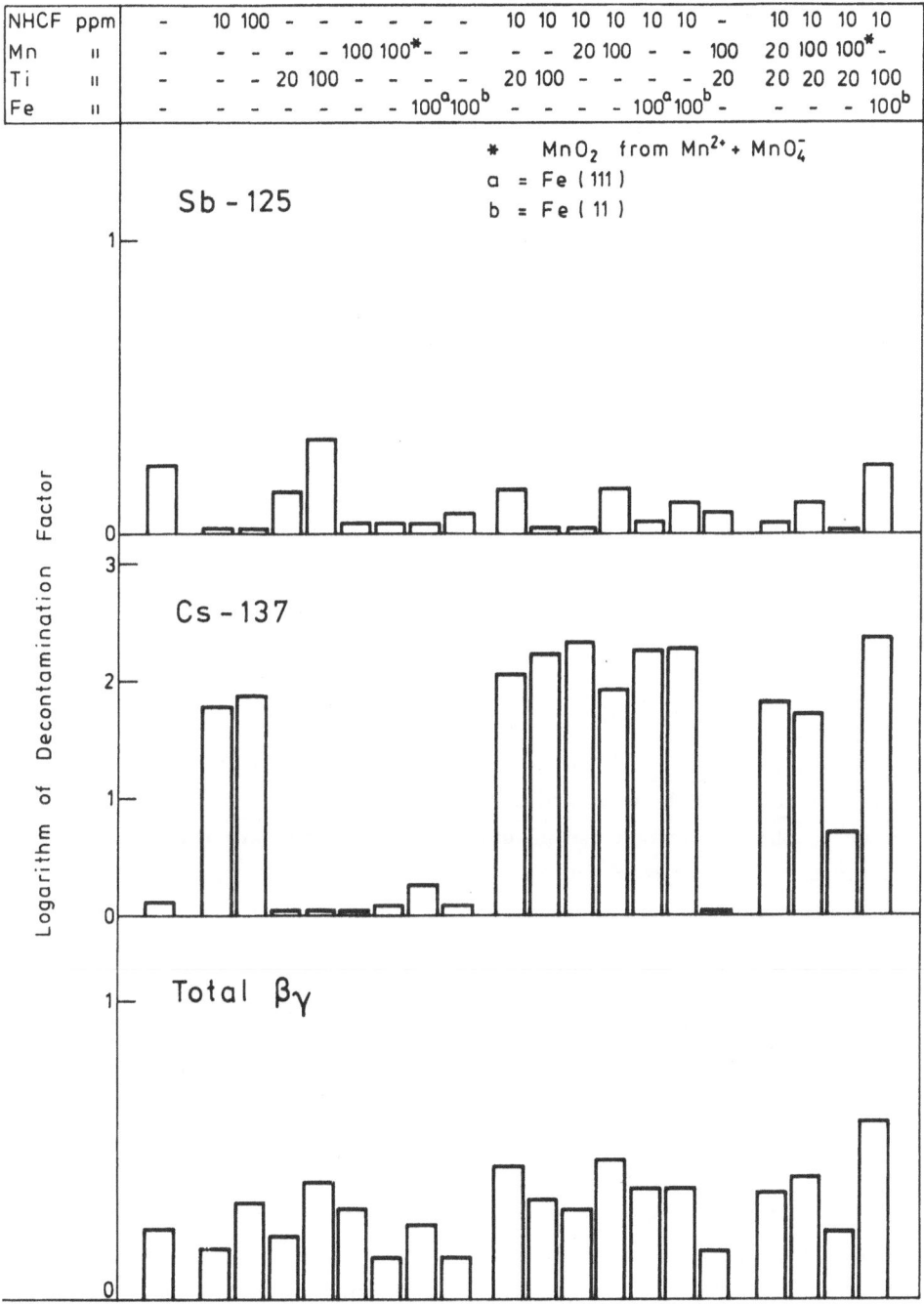

FIG. 4.

strong dependence of DF on concentration. In some instances (eg removal of $Fe^{59}$ by $Fe(OH)_3$) it is necessary to precipitate the seed in the waste solution (in situ generation) to bring about enhanced radionuclide removal. In such cases activity removal is presumably by a co-precipitation mechanism rather than ion exchange. We have not tested this point specifically but it seems reasonable to conclude that the order in which the seeds are introduced into the solution is important - an in situ generated seed would need to come first, followed by pre-formed absorbers. The incorporation of in situ generated additives introduces, therefore, an extra element of complexity and this is not desirable from a waste treatment process point of view.

From a radiological standpoint (that is in terms of dose to critical groups) the most important radionuclides likely to occur in nuclear power station liquid effluents are $Cs^{137}$ and $Co^{60}$. No single seed can reduce both these, but a mixture of, for instance, NHCF and HTiO, readily provides reductions of at least 98% (DF >50). Incorporation of $MnO_2$ prepared by $Mn^{2+}$ and $MnO_4^-$ provided reductions in a broader spectrum of radionuclides, but several important ones are only reduced by a small amount and overall DFs for βγ activity of only ca.3 can be achieved with the simulant we have investigated. Further novel absorbers need to be identified especially if large reductions (eg DF >10) in total activity are to be achieved. The radionuclides which have proved the most difficult to remove in our tests are $Cr^{51}$, $Sb^{125}$ and to a lesser extent $Ru^{106}$: these are present in our solutions in anionic form (as they would be in real effluents). Absorbers for anionic species need therefore to be identified or the possibility of converting these species to cationic form, for instance by reduction, needs to be considered.

This study has been concerned specifically with the reduction of already very low concentrations of radionuclides from nuclear power station effluents but there is no reason why aqueous wastes contaminated with even quite low levels of other materials, such as transition metal ions, should not be treated in a similar manner. The correct choice of seeds is essential if a variety of different species are to be treated simultaneously. In effect seeded ultrafiltration offers a way of tailoring ion exchange to the treatment of complex mixtures of chemicals. Although not considered here it is generally possible to achieve higher absorption efficiencies (amount of material absorbed per unit of absorber) with UF seeds than in conventional ion exchange beds. Seeded UF also does away with the need for the material to be granular in form and of sufficient strength that it could be used in a conventional ion exchange bed (often a problem with inorganic absorbers), and in principle allows a wider range of absorbers to be considered.

## ACKNOWLEDGEMENT

This paper is published by permission of the United Kingdom Atomic Energy Authority and the Central Electricity Generating Board.

## REFERENCES
1.  Reed, I M, UKAEA Report No AERE-R 11096, 1984
2.  Cross, J E and Hooper, E W, UKAEA Report No AERE-R 12559, 1987

# RECOVERY OF FISSION PRODUCTS BY FIXATION ON INORGANIC ADSORBENTS

S. ZOUAD, C. LOOS-NESKOVIC
Laboratoire P. Sue, C. E. N. Saclay
91191 Gif-sur-Yvette, France

E. GARNIER
Laboratoire de Cristallographie Minérale, Université de Poitiers
40, avenue du Recteur Pineau
86022 Poitiers Cedex, France

J. JEANJEAN, M. FEDOROFF
Centre d'Etudes de Chimie Métallurgique
15, rue Georges Urbain
94407 Vitry-sur-Seine, France

## ABSTRACT

The sorption of cesium and strontium from radioactive waste was studied on several inorganic absorbents. Cesium can be retained with a high efficiency on zinc or nickel ferrocyanides. The best results on columns were obtained for products with a high kinetics of fixation: mainly mixed alkaline-zinc ferrocyanides and mixed alkaline-nickel ferrocyanides with small crystals. The fixation of strontium was studied on polyantimonic acid and on two types of phosphatoantimonic acids. At low levels of strontium the best efficiency is achieved with the first product. One type of phosphatoantimonic acid with a layered structure is attractive, if a larger quantity of strontium has to be retained.

## INTRODUCTION

Cesium and strontium with half-lifes close to 30 years are the main fission products in radioactive wastes from the nuclear industry and nuclear research centres after a cooling time of some years. It seems attractive to separate these elements from primary wastes in a way suitable for further storage in safe conditions. Fixation on inorganic absorbents from an aqueous solution is a promising way of separation as such products are generally radiation resistant and are stable enough for long term storage. Among possible products, we chose insoluble ferrocyanides known for their affinity for cesium (1-13) and antimonic acid known for its affinity for strontium (14-20). The aim of this study was to verify the efficiency of such products and to develop our knowledge on the fixation mechanisms in order to determine the optimum conditions of separation. For that purpose, fixation kinetics and sorption isotherms were studied by batch experiments and connected to the composition and crystalline structure of the products. Several types of zinc and nickel ferrocyanides were studied for cesium sorption. Antimonic acid was compared to two types of phosphatoantimonic acids for strontium sorption. The results of batch experiments were then applied to separations on columns.

## MATERIALS AND METHODS

### Ferrocyanides

One nickel ferrocyanide was supplied from Recherche Apliquée du Nord (59330 Hautmont, France). All other products including nickel and zinc ferrocyanides were prepared using various precipitation methods (21,22) or by growth localized on alkaline ferrocyanide crystals (23,24). For column utilizations, products were mainly prepared by this last method. In this procedure, sodium or potassium ferrocyanide crystals are placed in a solution of nickel or zinc salt. At a selected concentration and temperature, a sphere of insoluble ferrocyanide film is formed around each crystal. This film grows and solidifies, while the initial crystal is consumed. The size of formed beads is related to the temperature, initial concentration of solution and size of crystals (24). This method leads with a good yield to products with a controlled granulometry suitable for using in chromatographic columns.

All products were analysed for their composition by non-destructive activation analysis, atomic absorption spectrometry, inductively coupled plasma atomic emission spectrometry and complexometry. Examples of compositions of products used in this study are given in Table 1. In many cases, the apparent stoichiometry is not fulfilled. The excess of cations is generally balanced if a complete determination of the anions, coming from the starting salts and adsorbed in the products, is performed. A deficiency of measurable cations may be attributed to the presence of $H^+$ ions.

TABLE 1

Compositions[a] and structures of ferrocyanides prepared
by classical precipitation (A) and by localized growth on alkaline ferrocyanide (B)

| Method | | Zn | Ni | Na | K | H[b] | $H_2O$ | Structure | $C_{max}$[e] |
|---|---|---|---|---|---|---|---|---|---|
| 1 | A | 2.02 | - | <0.1 | - | - | 5.5 | trigonal (27)[d] | 1.6 |
| 2 | A | 1.55 | - | 0.89 | - | - | 4.9 | rhomb (28)[d] | 1.0 |
| 3 | B | 1.60 | - | 0.03 | 0.96 | - | 3.4 | rhomb (25,28)[d] | 0.95 |
| 4 | A | - | 1.55 | 0.77 | - | 0.13 | 7.5 | f.c.c. | 0.83 |
| 5 | A | - | 1.76 | - | - | 0.48 | 9 | f.c.c. | 0.46 |
| 6[c] | A | - | 1.28 | - | 1.13 | 0.31 | 2.65 | f.c.c. | 0.96 |
| 7 | B | - | 0.90 | 1.82 | - | - | 3.9 | cubic (?) | 0.40 |

a : concentrations are given in atoms per iron atom
b : determined by difference in order to achieve ionic neutrality
c : purchased from Recherche Appliquée du Nord
d : structure described in the given references
e : retention capacity for Cs in atoms per iron atom of the solid

Nickel ferrocyanides can be represented by the general formula $M^I_{2x}Ni_{2-x}Fe(CN)_6$ $yH_2O$, where $M^I$ is an alkaline element, a proton or a mixture of them. The values of x vary from 0.2 to 1. A composition close to the $Na_2NiFe(CN)_6$ can be achieved by the localized growth method. All the X-ray diffraction diagrams can be indexed by a f.c.c. structure with a mean parameter of 10.2 Å except for some products prepared by localized growth. Almost all studied zinc ferrocyanides belong to one of the two following types. Mixed compounds have a $M^I_2Zn_3[Fe(CN)_6]_2$ $yH_2O$ composition and a rhomboedric structure (25,26). The simple ferrocyanide has a $Zn_2Fe(CN)_6$ $2H_2O$ composition and a trigonal structure (27). However a product with the same zinc to iron ratio but with a new cubic structure was observed (28). Precipitated products are composed of small crystals with mean diameters of 0.06 µm for nickel ferrocyanides and 0.2 µm for zinc ferrocyanides. Beads prepared by localized growth are composed of crystals of various shapes and sizes. In the outer part of the beads, crystals cannot be distinguished by scanning electron microscopy; the size of crystals is larger as the distance to the surface increases. Some of them reach 10 µm in their larger dimensions.

## Polyantimonic and phosphatoantimonic acids

Hydrated antimony pentoxide (HAP) was purchased from Carlo Erba, Milano, Italy. Analyses performed by nuclear activation and thermogravimetry led to a $Sb_2O_5, 2.54 \pm 0.02$ $H_2O$ composition. The crystallographic structure belongs to the pyrochlore type with a Fd3m space group and a 10.35 Å parameter. The crystallographic positions of the atoms before and after strontium fixation were determined from the diffraction line intensities. The reliability factor R $= \Sigma |Fo-Fc| / \Sigma Fo$, where Fo and Fc are the observed and calculated structure factors, was obtained after least squares refinement of the crystallographic parameters (program AFFINE ). Optical and electron microscopy showed aggregates of approximately 100 µm diameter, consisting of small particles (0.08 to 0.18 µm).

Two types of phosphatoantimonic acids, prepared at the Laboratoire de Chimie des Solides (29,30), Université de Nantes, France, were studied: $H_3Sb_3P_2O_{14}$, n $H_2O$ (RH3) and $H_5Sb_5P_2O_{20}$, 7.3 $H_2O$ (RH5). RH3 has a rhomboedric structure $R\overline{3}m$ consisting of lamellae of $SbO_6$ octrahedra connected to $PO_4$ tetrahedra and including large channels. Electron microscopy shows small crystals (<0.2 µm) aggregated in particles (1 µm). RH5 has a tridimensional orthorhombic structure Pnnm or Pnn2. Electron micoscopy shows sticks (length : 2 µm, thickness 0.7 µm) consisting of a few crystals.

## Batch experiments

Radioactive $^{134}Cs$ and $^{25}Sr$ were obtained by irradiating cesium chloride or strontium nitrate in the neutron flux of a nuclear reactor using the facilities of the Pierre Sue Laboratory of Saclay. A known amount of absorbent was shaken with a solution containing one of the radioactive tracers. After a certain time, the solid was separated from the solution by filtration. The quantity of element sorbed in the solid was determined by measuring the radioactivity of an aliquot of the solution. The kinetics of the fixation were determined by varying the time of shaking. The fixation equilibrium and capacity of retention were determined by varying the quantity of non-radioactive cesium or strontium introduced in the solution. The release of cations from the solid into the solution was measured by atomic absorption spectroscopy (11).

## Column experiments

Sorption was studied by passing a solution containing the radioactive tracer through a column of the absorbent and by measuring the radioactivity of the solution as a function of the eluted volume (31). The decontamination factor was defined as the ratio of the radioactivity of an element in the same volume of solution before and after passing through a column.

## RESULTS AND DISCUSSION

### Fixation of cesium

1. Kinetics: The equilibrium of sorption is achieved within a few minutes for sodium or potassium-zinc ferrocyanides (Fig. 1, curve A) or for precipitated nickel ferrocyanides. In the case of mixed zinc ferrocyanides, the fast kinetics may be explained by the zeolitic character of

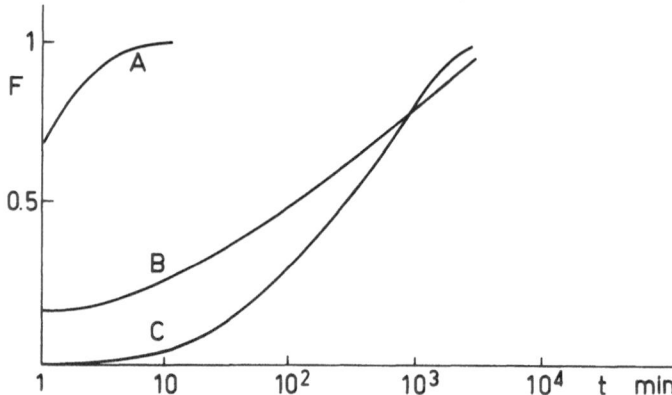

Figure 1. Examples of kinetics of cesium fixation on ferrocyanides. Variation of the fixation achievement factor F as a function of time t for three types of ferrocyanides. A: precipitated $Na_2Zn_3[Fe(CN)_6]_2$; B: $Na_2NiFe(CN)_6$ prepared by localized growth; C: precipitated $Zn_2Fe(CN)_6\ 2H_2O$

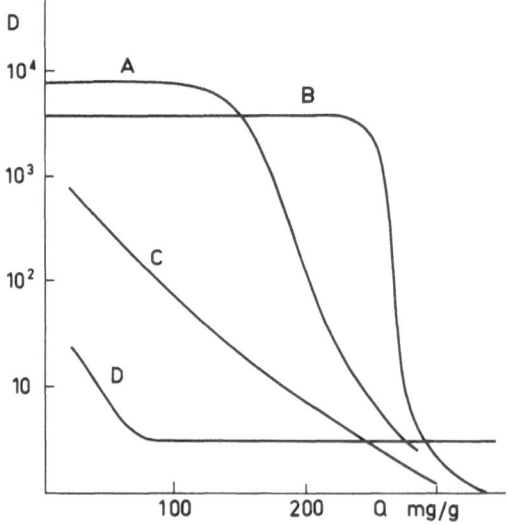

Figure 2. Variation of the decontamination factor D as a function of the cesium quantity Q per gram of product poured on a column of various zinc ferrocyanides prepared by localized growth. A: mixed sodium-zinc ferrocyanide (mixture of phases); B: $K_2Zn_3[Fe(CN)_6]_2$; C: mixed sodium-zinc ferrocyanide with a higher Zn concentration than A; D: zinc ferrocyanide with composition close to $Zn_2Fe(CN)_6$.

the structure leading to a high ionic conductivity (28). The fast kinetics in precipitated nickel ferrocyanide may be explained by diffusion in the cubic structure with partial occupancy of the iron sites (22). On simple zinc ferrocyanide or nickel ferrocyanides prepared by localized growth (Fig. 1, curves B and C), equilibrium achievement requires several days. For simple zinc ferrocyanides, the kinetics seems to be controlled by a slow diffusion within the new cubic phase formed after exchange, while in nickel ferrocyanides it seems to be limited by the size of the crystals.

2 Fixation at trace level: The fixation of cesium was studied over a large pH range. For $1 < pH < 10$, there is no obvious variation of the fixation rate. The distribution coefficient Kd is over $10^3$ $cm^3.g^{-1}$ in HCl solution from 0.01 to 2 M, as well as in neutral or basic solutions until pH 12 for nickel ferrocyanides and pH 10 for zinc ferrocyanides.

3. Sorption isotherms and capacity: In the case of the $M^I_2Zn_3[Fe(CN)_6]_2$ type, the capacity of fixation is close to the theoretical value corresponding to the alkaline content (one Cs per one Fe). For $Zn_2Fe(CN)_6$ $2H_2O$, the maximum uptake is 1.6 Cs per Fe atom and the exchange leads to a structural change. Cesium is fixed on mixed alkaline-nickel ferrocyanide without structural change. The capacity of Cs retention never reaches the theoretical value corresponding to a total release of the monovalent ions of the solid. Furthermore, the fixation of cesium is not balanced by the release of these ions. The capacity of the product with the highest sodium content is 0.4 Cs per Fe atom, with a sodium release of 0.26 ions. Higher capacities have been observed for some other nickel ferrocyanides with lower alkaline contents (Table 1).

TABLE 2

Cesium decontamination from PWR effluents
by fixation on mixed potassium-zinc ferrocyanide[a].

| column characteristics | | | |
|---|---|---|---|
| internal diameter (cm) | 0.85 | 3.2 | 0.85 |
| column length (cm) | 17.5 | 2 | 1.75 |
| column volume (mL) | 10 | 20 | 1 |
| solution flow rate (mL.cm$^{-2}$.h$^{-1}$) | 596 | 35 | 596 |
| total eluted solution (vol/vol) | 720 | 2350 | 40,000 |
| decontamination factor D[b] | > 6000 | 1000 | 200 |

a : product 3 of table 1

b : mean decontamination factor for $^{134}$Cs and $^{137}$Cs

4. fixation on columns: Examples of elution curves from zinc ferrocyanide columns are given in Fig.2. The more favourable case is observed for rhombohedral mixed alkaline-zinc ferrocyanides (curves B and A): we observe a plateau with a high value of the decontamination factor D, followed by a steep fall at the retention capacity. These conditions are fulfilled when the kinetics, as measured in the batch experiments, is fast. On the contrary a slow kinetics of fixation leads to a continuous decrease of D (curves C and D). In the case of mixed alkaline-

nickel ferrocyanides, a plateau is observed for products with small crystals. It is also connected to a fast kinetics of fixation.

Effluents from the primary circuit of a PWR reactor were tested. The pH of the solution is fixed to 7 by the presence of large quantities of lithium borate (7 g.l$^{-1}$). The results are summarised in table 2. Mixed potassium-zinc ferrocyanide was used. We noticed high values of the decontamination coefficient. Even after the passage of 40000 column volumes, this coefficient is equal to 200. However a detectable amount of potassium and zinc was released. This drawback hinders the use of ferrocyanides for the continuous treatment of PWR circuits, but these products are well adapted to the treatment of all effluents before final disposal. Another application is the selective removal of radioactive cesium from ion exchanger resins used in the primary circuits (32). A volume reduction factor superior to one order of magnitude can be achieved by transferring the radioactivity from the resin onto ferrocyanide.

### Fixation of strontium

1. kinetics: For RH3, the equilibrium is reached within 15 minutes even when the initial concentration of strontium is high (Fig. 3). This feature can be explained by the lamellar structure which allows a fast diffusion of strontium into the particle. For HAP and RH5, the equilibrium is achieved after several days for concentrations of some 10$^{-2}$ M. This slow kinetics is related to the diffusion in the tridimensional structure of these compounds. However, a fast step related to a superficial fixation is observed.

2. Fixation at trace level: The distribution coefficient Kd of strontium decreases when the concentration of H$^+$ ions increases in the solution. The observed law of variation is explained by a H$^+$-Sr$^{2+}$ exchange. The distribution coefficient varies in the order: HAP > RH5 > RH3. A competition due to the Na$^+$-Sr$^{2+}$ exchange is also observed for high concentration of sodium in the solution.

TABLE 3

Fixation capacity for strontium on polyantimonic acid HAP and two phosphatoantimonic acids RH3 and RH5 as a function of pH in equivalent per mol of product.

|  | pH 3 | pH 5.3 | pH 12 |
|---|---|---|---|
| HAP | 0.62 | 0.62 | 2.0 |
| RH3 | 2.1 | 2.4 | 4.9 |
| RH5 | 1.9 | 1.2 | 5.6 |

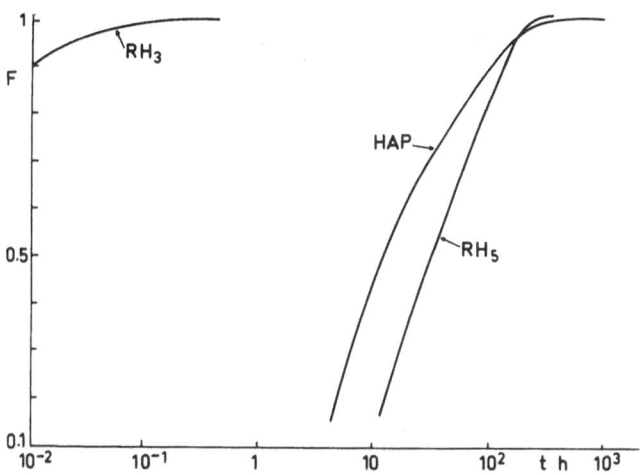

Figure 3. Examples of kinetics of strontium fixation. Variation of the fixation achievement factor F as a function of time t for polyantimonic acid HAP (initial Sr concentration 2.8 10$^{-2}$ M) and two phosphatoantimonic acids: RH3 (Sr concentration 0.33 M) and RH5 (Sr concentration 2.8 10$^{-2}$ M)

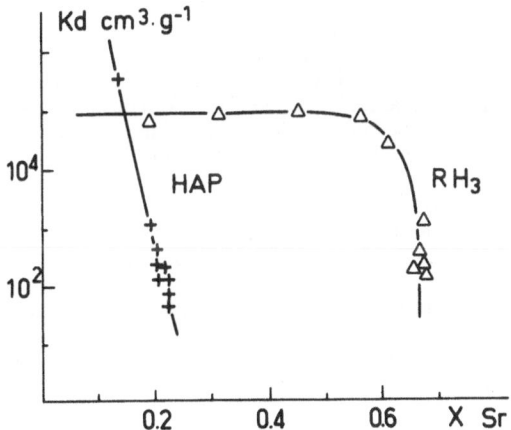

Figure 4. Variation of the distribution coefficient Kd of strontium as a function of the exchanged fraction $X_{Sr}$ in the solid for HAP ($X_{Sr} = 1$ for theoretical capacity of 2 Sr equivalent per mol) and RH3 ($X_{Sr} = 1$ for the theoretical capacity of 3 Sr equivalent per mol) in acid medium.

3. sorption isotherms and capacity: The capacity for strontium increases with pH (Table 3). The theoretical capacity of HAP is only achieved in basic media . A detailed study of the crystallographic structure showed the presence of two fixation sites in the lattice: $S_I$ sites are situated in the hexagonal windows while $S_{II}$ sites are located in the cavities. In acid or basic media : only the $S_I$ sites can be occupied by strontium, while in basic medium both sites are exchanged. For RH3 and RH5, the capacity does not reach the theoretical value in acid and neutral medium, but it exceeds this value in basic medium. The exchange isotherms of strontium on HAP and RH3 were studied. They showed an inversion of selectivity as the concentration of strontium in the solid increases (Fig. 4). HAP has a very high distribution coefficient for strontium ions at low concentrations. Owing to the rigidity of the tridimensional lattice and to the interaction between ions, the distribution coefficient decreases with the increase of the quantity of Sr ions in the solid. In the case of RH3, the distribution coefficient remains constant in a large domain, probably because the lamellar character of the structure allows a large change in the crystallographic parameter.

4. fixation on columns: The highest values of the distribution coefficient of strontium are reached on HAP, but the slow kinetics of diffusion does not seem to be favourable for a fixation of this element on a column. However for radioactive strontium produced by fission reactions, whose residual concentration in wastes is very low, the capacity corresponding to the fast step of the kinetics ($0.1$ eq mol$^{-1}$) is large enough. In column experiments, with synthetic radioactive strontium solutions, we observe a plateau with a value of $7 \times 10^3$ for the decontamination factor until a fixed quantity of $0.15$ eq mol$^{-1}$. In the case of RH3 and RH5, the particle size is too small for column utilization.

## CONCLUSION

Two types of products were studied for the fixation of cesium and strontium from radioactive waste: respectively insoluble ferrocyanides and antimonic or phosphatoantimonic acids. In order to select the best products three types of measurements had to be performed: distribution coefficients at trace level, variation of the exchange equilibrium as a function of fixed quantity and capacity, kinetics of fixation. If we consider the two first types of measurements, all studied ferrocyanides have a very high affinity for cesium. But, for good retention properties on column, a fast kinetics of fixation is needed. This excludes products on which the kinetics is limited by structural change or crystal size. Mixed alkaline-zinc ferrocyanides seem to be the most suitable absorbents for cesium. In the case of strontium, the best absorbent is polyantimonic acid as it shows the highest distribution coefficient for low concentrations and a fast step of kinetics with a non negligible capacity. For a larger quantity of strontium, a lamellar phosphatoantimonic acid could be used, but the granulometry of this product is not suitable for column utilization. In all cases an important factor must be taken into account: the grain size and the stability of the products when they are used in columns. We developed a new method of preparation by localized growth, which lead to ferrocyanides suitable for column utilization. This method was not yet applied to antimonic acid, which can be used only in small columns. Considerable effort is needed in the future for the preparation of inorganic exchangers with suitable grain size and long term stability in columns.

## REFERENCES

1. Loewenschuss, H., Radioactive Waste Manag.,1982, 2, 327.
2. Barton, G.B., Hepworth, J.L.,MacClanahan, E.D. Jr, Moore, R.L. and Van Tuyl, H.H., Ind. Eng. Chem., 1958, 102, 212.
3. Hendrickson, W.F. and Riel, G.K., Health Physics, 1975, 28, 17.
4. Bilewicz, A.and Narbutt, J., Isotopenpraxis, 1984, 20, 141.
5. Hooper, E.W., Phillips, B.A. Dagnall, S.P. and Monckton, N.P., AERE-R 11088, Harwell, G. B., 1984.
6. Faubel, W. and Ali, S.A., Radiochimica Acta, 1986, 40, 49.

7.  Ganzerli-Valentini, M.T., Stella, R. and Cola, M., J. Radioanal. Nucl. Chem., 1986, **102**, 99.
8.  Harjula, R., Lehto, J. and Walace, J., J. Radioanal. Nucl. Chem., 1987, **111**, 297.
9.  Nielsen, P., Dresow, B. and Heinrich, H.C., Z. Naturf., 1987, **42b**, 1451.
10. Loos-Neskovic, C., Fedoroff, M. and Revel, G., J. Radioanal. Nucl. Chem., 1976, **30**, 533.
11. Loos-Neskovic, C.and Fedoroff, M., Solv. Extr. Ion Exch., 1989, **7**, 131.
12. Loos-Neskovic, C.and Fedoroff, M., Reactive Polymers, 1988, **7**, 173.
13. Loos-Neskovic, C., Thesis, Paris, 1986.
14. Abe, M. and Uno, K., Sep.Sci. Techn., 1979, **14**, 355.
15. Beatsle, L.H. and Huys, D., J. Inorg. Nucl. Chem., 1968, **30**, 639.
16. Abe, M. and Ito, T., Bull. Chem. Soc. Japan, 1968, **41**, 333.
17. Beatsle, L.H., Van Deyck, D., Huys, D.and Genry, A., EUR 2497e, 1965.
18. Murphy, T.S., Balasubramanian, K.R. and Narasima Rao, K.L., Proc. Nucl. Radiochem. Symp. Andhra Univ., Waltair, 1980.
19. Konecny, C. and Kourim, V., Radiochem. Radioanal. Lett., 1969, **2**, 47.
20. Zouad, S., Loos-Neskovic, C.and Fedoroff, M., J. Radioanal. Nucl. Chem., 1987, **111**, 337.
21. Loos-Neskovic, C., Fedoroff, M., Garnier, E. and Gravereau, P., Talanta, 1984, **31**, 1133.
22. Loos-Neskovic, C., Fedoroff, M.and Garnier, E., Talanta, 1989, **36**, 749.
23. Fedoroff, M. and Loos-Neskovic, French Patent, 1984, 84-12139.
24. Loos-Neskovic, C., Abousahl, S. and Fedoroff, M., J. Mater. Sci., 1989, **24**, in press.
25. Garnier, E., Gravereau, P. and Hardy, A., Acta Cryst., 1982, **B38**, 1401.
26. Gravereau, P., Garnier, E. and Hardy, A., Acta Cryst., 1979, **B35**, 2843.
27. Siebert, H., Nuber, B. and Jentsch W., Z. Anorg. Allg. Chem., 1981, **474**, 96.
28. Garnier, E., Thesis, Poitiers, France, 1985.
29. Piffard, Y., Laghdar, A. and Tournoux, M., J. Solid State Chem., 1985, **58**, 253.
30. Piffard, Y., Laghdar, A. and Tournoux, M., Rev. Chim. Min., 1985, **22**, 101.
31. Loos-Neskovic, C.and Fedoroff, M., Radioactive Waste Manag., 1989, **11**, 43.
32  Nott, B.R, Int. Symp. Water Chem. Corrosion Problems Nucl. React. Syst. Comp., Vienna, Austria, nov. 22-26, 1982, IAEA-SM-264, p.295, 1983.

**Simultaneous Elimination of Heavy Metals
and Chelating Agents from Waste Waters**

WOLFGANG H. HÖLL
Karlsruhe Nuclear Research Center
Institute for Radiochemistry, Water Technology Division
P.O. Box 3640, D-7500 Karlsruhe, Fed. Rep. of Germany

## ABSTRACT

The simultaneous elimination of heavy metals and chelating agents by means of strong-base anion exchange resins has been investigated. The resins are applied in the bicarbonate form and regenerated by means of magnesium bicarbonate, which is produced from magnesium oxide, water, and cabon dioxide under pressure. Laboratory scale experiments demonstrated the feasibility of the proposed process. Macroporous exchange resins showed the best performance.

## INTRODUCTION

In industrial metal finishing processes chelating agents are applied for various purposes. Objectives are either to dissolve metals or to keep them in solution at conditions under which they normally would precipitate [1]. The most important chelating substances are ethylene diamine tetraacetic acid (EDTA), quadrol, citrate, and tartrate [2].

Chelating agents strongly complicate the waste water treatment: The precipitation of heavy metals becomes difficult or even impossible [1], and after discharge of the waste water these substances can dissolve heavy metals from sediments of bodies of water and thus cause problems for drinking water supplies [3]. Therefore, to avoid such problems, heavy metals and chelating substances have to be eliminated simultaneously, e.g. by a combination of oxidative or electrolytical and precipitation processes [4, 5]. For such a treatment, however, dilute solutions should be concentrated. This can be achieved by ion exchange processes.

## THEORY

### Conventional proposals

A simultaneous elimination of heavy metals and chelating agents in the form of negatively charged metal complexes can be achieved by applying strong-base anion exchange resins in the OH⁻ form. Using copper and EDTA as an example this exchange follows the formal equation:

$$2\ \overline{R - N(CH_3)^+OH^-} + CuEDTA^{2-} \rightleftharpoons \overline{(R - N(CH_3)^+)_2CuEDTA^{2-}} + H_2O$$

This exchange process is coupled with a neutralization reaction and due to the disappearance of OH⁻ ions there is a very effective uptake of CuEDTA²⁻ ions. Difficulties arise, however, if the resin is to be regenerated by means of caustic. Due to the preference of divalent species large excess volumes have to be applied. Furthermore, at high pH values CuEDTA²⁻ complexes can be split. As a consequence tetravalent EDTA⁴⁻ anions are generated, the removal of which is practically impossible [1].

In a second proposal the application of strong-base resins in the chloride form is described [6]:

$$2\ \overline{R - N(CH_3)^+Cl^-} + CuEDTA^{2-} \rightleftharpoons \overline{(R - N(CH_3)^+)_2CuEDTA^{2-}} + 2\ HCl$$

In this case no reaction-coupled process takes place. As a consequence the elimination of metal chelates is slightly less effective than with OH⁻ loaded resins. Regeneration uses NaCl solutions in the appropriate excess. Since these excess quantities cannot be recovered, the overall salt concentration in the effluent is increased, which might influence further treatment steps.

### Application of strong-base resins in the bicarbonate form

More favourable conditions for regeneration as well as for the post-treatment exist, if the resins are applied in the bicarbonate form to which they are converted in the regeneration cycle by means of magnesium bicarbonate ($Mg(HCO_3)_2$) which can be generated from magnesium oxide (MgO), water, and carbon dioxide ($CO_2$) under pressure [7]:

$$MgO + 2\ CO_2 + H_2O \rightleftharpoons Mg(HCO_3)_2$$

Depending on $CO_2$ pressure in the system solutions of fairly high concentrations can be produced (see fig. 1), which allow an effective regeneration. Furthermore, magnesium bicarbonate solutions have pH values of less than 6.5. As a consequence, there is no splitting of metal chelates and no formation of tetravalent species. Part of the metal chelates is even converted to monovalent ones which improves regeneration efficiency. The particular advantage, however, is that $Mg(HCO_3)_2$ decomposes in the absence of carbon dioxide into magnesium carbonate compounds which do not contribute to the total salt content of the waste water and which might be recovered and reused.

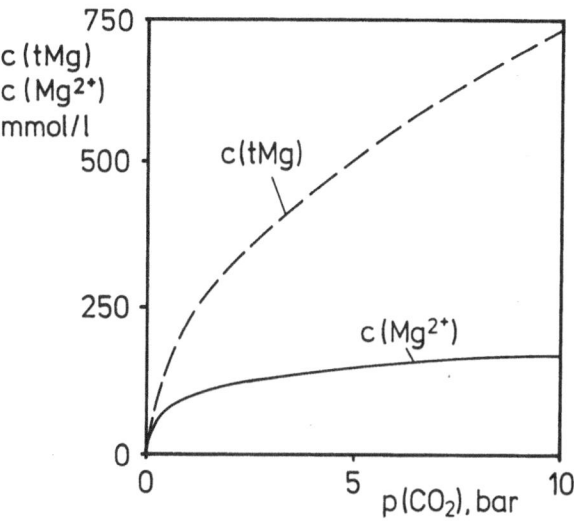

Figure 1. Concentration of total magnesium tMg ($=Mg^{2+}$ + magnesium bicarbonate complex species) as a function of carbon dioxide pressure

The development of the service cycle depends on both the heavy metal chelates and the ion exchange resin. Technical Cu-EDTA solutions are usually prepared from $CuCl_2$ and $Na_2H_2EDTA$. Thus, during the service cycle, $Cl^-$ and $CuEDTA^{2-}$ anions are competing for the fixed sites:

$$3 \overline{R - N(CH_3)^+HCO_3^-} + CuEDTA^{2-} + Cl^-$$
$$\rightleftharpoons \overline{(R - N(CH_3)^+)_3(CuEDTA^{2-},Cl^-)} + 3 HCO_3^-$$

According to their divalency $CuEDTA^{2-}$ should usually be preferred before $Cl^-$ ions. However, due to the size of the metal chelates this preference may be overcompensated by steric effects, mainly in gelular resins. If $CuEDTA^{2-}$ ions are preferred, they will replace chloride ions and an optimum column capacity is obtained. In the opposite case, if chloride ions have the highest affinity towards the resin, they will replace $CuEDTA^{2-}$ ions which will occur in the effluent in concentrations above the raw water values. The required selectivity sequence therefore is $CuEDTA^{2-} > Cl^- > HCO_3^-$.

## EXPERIMENTAL

For the experiments the resins LEWATIT MP 600 (BAYER AG), AMBERLITE IRA 458, and AMBERLITE IRA 958 (both from ROHM AND HAAS COMP.) were used. Characteristic data of these resins are summarized in table 1 [8, 9].

Table 1

Characteristic resin data

| Exchange resin | LEWATIT MP 600 | AMBERLITE IRA 458 | AMBERLITE IRA 958 |
|---|---|---|---|
| Matrix | Styrene/DVB | Acrylamide/DVB | Acrylamide/DVB |
| Funct. group | $-N((CH_3)_2C_2H_4OH)^+$ | $-NH(CH_2)_3N(CH_3)_3^+$ | $-NH(CH_2)_3N(CH_3)_3^+$ |
| Structure | macroporos | gelular | macroporos |
| Bead size | 0.3 - 1.0 mm | 0.3 - 1.0 mm | 0.3 - 1.0 mm |
| Capacity | 1.65 eq/l | 1.55 eq/l | 1.25 eq/l |

Elimination of heavy metal chelates was investigated using copper complexes of EDTA, nitrilotriacetic acid (NTA), and tartaric acid (TA). Regeneration experiments were carried out with $CuEDTA^{2-}$ and $CuNTA^-$ loaded resins.

Filter experiments were carried out using copper chelate solutions which were produced from $CuCl_2$ and the sodium salt of the relevant chelating agent and addition of NaOH for pH adjustments. For measurements of the binary equilibrium $CuEDTA^{2-}$ /Cl⁻ chloride-free solutions were prepared from the free acids, and solid $Cu_2CO_3(OH)_2$ (malachite). Since in these cases pH was below 4, $CO_3^{2-}$ ions are converted to $CO_2$ which can be stripped out.

## EXCHANGE EQUILIBRIA

### Binary equilibria of the system copper chelate-Cl⁻-HCO₃⁻

For evaluation of binary equilibria different amounts of resin material in the Cl⁻ form were contacted with constant volumes of copper chelate solutions with an initial concentration of 10 mmol/l. After equilibration, chloride and DOC concentrations were measured in the liquid phases. From these measurements it can be concluded that copper-EDTA chelate anions are adsorbed as divalent species, copper-NTA and copper-TA anions as monovalent species. Thus the corresponding exchange isotherms can be calculated. The results, plotted in figs. 2a and 2b demonstrate that each of the three resins prefers the chelate anions, even the gel-type AMBERLITE IRA 458.

The equilibrium of the binary exchange of Cl⁻ vs. HCO₃⁻ ions has been studied intensively in the past. Details and results have been reported in previous papers [10, 11].

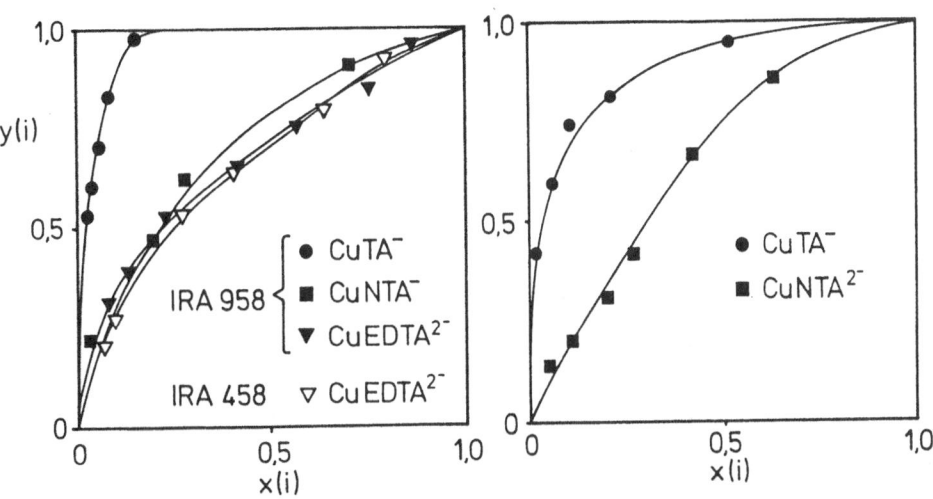

Figures 2a and 2b. Dimensionless isotherms of the binary exchange of copper chelate anions vs. Cl⁻ ions. Left: AMBERLITE IRA 958 and IRA 458, right: LEWATIT MP 600.

### Regeneration equilibria

Regeneration equilibria have been studied at a $CO_2$ pressure of 1 bar with AMBERLITE IRA 458. One of the results from experiments with $CuEDTA^{2-}$ loaded resin material is shown in fig. 3 as both the copper and $CuEDTA^{2-}$ loading of the resin vs. the corresponding concentrations in the liquid phase. The development of these isotherms demonstrates the very efficient removal of metal chelates by magnesium bicarbonate, even at such moderate conditions.

## FILTER EXPERIMENTS

### General

Experiments with respect to pure breakthrough behaviour were carried out in a small laboratory plant with resin columns of 2 cm diameter and various heights of 10 - 25 cm. Throughput (in upstream direction) amounted to 5 bed volumes (BV) per hour. For analyses samples were taken from the effluent at preset intervals. In addition pH value was measured and recorded.

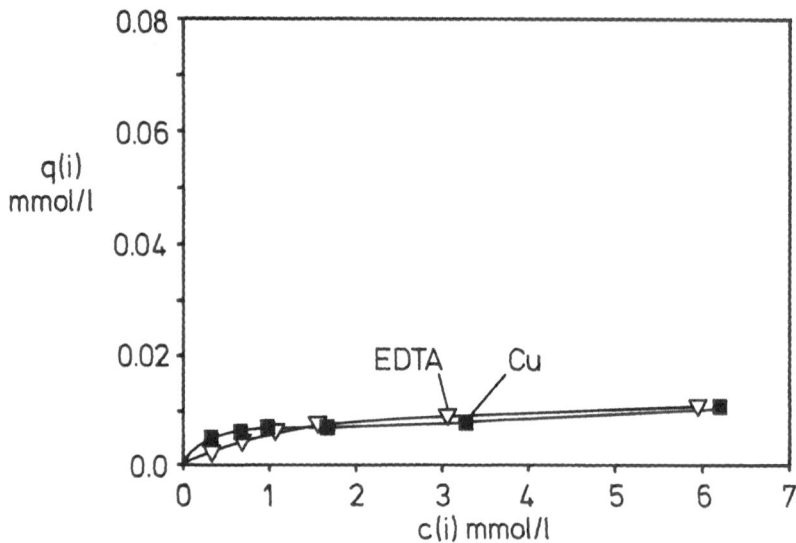

Figure 3. Isotherms of regeneration of AMBERLITE IRA 458 at a $CO_2$ pressure of 1 bar.

For experiments with periodical change of service and regeneration cycles another installation was used. It contained a stainless steel column of 50 mm diameter with a bed height of approximately 80 cm. The feed solution was pumped in downstream direction across the resin material at a throughput of 6.6 BV/h. For regeneration 10 l of a $Mg(HCO_3)_2$ solution, saturated at a $CO_2$ pressure of 6 bar were pumped in counterflow across the resin bed. Figure 4 schematically shows this plant.

**Breakthrough behaviour**

Typical results from experiments for elimination of $CuEDTA^{2-}$, using the resin AMBERLITE IRA 958 are plotted in fig. 5. In each case the raw water contained 10 mmol/l copper chelate and 20 mmol/l $Cl^-$ ions. The development of the breakthrough curves indicate that $CuEDTA^{2-}$ ions are strongly held by this resin whereas $Cl^-$ ions are replaced. Similar results were found during the elimination of copper tartrate complexes with the type II exchanger LEWATIT MP 600 (fig. 6).

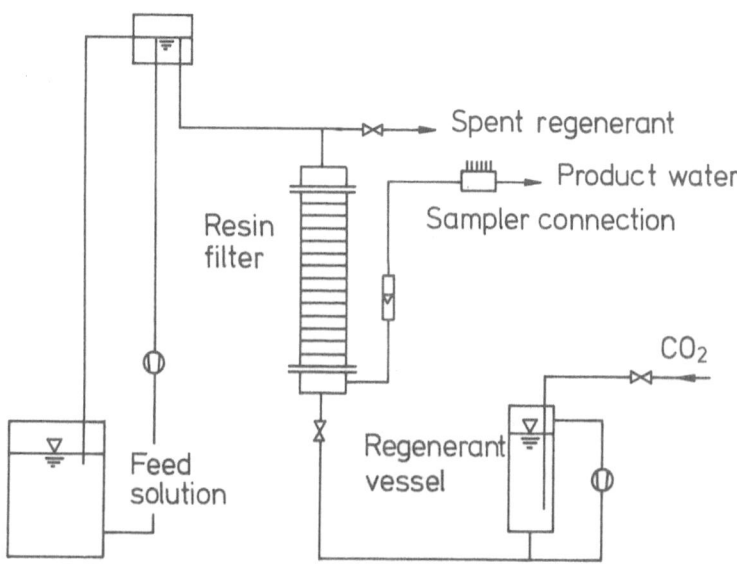

Figure 4. Schematic plot of the laboratory plant for periodical change of service and re-generation cycles.

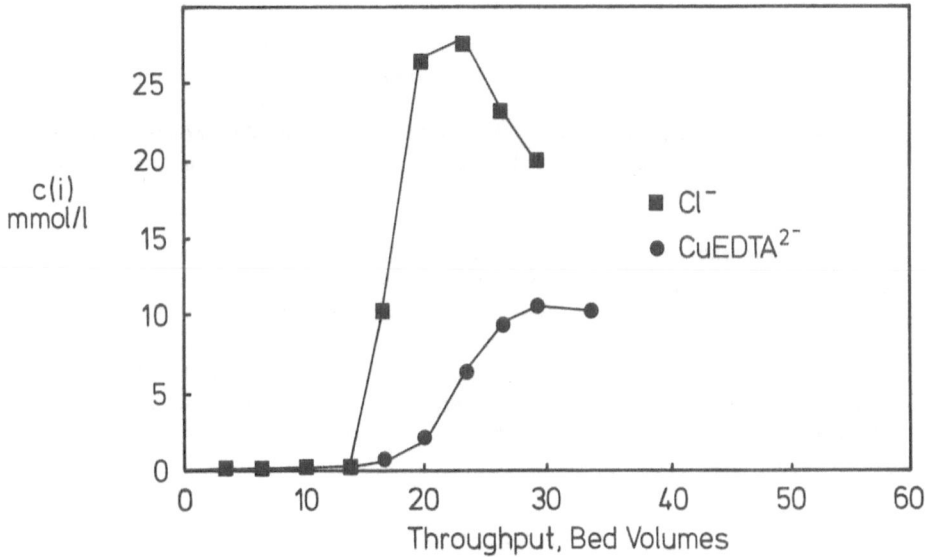

Figure 5. Breakthrough curves from the elimination of $CuEDTA^{2-}$ ions by AMBERLITE IRA 958 in the $HCO_3^-$ form, pH = 8.

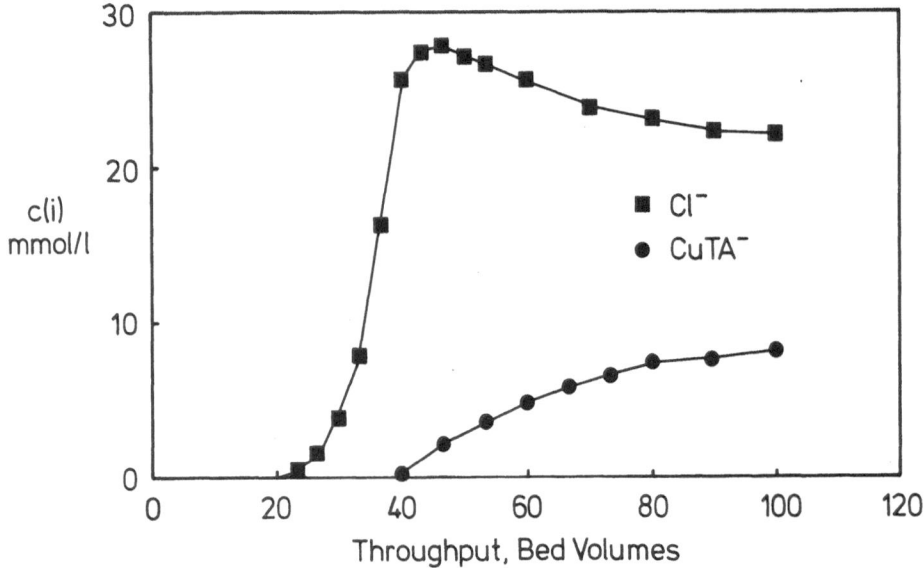

Figure 6. Breakthrough curves from the elimination of CuTA ions by LEWATIT MP 600 in the $HCO_3^-$ form, pH = 8.

In practical cases waste water usually contains the chelating agents in excess. As a consequence a third kind of species is present. Figure 7 shows results from corresponding experiments with 10 mmol/l $CuCl_2$ and 15 mmol/l $Na_2H_2EDTA$ in the raw water. At pH = 8, as adjusted in this experiment, a considerable part of free EDTA consists of trivalent anions which replace $CuEDTA^{2-}$ from the resin. The quick sequence of concentration waves in fig. 7 is due to the short bed height.

Results from experiments with cyclic change of regeneration and service cycles are plotted in figure 8 for the macroporous resin AMBERLITE IRA 958. According to the strong regeneration efficiency there is an excellent elimination of $CuEDTA^{2-}$ anions.

During regeneration of the resins, copper concentrations of up to 4000 mg/l were measured while magnesium carbonate precipitated. Thus the proposed process allows concentration of    solutions without increasing the total salt content.

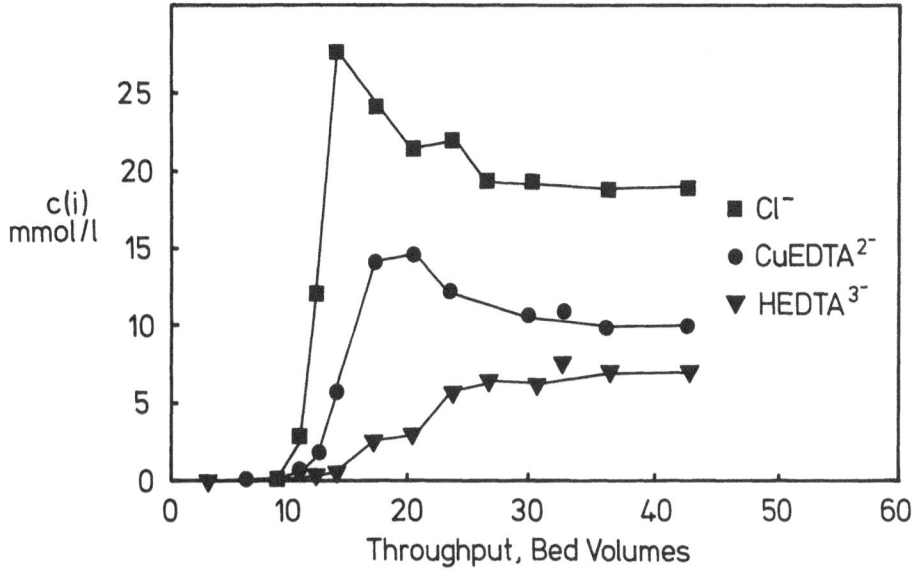

Figure 7. Breakthrough curves from the elimination of CuEDTA$^{2-}$ ions ions by AMBERLITE IRA 958 in the HCO$_3^-$ form in the presence of excess Na$_2$EDTA$^{2-}$, pH = 8.

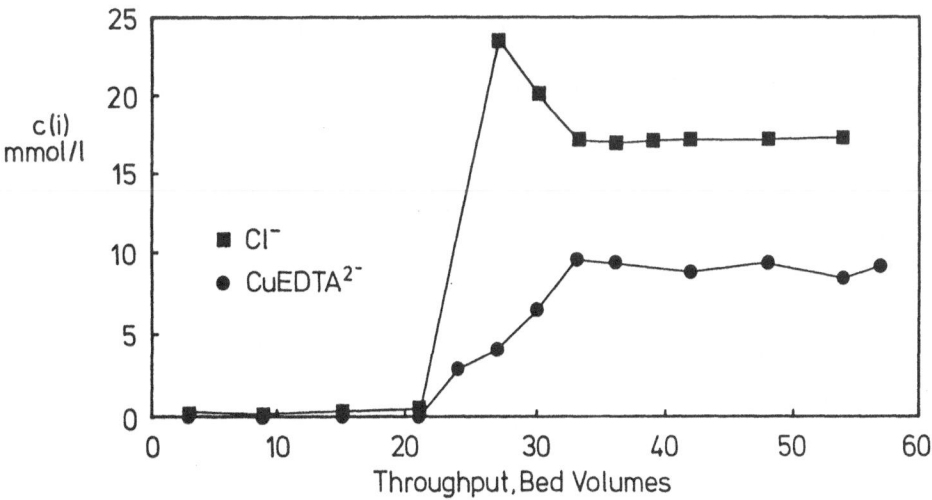

Figure 8. Breakthrough curves from the elimination of CuEDTA$^{2-}$ ions ions by AMBERLITE IRA 958 in the HCO$_3^-$ form, pH = 3.

## REFERENCES

1. Hartinger, L., Taschenbuch der Abwasserbehandlung, Carl Hanser-Verlag München, Wien, 1976.

2. Wing, R. E., Complexed and chelated copper-containing rinsewaters. Plating and Surface Finishing, 1973, **73**, 20 - 41.

3. Bernhardt, H. (editor), NTA: Studie über die Umweltverträglichkeit von Nitrilotriacetat. Verlag Hans Richarz, Sankt Augustin, 1984.

4. Morooka, S., Ikemizu, K., Kamano, H. and Kato, Y., Ozonation rate of water soluble chelates and related compounds. J. Chem. Eng. Japan, 1986, **19**, 294 - 299.

5. Müller, K.-J., Bolch, T. and Merz, K., Festbettelektrolyse zum Abbau bzw. zur Rückgewinnung von organischen Komplexbildnern im Abwasser. Galvanotechnik, 1988, **79**, 172 - 176.

6. Vignola, M., Verfahren zum Entfernen von Metallkomplexen aus Abwasserlösungen. Offenlegungsschrift DE 36 14 061, Oct. 30, 1986.

7. Höll, W. H., Horst, J., Nagel, S. and Ma Shishen, Conversion of cation exchange resins to the magnesium form using $Mg(OH)_2$ and $CO_2$. In Ion Exchange for Industry, ed. M. Streat, Ellis Horwood Limited, Chichester, 1988.

8. LEWATIT-LEWASORB Handbuch, BAYER AG, Leverkusen, 1974.

9. ROHM AND HAAS DEUTSCHLAND, AMBERLITE ion exchangers, 1985.

10. Höll, W. H., Entwicklung und Grundlagen eines neuen Verfahrenskonzepts zur Teilentsalzung von Wasser mit Ionenaustauschern unter Verwendung von Kohlendioxid als Regenerierchemikalie. Karlsruhe Nuclear Research Center, Report No. 4022, 1985.

SEPARATION OF RUTHENIUM AND MOLYBDENUM BY CATION EXCHANGE
CHROMATOGRAPHY ON AG 50W-X4 IN A NITRIC ACID-ETHYLENE=
DIAMINE MIXTURE

T N van der Walt and P J Fourie
Isotope Production Centre,
Atomic Energy Corporation of South Africa Ltd.,
P O Box 582, Pretoria, 0001,
Republic of South Africa.
and
P P Coetzee
Department of Chemistry,
Rand Afrikaans University
P O Box 524, Johannesburg, 2000,
Republic of South Africa.

## ABSTRACT

The $^{99}$Mo-$^{99m}$Tc-generator is widely used in nuclear medicine. $^{99}$Mo can be obtained from the fission of $^{235}$U in a nuclear reactor. $^{103}$Ru is one of the numerous fission products to be separated from $^{99}$Mo. $^{103}$Ru can easily be separated from $^{99}$Mo by cation exchange chromatography on AG 50W-X4 in a nitric acid-ethylenediamine mixture. Ruthenium is sorbed as a posi= tively charged ruthenium-ethylenediamine complex from a mixture containing 0,10M nitric acid and 0,01M ethylene= diamine or 1,0M nitric acid and 0,10M ethylenediamine. Molybdenum passes through the resin column and can be separated from the ethylenediamine by ion exchange chromatography on aluminium oxide or manganese dioxide.

## INTRODUCTION

Technetium-99m (half-life 6,0h) is the radioisotope which
has probably found the widest application in nuclear
medicine. It is used to label various labelling kits to
perform a variety of diagnostic studies (Table 1).

## TABLE 1

Examples of labelling kits for diagnostic studies.

| Kits for labelling with $^{99m}$Tc | Diagnostic study |
|---|---|
| HSA (human serum albumin) | Blood, vascular system, heart, brain |
| Pyrophosphate | Blood, vascular system, heart, brain, skeleton |
| RBC (red blood cells) | Blood, vascular system, heart, spleen |
| Human fibrinogen | Blood, vascular system, tumours |
| DTPA (diethylenetriamine pentaacetic acid) | Brain, kidneys, lungs |
| Gluconate | Brain, kidneys, tumours |
| MDP (methylene diphosphonic acid) | Joints, skeleton |
| DMSA (dimercaptosuccinic acid) | Kidneys |
| Colloidal sulphide | Liver, bone marrow, spleen |
| HSA macroaggregates/microspheres | Lungs |
| Colloidal (Re) sulphide | Lungs, lymphatic system |
| BIDA (p-butyl-IDA) | Hepatobiliary system |
| DISIDA (diisopropyl-IDA) | Hepatobiliary system |

Technetium-99m is obtained from the $^{99}$Mo-$^{99m}$Tc-gene=
rator, Figure 1 which usually consists of an ion exchange column
which is housed in a lead container. The ion exchange
column contains an inorganic ion exchanger such as
alumina[1,2,3], silica gel[4] manganese dioxide[5],
zirconium arsenate[6] etc. Molybdenum-99 (half-life
66,0h) is loaded onto the ion exchanger. Molybdenum-99
decays to technetium-99m, which is eluted every 24 hours
with sterile and pyrogen-free physiological saline
solution (0,9% m/v sodium chloride).

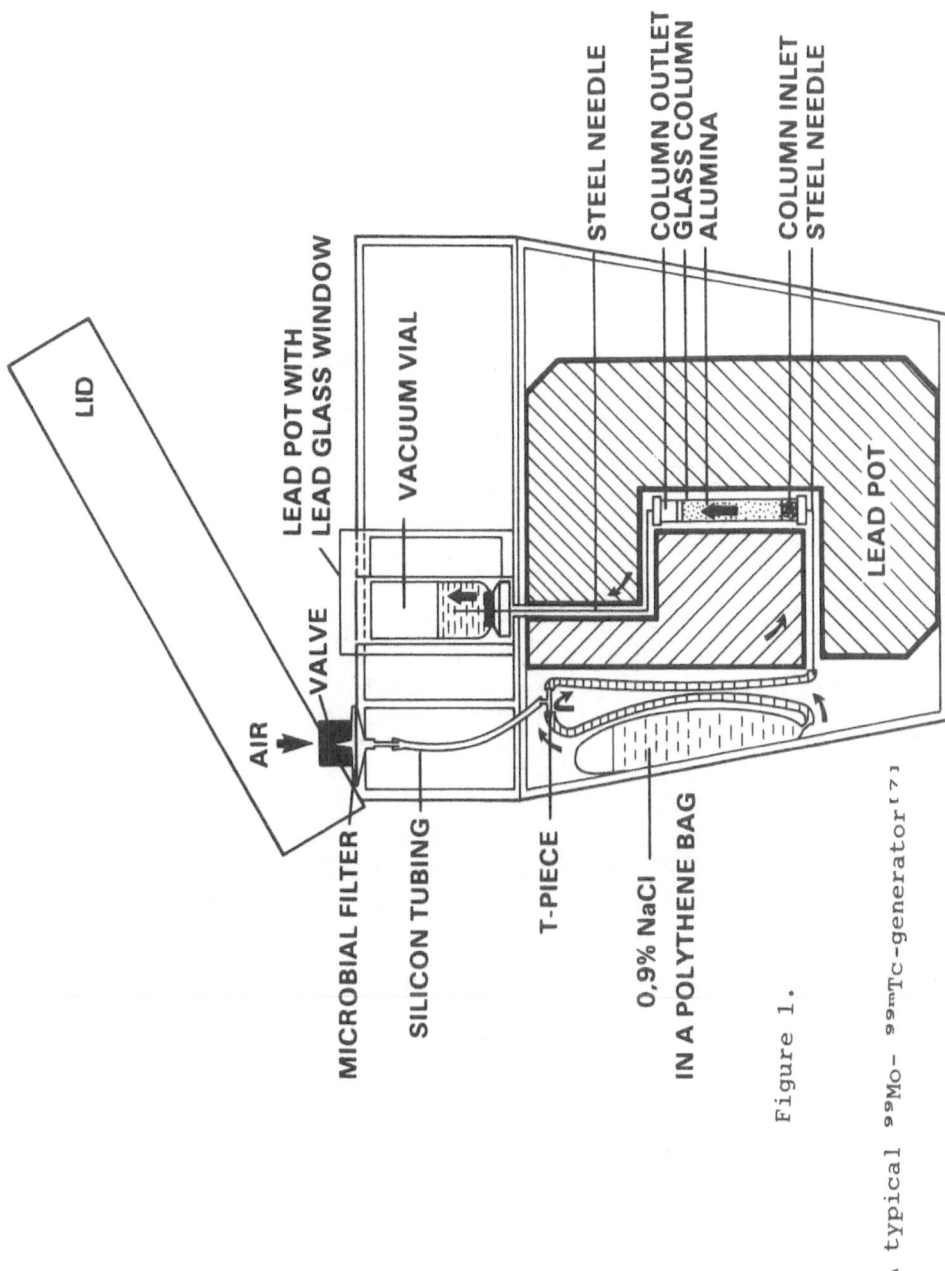

Figure 1.

A typical $^{99}Mo$- $^{99m}Tc$-generator[7]

$^{99}$Mo has to comply to the product specifications, which is a certificate of guarantee in the final quality report received from the manufacturer (Table 2).


**TABLE 2**

Product specifications for $^{99}$Mo[7]

---

Chemical form : sodium molybdate
Solution alkalinity : ca 0,2 M NaOH
Specific activity : carrier-free, > 185 000
  GBq/g Mo at calibration time
Radiochemical purity : $^{99}$Mo as molybdate > 99%
Radionuclidic purity : at calibration time,

Gamma:  $^{131}$I/$^{99}$Mo        < 5 x 10$^{-5}$
        $^{103}$Ru/$^{99}$Mo       < 5 x 10$^{-5}$
        $^{132}$Te/$^{99}$Mo       < 5 x 10$^{-5}$
Beta:   $^{89}$Sr/$^{99}$Mo        < 6 x 10$^{-7}$
        $^{90}$Sr/$^{99}$Mo        < 6 x 10$^{-8}$

All other $\beta, \gamma$ -emitters/$^{99}$Mo < 1 x 10$^{-4}$
                      ($^{99m}$Tc excluded)
Total $\alpha$-emitters/$^{99}$Mo          < 1 x 10$^{-9}$

---

$^{99}$Mo with a high specific radioactivity can be obtained by the fission of $^{235}$U in a nuclear reactor. Various methods have been described for the separation of $^{99}$Mo from the target material and other fission pro= ducts[8-14]. Problems are sometimes experienced with the separation of $^{99}$Mo from $^{103}$Ru.

This can probably be ascribed to the various oxidation states of ruthenium and the chemical behaviour of ruthenium in the various solutions during the separation steps.

Weinert et al [15-16] described the cation-exhange be= haviour of the noble metals in hydrobromic acid - acetone - thiourea mixtures and in nitric acid and hydrochloric acid. However, before this systems can be employed to separate

¹⁰³Ru from ⁹⁹Mo, two important factors must be borne in mind: Firstly, molybdenum(VI) is reduced by thiourea to the pentavalent form which also forms a cationic species, and it is also sorbed on the cation exchanger. Secondly, nitric acid has to be used preferably in a hot cell to pre= vent corrosion of the cell and equipment used in the cell. Ethylenediamine and 2,2¹-dispiridyl Fig.2 can also be used as bidentate ligands to form cationic ruthenium complexes in halogen acid solutions and in nitric acid solutions. These compounds do not reduce molybdenum (VI). Molybdenum re= mains as an anion in the acid solutions and can thus be separated from the ruthenium by cation exchange chromato= graphy. This paper describes the separation of ⁹⁹Mo from ¹⁰³Ru and the further purification of ⁹⁹Mo prior to sorption on the generator column.

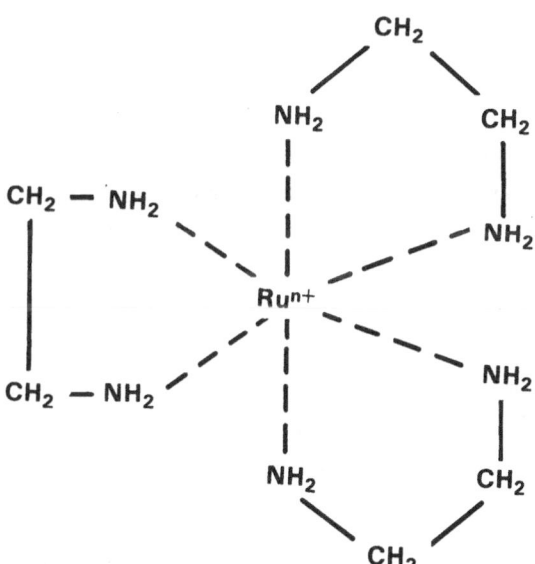

Figure 2. Structure of the ruthenium-ethylenediamine complex.

## EXPERIMENTAL

<u>Reagents</u>.    Analytical   reagent grade chemicals and demine=
ralized  water  were  used  throughout.  The cation exchange-
resin  AG 50W-X4, of 100 to 200 mesh particle size, was used
in  the  hydrogen  form.   The resin was supplied by BIO RAD
Laboratories,  Richmond,  California.   Hydrated  manganese
oxide  (MANOX  A)  was  obtained from Recherche Appliquee du
Nord,  Haut  Nord,  Hautmont,  France  and  alumina (neutral
Akt.1) from Woelm Pharma, Eschwege, Germany.

<u>Apparatus</u>.   Two  types  of ion exchange columns were used:
One  column  (type  A) consists of a borosilicate glass tube
(10  mm  bore  and 30 mm long), fitted with a no. 1 porosity
sintered-glass  disc  and a burette tap at the bottom, and a
B14  ground-glass  sleeve  at  the  top  to  hold a dropping
funnel  as  an  eluent  reservoir.  The other type of column
(type  B)  was  similarly  made  but with a 25 mm long boro=
silicate glass tube of 6 mm bore.

An  ion exchange column (type A) was filled with a slurry of
AG  50 W-X4 until the settled resin had a volume of  13,0 ml
($\equiv$  3,0 g of resin) in water.  The resin was equilibrated by
passage  of  50  ml  of  0,10M nitric acid - 0,01M ethylene=
diamine.

A  slurry  of  alumina in water was prepared and decanted to
remove  the fine particles.  An ion exchange column (type A)
was  filled  with  the  slurry until the settled alumina had
reached  a  mark  at  10  ml volume.  The alumina was condi=
tioned  by passage of 100 ml of 0,1M nitric acid followed by
100  ml  of 1mM nitric acid.  A slurry of hydrated manganese
dioxide  (HMD)  was similarly prepared and decanted.  An ion
exchange  column  (type  B) was filled with the HMD until it
had  reached  a  mark  at 2,0 ml volume.  The HMD was condi=
tioned  by  passage  of  solutions  in. the following order:

20 ml of 2,0M sodium hydroxide, 50 ml of water, 25 ml of 0,1M nitric acid and 50 ml of water.

A Canberra Series 30 Multichannel analyzer with a sodium iodide detector was used to identify the radioisotopes and to measure their activities.

## Elution curves

Mo-Ru. Sodium molybdate dihydrate (2,5 mg) and ru= thenium (III) chloride (2,1 mg) were dissolved in 2,0 ml of 5,0M nitric acid. Ten ml of water, 1,0 ml of 1,0M ethylenediamine (EDA) and 1,0 ml of 1% hydrogen peroxide were added. The solution was heated to boiling point, cooled and finally diluted to 100 ml volume with water. Known amounts of $^{99}$Mo and $^{103}$Ru were added and the solution was passed through the cation exchange column (type A, AG 50 W-X4, prepared and equilibrated as described above).

The elements were washed onto the resin with small por= tions of 0,1M nitric acid - 0,01 M ethylenediamine - 0,01% hydrogen peroxide (0,1M $HNO_3$ - 0,01M EDA - 0,01% $H_2O_2$) (I). Molybdenum was eluted with more of solution I (160 ml in total). A flow-rate of 5,5 ± 0,3 ml/min was maintained and 10 ml fractions were col= lected from the beginning of the sorption step. The amounts of molybdenum and ruthenium were determined in each fraction by measuring their activities with the multichannel analyser. The elution curve is shown in Figure 3.

In a similar experiment an elution curve was obtained for the elements but using 1,0M nitric acid - 0,10 M ethylenediamine - 0,06% hydrogen peroxide in stead of solution I. The elution curve is also shown in Figure 3.

178

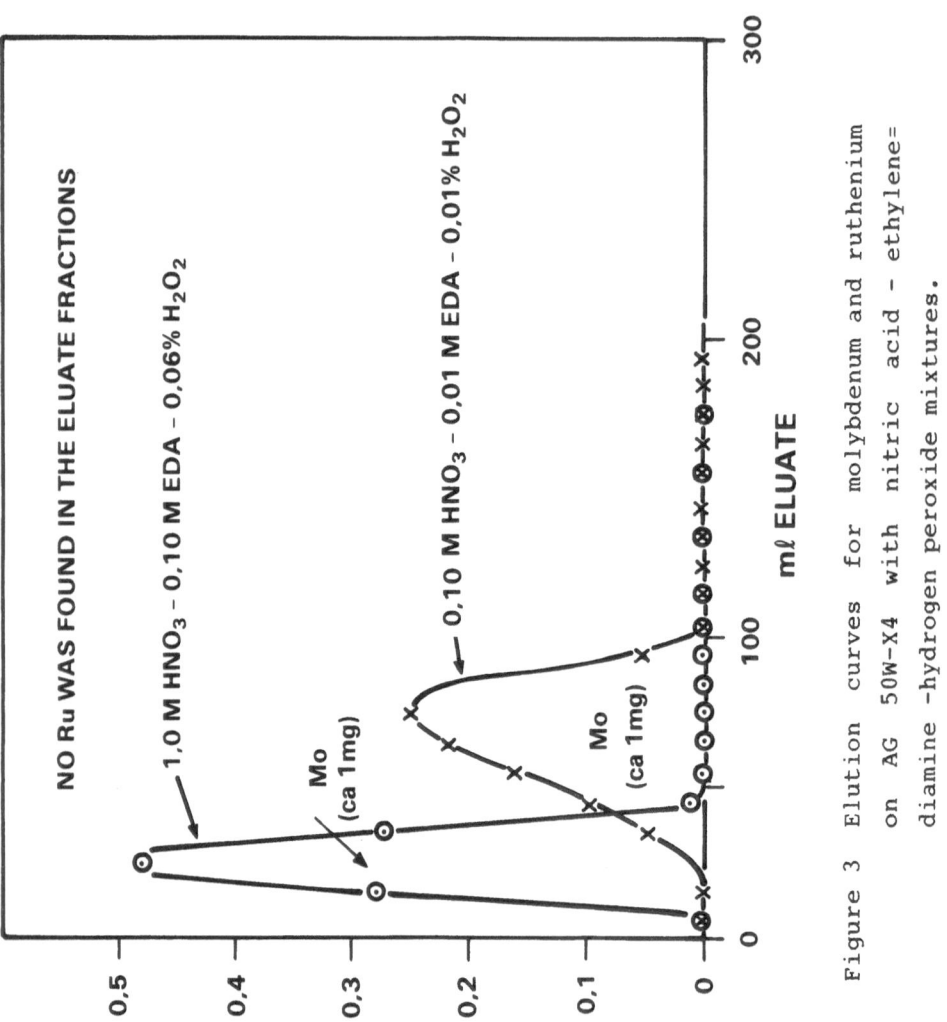

Figure 3 Elution curves for molybdenum and ruthenium on AG 50W-X4 with nitric acid - ethylene= diamine -hydrogen peroxide mixtures.

<u>Mo</u>. The eluate fractions, obtained from the first elu=
tion curve experiment and containing the molybdenum,
were combined. The solution was neutralized to pH 3,5
by the addition of 1,0M sodium acetate. The solution
was passed through the ion exchange column of type B,
containing the HMD (prepared and conditioned as de=
scribed above). The molybdenum was washed onto the
resin with small portions of 1mM nitric acid and the
ethylenediamine eluted with more 1mM nitric acid
(120 ml in total). The nitric acid was eluted with 20
ml water and the molybdenum with 100 ml of 2,0M ammo=
nium hydroxide. A flow-rate of 4,5 ± 0,3 ml/min was
maintained and 10 ml fractions were collected from the
beginning of the sorption step. The amounts of molyb=
denum were determined in each fraction as described
above and the elution curve is presented in Figure 4.

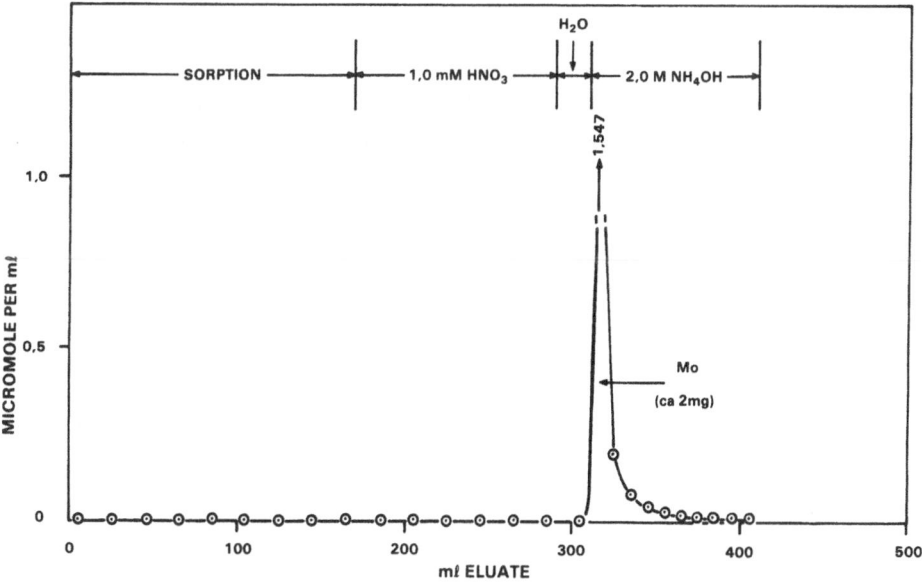

Figure 4 Elution curve for molybdenum on hydrated
manganese oxide, sorbed from a nitric acid -
ethylenediamine - hydrogen peroxide mixture.

An elution curve was obtained for molybdenum on alumina in a similar experiment. The elution curve is shown in Figure 5.

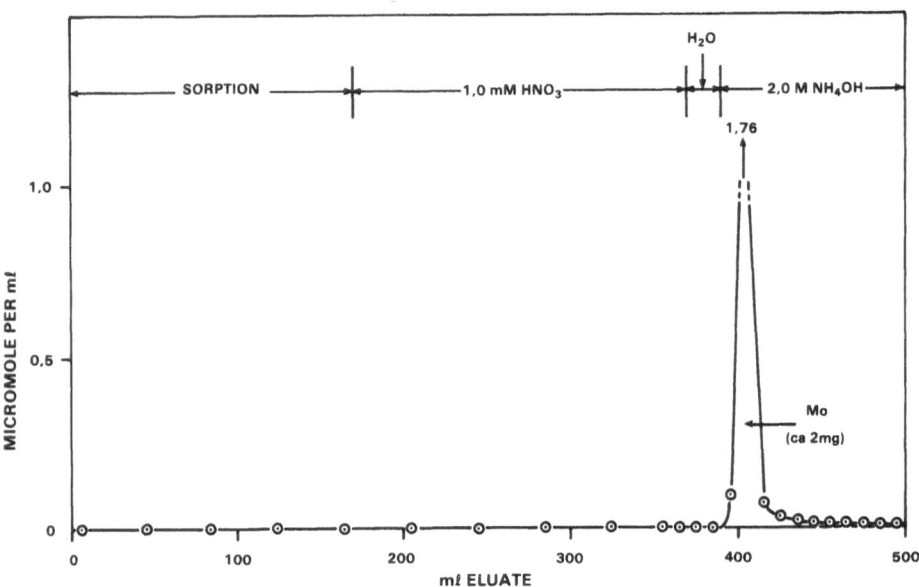

Figure 5 Elution curve for molybdenum on alumina, sorbed from a nitric acid - ethylenediamine - hydrogen peroxide mixture.

## RESULTS AND DISCUSSION

Figure 3 shows that a sharp separation of ruthenium from molybdenum was obtained on the cation exchange column con= taining 13,0 ml ($\equiv$ 3,0 g) of AG 50W-X4 with both the nitric acid - ethylenediamine - hydrogen peroxide mixtures as eluent. Molybdenum was completely eluted while ruthenium remained on the resin as a narrow brownish band, sorbed at the top of the resin column. No ruthenium was found in the eluate fractions. Molybdenum was eluted more rapidly with the 1,0M nitric acid 0,01 M ethylenediamine - 0,06% hydro= gen peroxide mixture. This can be ascribed to the higher nitric acid concentration. However, it is advantageously to use the 0,10M nitric acid - 0,01M ethylenediamine - 0,01% hydrogen peroxide mixture because a smaller volume of 1,0M sodium acetate is required for the neutralization step. The sorption of the ruthenium indicates that a stable cationic complex of ruthenium was formed with the bidentate ethylenediamine ligand.

Figures 4 and 5 show that molybdenum was quantitatively sorbed, respectively, on 2,0 ml of hydrated manganese dioxide and on 10 ml of alumina from the nitric acid - ethylenediamine - sodium acetate mixtures. The ethylene= diamine and sodium acetate were easily eluted from the HMD and alumina with 1mM nitric acid. The figures show that molybdenum can be eluted with 2,0M ammonium hydroxide. The molybdenum can also be sublimized from the HMD and alumina rendering a very pure product of high specific radio= activity.

## CONCLUSION

A rapid and sharp separation of $^{99}$Mo and $^{103}$Ru can be
obtained by cation exchange chromatography on AG 50W-X4 in
a nitric acid-ethylenediamine - hydrogen peroxide mixture.
The organic solvent can be removed by ion exchange chroma=
tography on hydrated manganese dioxide or alumina. $^{99}$Mo
can be eluted with ammonium hydroxide from the inorganic
ion exchangers or can be sublimized from them to obtain a
highly purified product.

## REFERENCES

1          Lavi, N., The study of conditions for the pre=
           paration and application of $^{99}$Mo - $^{99m}$Tc gene=
           rators starting from irradiated molybdenum metal.
           J. Radional. Chem., 1978, **42**, 25-34

2          Milenkovic, S. M., Vucina, J. L., Jacimovic, Lj.
           M., Karanfilov, E.S. and Memodivic, T. V.,
           Universal $^{99m}$Tc generator for human use.
           Isotopenpraxis, 1982, **18**, 85-87

3          Brown, J. L. and Harris, O. A. Technetium-99m
           generator. Ger. Offen. 2,242,395 (Cl. C Olg), 15
           Mar 1973, US Appl. 177,249, 2 Sep 1971, 25pp

4          Levin, V. I., Kozyreva - Alexandrova, L. S.,
           Sokolova, T. N. and Bagenova, T. L. A new $^{99m}$Tc
           generator of higher activity. Int. J. Appl.
           Radiat. Isot., 1979, **30**, 450-451

5          Meloni, S. and Brandone, A. a new technetium-99m
           generator using manganese dioxide. Int. J. Appl.
           Radiat. Isot., 1968, **19**, 164-166

6     Bhattacharyya, D. K. and De, A. Separation of carrier-free $^{99m}$Tc from $^{99}$Mo and $^{144}$Pr from $^{144}$Ce over a column of zirconium arsenate. Sep. Science Tehnology, 1982, 17, 925-933

7     Fourie, P. J. Progress in the development of technetium-99m technology in South Africa. In Technetium-99m Technology, ed. M. Beyers, Third Congress of the S A Society of Nuclear Medicine, Bloemfontein, 14-17 August 1988, pp 3-10

8     Ejaz, M. Separation of molybdenum (VI) from uranium (VI) and fission products. Radiochimica Acta, 1975, 22, 51-52

9     Salacz, J. Reprocessing of irradiated uranium-235 for the production of $^{99}$Mo, $^{131}$I, $^{133}$Xe radioisotopes. Revue IRE Tijdschrift, 1985, 9, 22-28

10    El-Garhy, M., Shehata, M. K. K. and El-Bayoumi, S. Selective separation of $^{99}$Mo from fission products in chloride media on activated alumina. J. Radioanal. Chem., 1972, 10, 35-40

11    Arino, H. and Kramer, H. H. Separation and purification of radiomolybdenum from a fission product mixture using silver-coated carbon granules. Int. J. Appl. Radiat. Isot., 1978, 29, 97-102

12    Cheng, W.L., Lee, C. S., Chen, C. C., Wang Y. M. and Ting, G. Study of the separation of molybdenum-99 and recycling of uranium to water boiler reactor. Appl. Radiat. Isot., 1989, 40, 315-324

13      Münze, R., Hladik, O., Bernard, G., Boessert, W. and Schwarzbach, R.   Large scale production of fission $^{99}$Mo by using fuel elements of a research reactor as starting material. Int. J. Appl. Radiat. Isot., 1984, 35, 749-754.

14      Sameh, A.      Production of molybdenum-99. PROCEEDINGS International Symposium on Isotope Applications, Taipei, Taiwan, Republic of China, December 4 - 5, 1986, 291-303.

15      Weinert, C. H. S. W. and Strelow, F. W. E. Cation exchange behaviour of the platinum group and some other rare elements in hydrobromic acid- thiourea-acetone media. Talanta, 1983, 30, 755-760.

16      Weinert, C. H. S. W. and Strelow, F. W. E. The influence of thiourea on the cation-exchange behaviour of various elements in dilute nitric and hydrochloric acids.   Talanta, 1986, 33, 481-487.

# Part 4

# THEORETICAL ASPECTS
# AND NEW ADVANCES

# CALCULATION OF RETENTION DATA IN ION EXCHANGE CHROMATOGRAPHY

JÁNOS INCZÉDY
Institute of Analytical Chemistry
University of Veszprém, P.O.Box 158,
Veszprém H-8201, Hungary

## ABSTRACT

Fundamental equations for the calculation of retention data for the separation of metal cations by ion exchange chromatography are introduced. For the separation of $Na^+$, $K^+$, $Ca^{2+}$ and $Mg^{2+}$ ions a histidine containing eluent was used. The values of the necessary equilibrium constants (e.g. selectivity ratios) were calculated by reiteration from retention time data obtained experimentally. The calculation proved to be useful also for prediction of the retention data of the caesium and strontium ions.

## INTRODUCTION

Comparing     ion exchange chromatography to other types chromatographic methods, one of its main advantage is, that the distribution of the solutes between the stationary and mobile phases, and therefore the retention times, can be controlled by chemical reactions. If the chemical reactions, and their equilibrium constants are known, the most relevant interrelation    between the distribution ratios of the components to be separated and the composition of the eluent can be computed so that   optimal conditions of the separation can also be predicted.

As a starting point the ion exchange reaction taking place between sodium and hydrogen ions may be considered, using a $H^+$-form strong acid cation exchanger as stationary    phase with sodium chloride solution as a mobile phase.

$$\overline{H^+} + Na^+ \rightleftarrows H^+ + \overline{Na^+} \tag{1}$$

The bars refer to the ion exchanger phase. At equilibrium the concentrations are interrelated   according to the following relation:

$$K^*_{Na/H} = \frac{(Na^+)[H^+]}{[Na^+](Na^+)} \qquad (p, T, I \text{ const.}) \qquad (2)$$

Round brackets are for the concentrations in the solid and the squares for that of the solution phases. $K^*$ is the selectivity ratio, which at certain conditions (the pressure, temperature, ionic strength are constant, and the resin phase is almost completely in $H^+$-form, the sodium ion is present in very low amount) can be considered as constant (1).

Since the concentration of the hydrogen ion in the resin phase is close to the capacity of the resin, the distribution ratio can expressed as follows:

$$D_{Na} = \frac{(Na^+)}{[Na^+]} = K^*_{Na/H} Q[H^+]^{-1} \qquad (3)$$

where Q is the total capacity of the resin (equ/l). In log-arithmic form:

$$\log D_{Na} = \log K^X + \log Q - \log [H^+] \qquad (4)$$

The equation can be generalised also for the cases where cations of different types are exchanged with eluent ions of various charges.

$$\log D_M = \frac{1}{x} \log K^X_{M/A} + \frac{y}{x} \log Q - \frac{y}{x} \log [A] \qquad (5)$$

where M is the cation, with charge y and A is the ion with charge x.

It is much more interesting, and leads to valuable con-clusions, that the relation is valid also in those cases where the capacity of the ion exchanger is varied, but the concentration of the solute and eluent remains the same.

Figures 1 and 2 show that there are linear relationships between $\log D$ and $\log [HNO_3]$ and $\log Q$.

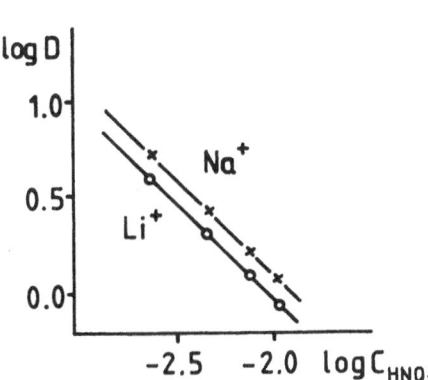

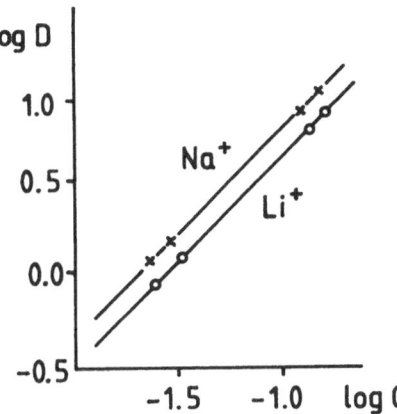

Fig.1.Logarithm of the distribution ratio of metal cations
plotted against the logarithm
of the eluent concentration.
Q = 0,019 equ/l

Fig.2.Logarithm of the distribution ratio of metal cations
plotted against the logarithm
of the exchange capacity of
the resin. Eluent concentration: $10^{-2}$ M $HNO_3$.

This experimental observation is an important fact because it means, that for informatory calculations, the ion
exchange selectivity ratio given for conventional ion exchange resins of high capacity, can also be used for low capacity packing materials used in the ion chromatography.
Assuming, of course, that the composition and structure of
the surface coating resin layer are similar to those of the
traditional resins (2).

If the distribution ratio is constant during the elution,
a very simple relation is held between the retention volume
$V_R$ and the distribution ratio D:

$$V_R = V_C D + V_0 \qquad (6)$$

Using constant flow rate of the eluent $q_v$, the retention time
can be expressed with the following equation:

$$t_R = \frac{V_R}{q_v} = t_0(k+1) \qquad (7)$$

$k = D \dfrac{V_C}{V_0}$ is the capacity ratio, and $t_0 = \dfrac{V_0}{q_v}$ the dead
time of the column. $V_C$ is the volume of the column.

The introduced equations (2-5) with some modifications are applicable also in those cases where metal ions are separated and for improvement of the selectivity complex forming agents are used in the eluents. The derived equations proved to be useful also in such complicated cases, where the metal ions were eluted with complexing agent (oxalic acid) containing eluent and with a gradually increasing pH (3).

For the ion chromatographic separation of the mono and divalent metal ions it seemed to be advantageous to use basic amino acid containing eluents. The basic amino acids can form mono- and also divalent cations depending on the pH of the solution, and by selection of a proper pH, the elution strength can be increased also at low eluent concentration much more effectively, than at the use of the customary hydrochloric or nitric acid eluents.

In the presence of the basic amino acid, however, simultaneous ion exchange equilibria will take place.

For the monovalent metal ion the distribution ratios are expressed as follows:

$$D_{M1} = K^x_{M1/A1}(x_{A1}Q) [A1]^{-1} + K^x_{M1/A2}(\frac{1}{2} x_{A2}Q)^{\frac{1}{2}} [A2]^{-\frac{1}{2}} + \tag{8}$$

$$+ K^x_{M/H}(x_H Q) [H^+]^{-1}$$

$$D_{M2} = K^x_{M2/A1} = (x_{A1}Q)^2 [A1]^{-2} + K^x_{M2/A2}(\frac{1}{2} x_{A2}Q) [A2]^{-1} + \tag{9}$$

$$+ K^x_{M2/H}(x_H Q)^2 [H^+]^{-2}$$

$K^x_{M1/A1}$, $K^x_{M1/A2}$ and $K^x_{M1/H}$ are the selectivity ratios of the monovalent metal ion referred to the monovalent amino acid cation A1, to the divalent amino acid, cation A2 and to the hydrogen ion respectively. Similarly are denoted the selectivity ratios for the divalent metal ions. With $x_{A1}$, $x_{A2}$ and $x_H$ are denoted the fractions of the total capacity of the ion exchanger phase occupied by the A1, A2 and $H^+$ ions respectively.

$$x_{A1} + x_{A2} + x_H = 1 \tag{10}$$

Considering the ion exchange equilibria taking place among the three eluent ions, the values of the above fractions i.e. the concentrations of the different eluent ions in the stationary phase can be expressed using the corresponding selectivity ratios $K_{A2/A1}$, and $K_{H/A1}$ and the actual concentrations of the A2, A1, and $H^+$ ions in the solution.

Since we found that the computed values are more consistent if we use activities instead of concentrations in the solution, the following formulations were used:

$$A1 = f_1 \phi_1 C_A; \qquad (11)$$

$$A2 = f_2 \phi_2 C_A \qquad (12)$$

$$H^+ = 10^{-pH} \qquad (13)$$

where $f_1$ and $f_2$ are the activity coefficients for the mono and divalent ions, and $\phi_1$, $\phi_2$ the molefractions of the species respectively. $C_A$ is the total concentration of the amino acid.

The mole fractions of the species A1 and A2 can be easily calculated using the protonation constants of the amino acid base and the actual pH of the solution. The final form of the equations of the distribution ratios are as follows:

$$D_{M1} = \frac{K^x_{M1/A1}}{K^x_{A2/A1}} \; \frac{([A1] + K^x_{H/A1} [H^+])}{4 [A2]} \left[ (1+m)^{\frac{1}{2}} - 1 \right] \qquad (14)$$

$$D_{M2} = \frac{K^x_{M2/A1}}{K^x_{A2/A1}} \; \frac{([A1] + K^x_{H/A1} [H^+])}{8 [A2]^2} \left[ 1 + \frac{m}{2} - (1+m)^{\frac{1}{2}} \right] \qquad (15)$$

$$m = 8 K_{A2/A1} Q [A2] ([A1] + K_{H/A1}[H^+])^{-2}$$

## EXPERIMENTAL

Experiments were carried out using $Na^+$, $K^+$, $Mg^{2+}$ and $Ca^{2+}$ containing sample solutions (0,1 mM), for preparation of the eluent solution ($C_A$ = 3-5 mM) L-hystidine monochloride and hydrochloric acid was used.

For the elution experiments Dionex-2010 (Dionex Co., USA) ion chromatography, Dionex CG3 (50x4) precolumn and Dionex CS3 (250x4mm) cation exchanger separation column, and Dionex CMMS-1 supressor columns were used.

From the obtained retention time data, the distribution ratios were calculated using equations (6) and (7). The necessary data for the calculations are as follows:

$$t_0 = 1,93 \text{ min}$$
$$V_C = 2,33 \text{ ml (stationary phase)}$$
$$q_v = 1 \text{ ml/min}$$
$$Q = 16,58 \text{ } \mu equ/ml$$

## RESULTS AND DISCUSSION

The calculated selectivity ratio data obtained by reiteration process from the $t_r$ values found by the experiments, and from the data of the actual concentrations and pH, are summarized in table 1. The obtained selectivity ratio data seem to be reliable and independent on the experimental conditions. It is maybe noteworthy to mention that three equilibrium constants are enough to determine the position of the five simultaneous ion exchange processes.

TABLE 1
Calculated selectivity ratios

| $C_{hys}$ mM | $Na^+$ | | | $K^+$ | | |
|---|---|---|---|---|---|---|
| | $K^x_{Na/Al}$ | $K_{A_2/Al}$ | $K_{H/Al}$ | $K^x_{K/Al}$ | $K_{A2/Al}$ | $K_{H/Al}$ |
| 3 | 0,29 | 10,7 | 0,45 | 0,56 | 10,6 | 0,45 |
| 3,8 | 0,30 | 10,7 | 0,47 | 0,59 | 11,0 | 0,52 |
| 4,5 | 0,28 | 11,0 | 0,45 | 0,56 | 11,0 | 0,50 |
| 5,2 | 0,29 | 10,7 | 0,55 | 0,55 | 10,0 | 0,55 |
| 6 | 0,29 | 11,0 | 0,48 | 0,55 | 10,8 | 0,55 |

| $C_{hys}$ mM | $Mg^{2+}$ | | | $Ca^{2+}$ | | |
|---|---|---|---|---|---|---|
| | $K^x_{Mg/Al}$ | $K_{A2/Al}$ | $K_{H/Al}$ | $K^x_{Ca/Al}$ | $K_{A2/Al}$ | $K_{H/Al}$ |
| 3 | 7,0 | 10,7 | 0,45 | 15,0 | 11,0 | 0,45 |
| 3,8 | 7,0 | 10,6 | 0,49 | 14,9 | 10,7 | 0,51 |
| 4,5 | 7,5 | 10,7 | 0,55 | 15,3 | 10,1 | 0,55 |
| 5,2 | 7,5 | 10,6 | 0,55 | 15,1 | 10,0 | 0,55 |
| 6 | 7,4 | 10,0 | 0,55 | 15,4 | 10,0 | 0,55 |

The selectivity ratios for thé metal cations referred to hydrogen, to the mono- and divalent hystidine cation were calculated. The trend of the selectivity ratio values correspond to the trend of the known ion exchange properties of the metal cations investigated.

For the $K^+$, $Cs^{2+}$, $Mg^{2+}$ and $Sr^{2+}$ ions there are available selectivity ratio data referred to the hydrogen ion for sulphonated cation exchangers in the literature. [1]

Using these data, and the selectivity ratio data for $K^+$ and $Mg^{2+}$ referred to monovalent hystidine ion, the values of $K^x_{Cs/Al}$ and $K^x_{Sr/Al}$ were calculated and found as 0,83 and 3,21 respectively. The $t_r$ values in minutes are listed in Table 2.

193

TABLE 2

| Ion | Calculated | found exp. |
|---|---|---|
| $Cs^+$ | 3,0 | 2,9 |
| $Sr^{2+}$ | 16,7 | 16,2 |

The agreement between the calculated retention time values obtained by calculation from the data of literature sources, and those of found experimentally is good.

ACKNOWLEDGEMENTS

Thanks are expressed to P. Hajós and M. Magyari, who carried out the experiments and also the calculations. The work with all details and results will be published elswhere (4).

REFERENCES

1. Inczédy, J.: Analytical Applications of Complex Equilibria Ellis Horwood, Chichester 1976.

2. Hajós, P., Inczédy, J.: Preparation, examination and parameter optimisation of cation exchangers in ion chromatography, in Ion exchange technology (Ed.D.Naden, M. Streat) p 450-457. Ellis Horwood, Chichester 1985.

3. Inczédy, J.: J. Chrom. 154, 175-181 (1978)

4. Hajós, P., Magyari, M., Inczédy, J.: Multiple species model for prediction of retention using amino acid eluents in ion-chromatography 14th Internat. Symp. HPLC '90, Boston, May 20-25 1990.

# LIGAND EXCHANGE PRINCIPLES FOR TRACE ENRICHMENT AND SELECTIVE DETECTION OF IONIC COMPOUNDS

U.A.TH. BRINKMAN AND H. IRTH

Free University, Department of Analytical Chemistry, de Boelelaan 1083, 1081 HV
Amsterdam, the Netherlands

## ABSTRACT

Ligand exchange principles are well suited to improve the pretreatment (trace enrichment and clean-up), separation as well as detection procedures in the trace-level determination of cationic and anionic compounds by means of reversed-phase column liquid chromatography. The pertinent strategies are outlined and several relevant applications are shown.

## INTRODUCTION

In the past fifteen years, high-performance liquid chromatography (HPLC) has become a versatile and widely applied method of analysis for a variety of, e.g., environmental, biomedical and food samples. In a large majority of all cases, such analyses deal with the determination of organic compounds which, normally, have to be determined at trace level, i.e., in the low ppm to high ppt range. In other words, satisfactory performance - i.e., satisfactory selectivity as well as sensitivity - has to be found by combining adequate separation with efficient trace enrichment and clean-up, and sophisticated pre- or post-column derivatization or reaction detection of the analyte(s) of interest. Besides, since large series of samples often have to be processed, on-line procedures are generally preferred to off-line analyses, in order to make automation more easily accessible.

Since HPLC is conventionally carried out in the reversed-phase mode - i.e., on hydrophobic C18- or C8-modified silica - with normal-phase HPLC with its non-aqueous eluents lagging far behind, the technique have never been highly attractive for studies dealing with inorganic ions and related ionized organic compounds. Still, in recent years an increasing number of papers has been published on the HPLC separation of heavy metal ions, generally as their dialkyl-dithiocarbamate (DAlkTC) complexes [1-4]. For both normal-phase and reversed-phase separations, UV/vis absorbance detection is the most frequently employed mode of detection with, for reversed-phase separations, electrochemical detection coming in second place [5]. With most published methods, however, there are rather serious drawbacks. For example, absorbance detection is only possible at wavelengths higher than 310 nm, because the excess of DAlkTC present in the HPLC eluent strongly absorbs in the low UV range. Electrochemical detection requires the removal of the DAlkTCs in order to avoid a high background signal. In several cases, on-line trace enrichment also becomes rather complicated [6].

In the present review, it is our intention to outline how earlier work on the trace-level determination of organic ligands - where reaction with a metal was used to obtain selectivity - led us to study the HPLC of inorganic ions, now using rather similar ligands for the trace enrichment and clean-up and/or reaction detection of such ions. The applicability of the various procedures will be demonstrated.

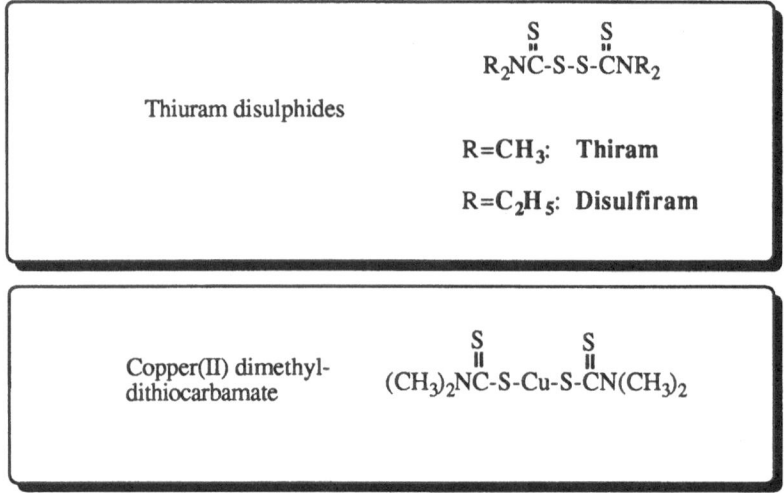

Figure 1. Thiuram disulphides and copper(II) dimethyldithiocarbamate.

## THIRAM AND DISULFIRAM

Thiram is a widely used protective fungicide in agriculture. It is a so-called thiuram disulphide, and has a structure closely related to that of the dithiocarbamates already mentioned in the introduction to this paper. It is also worthy of note that the tetraethyl analogue of thiram, disulfiram, is used as a drug against alcohol abuse (see Fig. 1). The HPLC methods for these compounds reported in the literature generally are rather non-selective [7,8], while the gas chromatographic methods are mostly based on the determination of carbon disulphide produced upon acid digestion of the dithiocarbamates and/or the thiuram disulphides [9]. A characteristic property of the two classes of compounds is that they form strong complexes with a large variety of metal ions. This characteristic was used to develop a selective trace-level determination of thiram, to allow its analysis in environmental and biological samples without excessive clean-up [10].

Thiram can easily be chromatographed, and separated from one of its degradation products, $Cu(DMeTC)_2$, in a reversed-phase C18-bonded silica/acetonitrile-aqueous acetate buffer (pH 5) system, provided the sample is stabilized with 10 mM EDTA/10 mM citrate to prevent breakdown of the fungicide due to complexation with metal ions. Besides, since thiram is a rather hydrophobic compound, it can easily be preconcentrated from an aqueous sample solution onto a short (4 x 2.1 mm ID) precolumn packed with C18-bonded silica. Actually, the breakthrough volume of thiram on such a column is over 50 ml, which means that - compared with a conventional 50 µl loop injection - an enrichment factor of 1,000 can readily be achieved. However, trace enrichment on a strongly hydrophobic surface will not do much to increase selectivity, and neither will UV detection at 262 nm, which is the wavelength of maximum absorption of thiram. We therefore tried to enhance selectivity by means of ligand exchange, i.e., by using the thiram-copper reaction.

Early experiments showed that neither on-line precolumn derivatization with copper(I) chloride nor post-column reaction in a solid-phase reactor with either copper(I) chloride mixed with C18-bonded silica or red metallic copper gave a reliable and kinetically fast solution of the problem in hand. If, however, a post-column reactor was filled with black metallic copper - prepared by the reduction of a copper(I) salt with sodium borohydride - an almost instantaneous and quantitative conversion of thiram into $Cu(DMeTC)_2$ was achieved. Actually, another 4 x 2.1 mm ID precolumn - now used in the post-column mode - filled with the reactive copper modification sufficed to effect an over 90% yield of the reaction product. This was then selectively monitored at $\lambda = 435$ nm, the absolute detection limit being 3 ng of thiram. In other words, with preconcentration from a 40-50 ml sample volume, sub-ppb detection limits can easily be obtained. Today, the method has successfully been used for the determination of thiram in tap and surface water, fruit and vegetables, and for disulfiram in urine [10-12].

The main conclusion from this work for the actual topic of the present paper is that organic ligand-metal ion exchange and complexation phenomena have a good potential in HPLC, and that they may well be useful for metal ion determination.

## HEAVY METAL IONS

### Transition metal ions

As was already stated in the introduction, the use of organic ligands, notably dialkyldithiocarbamates (DAlkTCs=alkyl; Me=methyl; Et=ethyl), to facilitate HPLC of heavy metal ions is rather well known, but most methods lack the sensitivity and/or selectivity required for trace-level studies. As an alternative solution to the problem, we tried to achieve simultaneous metal dithiocarbamate formation - using DEtTC as reagent - and preconcentration on a C18-bonded silica column in an on-line mode [13].

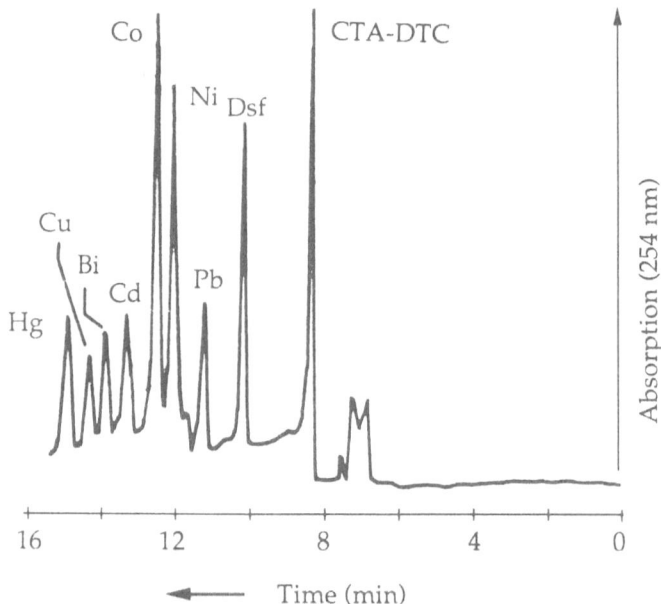

Figure 2. On-pre-column derivatization and separation of seven metal ions as their dithiocarbamates using CTA-DEtTC as derivatization reagent. LC conditions: analytical column, 250 x 4.6 mm, 5 μm Spherisorb ODS; eluent: solvent A: acetonitrile containing 10 mM CTAB; solvent B: 10 mM phosphate buffer, pH 6.8, containing 2 mM CTAB; gradient: from 40% to 75% solvent A in 8 min; flow-rate, 1.5 ml/min; UV detection at 254 nm (0.1 AUFS); derivatization pre-column, 2.0 x 4.6 mm packet with 5 μm Spherisorb ODS, loaded with 0.4 μmole CTA-DEtTC; injection, 100 μl of 50 μM Bi(III), Cd(II), Pb(II), Ni(II), Co(II), Hg(II) and Cu(II) (Dsf = Disulfiram).

First attempts were made with Zn(DEtTC)$_2$ immobilized on the C18 phase. The zinc complex was selected because it has a rather low formation constant, so that it can be expected to react with many heavy metal ions via ligand exchange to form the corresponding metal dithiocarbamates. Good results were indeed obtained for Co(II), Cu(II), Ni(II) and Hg(II), but ions such as Bi(III), Cd(II) and Pb(II) could not be determined, probably as a result of slow kinetics of the ligand-exchange reaction. To overcome this problem, Zn(DEtTC)$_2$ was exchanged for the ion-pair formed between DEtTC and cetyltrimethylammonium (cetrimide; CTA) as the cation, viz. CTA-DEtTC. With the latter reagent immobilized on a C18 phase in a 2 x 4.6 mm ID precolumn, a rapid conversion was observed for all seven metal ions mentioned above, with RSD values of 1-5% at the 1-3 ppm level. A typical result is shown in Fig. 2. Here, one should realize that the disulfiram peak is caused by the oxidation of a small amount of the excess of DEtTC used as reagent.

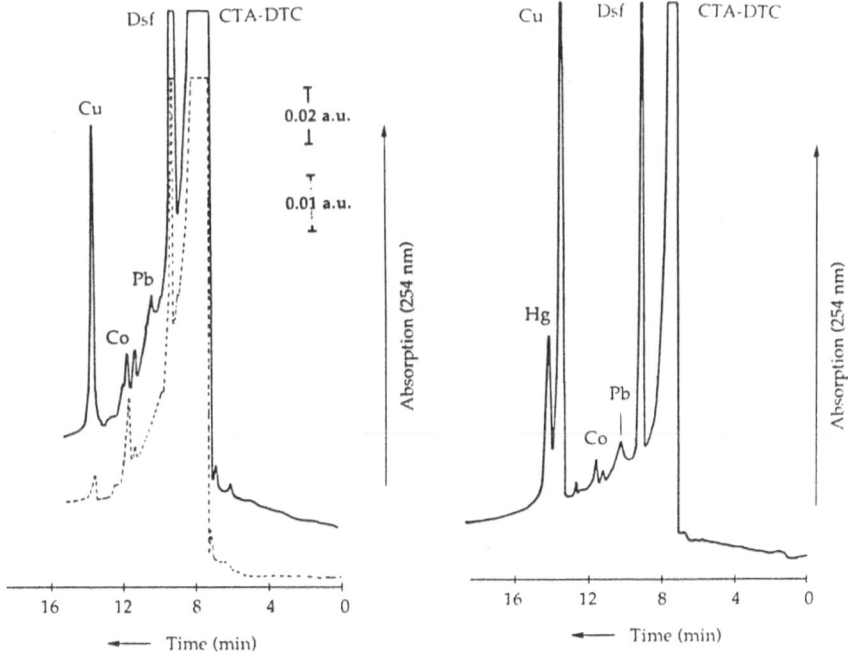

Figure 3. Determination of metal ions in drinking water with on-column formation/preconcentration using CTA-DEtTC as derivatization reagent. *Left,* preconcentration of 10 ml fresh drinking water and 10 ml doubly distilled water (dashed line); *right,* 10 ml water which stood for 12 h in a copper main, spiked with 80 ppb Hg(II).

The CTA-DEtTC-loaded precolumn did not show breakthrough of the reagent even after processing 25 ml of sample which, therefore, is a safe upper limit for trace-enrichment studies. The only precaution required is that the sample solution should be buffered at pH = 5.5-6.0, preferably with an acetate buffer, to prevent decomposition of CTA-DEtTC. As a practical example, the determination of Cu(II) in drinking water, distilled water, and drinking water that stood in a copper main for 12 h, is shown in Fig. 3. From the 64 ppb (Fig. 3a) and 130 ppb (Fig. 3b) contributions found for the 10-ml samples, one readily sees that trace-level determinations can easily be carried out at the 1 ppb level, with metal parts of the HPLC system rather than the inherent sensitivity of the total set-up determining the actual detection limit.

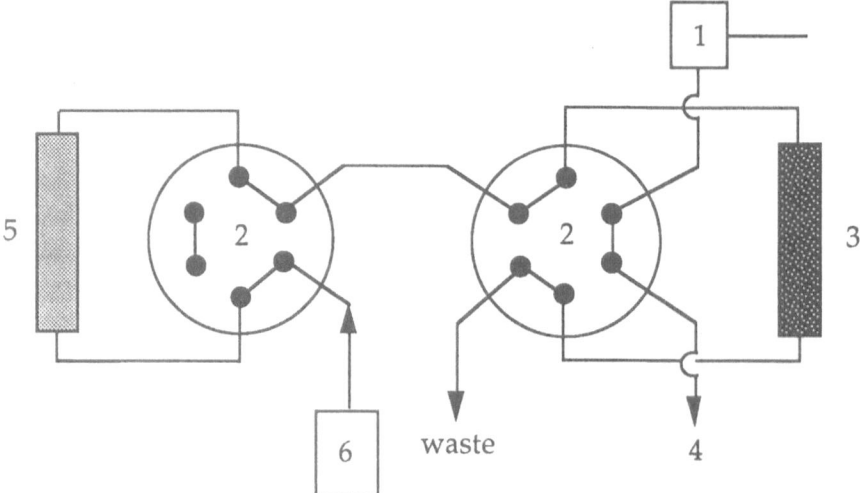

Figure 4. Scheme of the loading and elution procedure. 1. LC pump; 2, six-port injection valve; 3, clean-up pre-column; 4 derivatization pre-column; 5, preconcentration pump; 6, to analytical column and UV detector. Preconcentration of the metal ions as metal DEtTP complexes is shown.

### Trivalent As, Sb and Bi

A drawback of the system described in the previous section is that only a limited number of metal ions can be determined as their DAlkTC complexes by means of reversed-phase HPLC. For example, As(III) can not be determined because it forms only weak complexes with the various DAlkTCs, which probably dissociate or undergo ligand-exchange reactions. Working under acidic conditions (pH = 1-4) may well eliminate this problem; unfortunately, however, the DAlkTC ligands are not stable at low pH. The use of a ligand which forms strong, stable and neutral complexes with As(III) at pH = 1-4 should be a useful alternative. A dithio-type chelating agent which meets these requirements is diethyldithiophosphoric acid (DEtTP) [13-16]. Although the liquid-liquid extraction of metal-DEtTP complexes has been reported in some detail, and their use for normal-phase HPLC has been described, DEtTP has received much less attention than the DAlkTCs regarding the reversed-phase separation of metal ions. This may well be due to the fact that DEtTP forms weaker complexes with most metal ions than do the DAlkTCs. The stability under strongly acidic conditions is, however, of paramount importance when studying trivalent As, Sb and Bi, and this stability is much higher for DEtTP [16]. To quote one example, the half-life NaDEtTP in 1 M HCl has been reported to be about 250 h, while - under similar conditions - NaDAlkTCs have half-lives of a few seconds.

In this study, the same on-line on-column derivatization/trace enrichment set-up was used as in the earlier work on heavy metal ions (cf. above and Fig. 4) [17]. However, it was necessary to optimize the HPLC separation of the three DEtTP complexes before studying the precolumn part of the system. Separation of the complexes in a C18-bonded silica/acetonitrile-aqueous buffer HPLC system, as used for the heavy metal ion-DEtTC complexes was completely unsuccessful: the DEtTP complexes adhered strongly to the top of the analytical column and no signal was observed at all. Lowering the eluent pH to about 3 caused $Bi(DEtTP)_3$ to elute with good peak shape. This procedure did not help, however, in the case of arsenic and antimony - not even if the pH was further decreased to about 1.5. When working at a pH of 2, the addition of 1 mM (antimony) or even 10 mM (arsenic) of DEtTP to the HPLC eluent was required to obtain elution - with good peak shapes - of the remaining two DEtTP complexes. Under these optimized conditions, the separation took about 10 min. It may seem a distinct drawback that the organic ligand has to be added to the mobile phase - and at a rather high concentration - because with the DAlkTCs, this leads to a fairly high UV background and, consequently, reduced sensitivity. Fortunately, however, NaDEtTP has a rather low UV absorption over the whole 250-300 nm region and monitoring at 280 nm gave detection limits as low as 2 ng for all three elements (3-order linearity).

Returning to the precolumn set-up, several small changes were made compared with the system used for the heavy metal ions. Because of the low pH value required for the reaction between CTA-DEtTP and the trivalent ions, a styrene-divinylbenzene copolymer phase - which is stable between pH 1 and 13 -

was used instead of a C18 phase. The strongly hydrophobic polymer phase facilitates both the retention of the CTA-DEtTP reagent and the trace enrichment of the $M(DEtTP)_3$ complexes. Desorption, obviously, will require a high modifier content of the eluent, i.e., 85% acetonitrile. It is also interesting to note that in some cases another polymer-packed precolumn was inserted in front of the CTA-DEtTP-loaded precolumn (cf. Fig. 4). This helps to remove non-polar sample constituents, but does not influence the behaviour of the trivalent elements which, at this stage, have not yet been converted into hydrophobic DEtTP complexes!

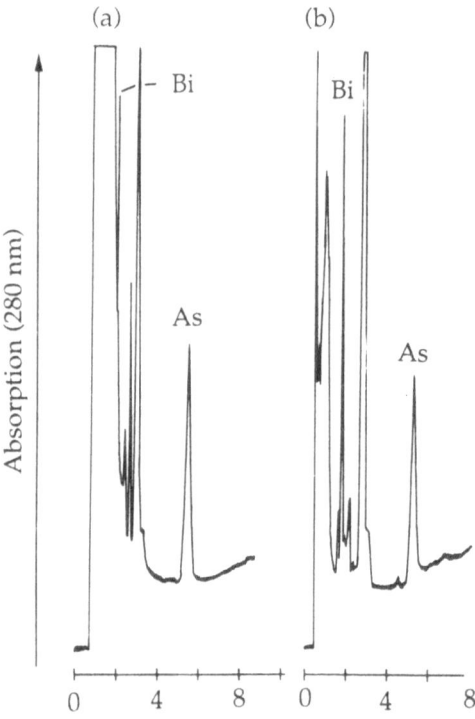

Figure 5. Chromatogram of a urine sample spiked with 19 ppb As(III) and 57 ppb Bi(III) in urine (1 ml preconcentration). LC conditions: On-pre-column formation of the metal DEtTP complexes occurred (a) without or (b) with a PRP$_1$ clean-up pre-column.

As an illustration of the practicality of the procedure, As(III) and Bi(III) were determined in spiked urine (10-60 ppb). Results for a 1-ml sample are shown in Fig. 5, which also nicely demonstrates the beneficial influence of the second polymer-packed precolumn on the clean-up. In further studies with 10-ml instead of 1-ml urine samples, the detection limit for arsenic and bismuth was found to be 1 ppb.

## POST-COLUMN REACTION DETECTION OF CATIONS

In the studies reported in earlier parts of this paper, the selective trace enrichment of metal ions as their DAlkTC or complexes was discussed. In those cases, selectivity was introduced via a dual precolumn system which eliminates most of the interferring hydrophobic as well as polar compounds. As an alternative, the performance of the total analytical system can be improved by using more selective detection.

In the section on the trace-level determination of thiram and disulfiram, it has been shown that these compounds can be rapidly converted into $Cu(DAlkTC)_2$ complexes by reacting them with black metallic copper in a short solid-state reactor. Other experiments showed that insoluble salts such as copper(II) sulphide and copper(II) phosphate also react rapidly with thiram, to form $Cu(DMeTC)_2$. This can be explained by the rather high complex stability of $Cu(DMeTC)_2$, which has a log $\beta_2$ of 21.8. The formation of $Cu(DAlkTC)_2$ complexes was also observed when copper(II) phosphate was added to a solution of lead or cadmium dialkyldithiocarbamate, which possess a considerably lower complex stability than the copper complex. Since the ligand-exchange reaction appears to proceed rather fast, an attempt was made to use the system for the selective detection of Bi(III), Cd(II) and Pb(II) by means of post-column reaction detection comparing, moreover, the behaviour of copper(II) phosphate with that of nickel phosphate as the solid-phase post-column reagent.

The schematic of the analytical system is shown in Fig. 6. It contains the solid-state reactor to be discussed below as well as a short precolumn that can be used for the on-line trace enrichment of the rather non-polar M(DAlkTC) complexes on a C18-bonded stationary phase. The metal chelates were formed off-line by adding 0.5 ml of an aqueous buffer (pH 6) and 0.5 ml of a NaDEtTC solution to an appropriate amount of the sample solution to be analyzed which, in this case, was urine. After loading the precolumn and flushing it with 5 ml of the aqueous pH 6 buffer to effect further clean-up, the chelates of interest were desorbed to the C18 analytical column with the HPLC eluent. This was a 70:30 mixture of acetonitrile and an aqueous pH 6 buffer containing 10 mM cetrimide to improve the peak shape of Cd(DEtTC)$_2$.

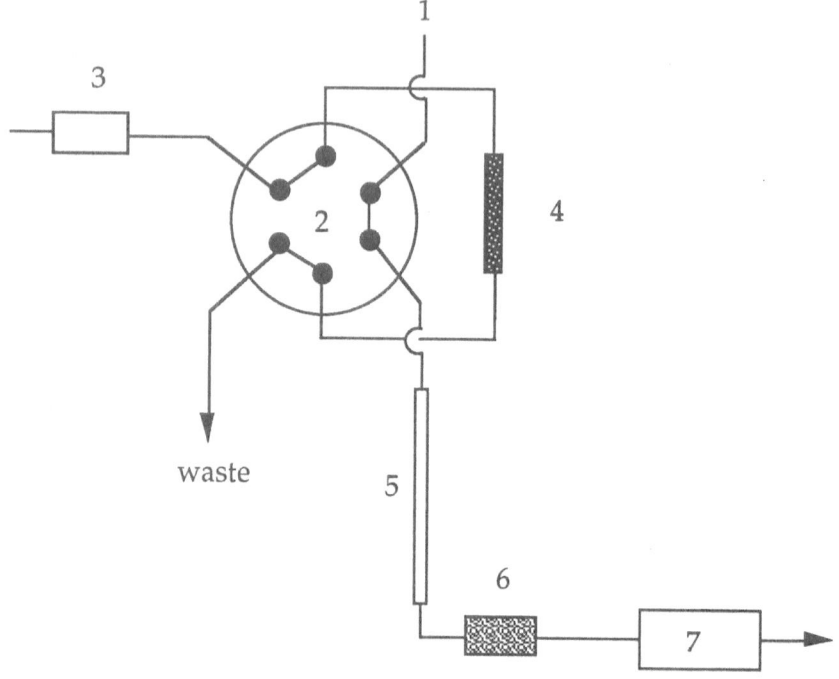

Figure 6. Scheme of the chromatographic system. 1: HPLC pump; 2: six-port injection valve; 3: preconcentration pump; 4: C$_{18}$ pre-column; 5: analytical column; 6: metal phosphate post-column reactor; 7: detector.

As regards the post-column reaction detection system, Cu(DEtTC)$_2$ and Ni(DEtTC)$_2$ have rather different absorption characteristics. The former complex has a $\lambda_{max}$ of 435 nm with a corresponding $\varepsilon_{max}$ of 13,000, while the nickel complex has a shorter wavelength maximum at 325 nm with $\varepsilon_{max}$= 38,000. The main parameters that will affect the conversion of the lead, cadmium and bismuth complexes into the corresponding copper or nickel salts will be the reaction time and temperature, and the composition and pH of the HPLC eluent. Using the copper(II) phosphate reactor with reactor residence times of between about 1 and 4 sec, it was observed that all three metal ions studied were quantitatively converted into Cu(DEtTC)$_2$ already at the shortest reaction time tested of 1.3 sec and at ambient temperature (20$^\circ$C). With the reactor containing nickel(II) phosphate, which produces the considerably less stable Ni(DEtTC)$_2$, rather different results were obtained. Quantitative conversion within 4 sec required a temperature of at least 60°C. Typical results for the slowest reacting complex, Bi(DEtTC)₃, are shown in Fig. 7. Within the range of conditions tested, the nature of the organic modifier used neither influenced the ligand-exchange reaction nor the lifetime of the solid-state reactor. However, in order to prevent dissolution of the metal phosphate in the reactor, the pH of the HPLC eluent has to be higher than 5.

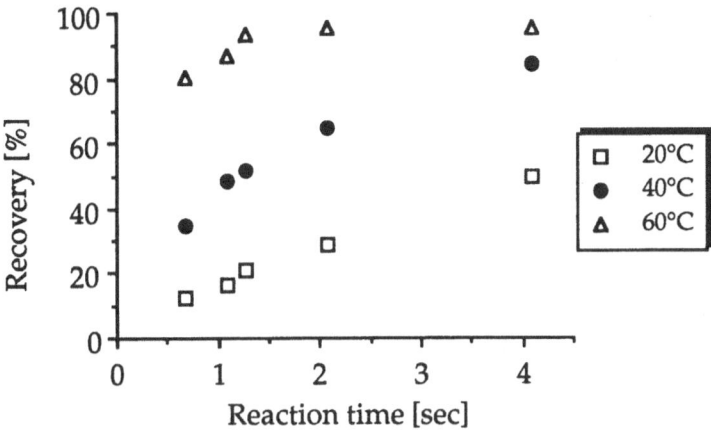

Figure 7. Conversion rates for Bi(DEtTC)₂ at 20, 40 and 60°C in dependence of the residence time using a nickel(II) phosphate reactor.

In order to prepare a packed reactor, the appropriate metal phosphate was precipitated from an aqueous solution, washed with water and methanol, then ultrasonically treated as a suspension in methanol and finally dried on tissue paper. The metal phosphate was pressed as densely as is possible into a 2- or 4-mm long reactor column using a micro spatula. A single reactor could be used for at least a week. Even with such a rather crudely prepared reactor, the contribution to band broadening was negligible; peak asymmetry typically increased from 1.2 to 1.3-1.4.

The detection limits for bismuth, cadmium and lead were 7 and 3 ng with the copper(II) phosphate and the nickel phosphate reactor, respectively, the better sensitivity of the latter reactor being explained by the higher $\varepsilon_{max}$ of the nickel complex (see above). Linearity was observed over three orders of magnitude. If, after separation, the metal diethyldithiocarbamate complexes themselves are detected at 254 nm, quantitation is sometimes difficult because of a large interferring peak due to disulfiram - which is invariably formed by partial oxidation of diethyldithiocarbamate (see above) - and a peak to be assigned to Ni(DEtTC)$_2$, the nickel originating from the stainless-steel parts of the HPLC system. The problem is especially pressing for Pb(II) since its DEtTC complex elutes in between disulfiram and the nickel complex. Fortunately, both compounds react rather sluggishly with the copper (II) phosphate in the post-column reactor, and interferences are now completely circumvented (Fig. 8). Using the present method for the analysis of spiked urine samples, even with a 1-ml pre-concentration step only, detection limits of 10 ppb were obtained for Pb(II) and Cd(II) with the copper(II) phosphate reactor. Despite its inherently higher sensitivity, under these real-life conditions, the performance of the nickel phosphate reactor was less good: many interferences showed up in the chromatogram at the rather low detection wavelength of 325 nm.

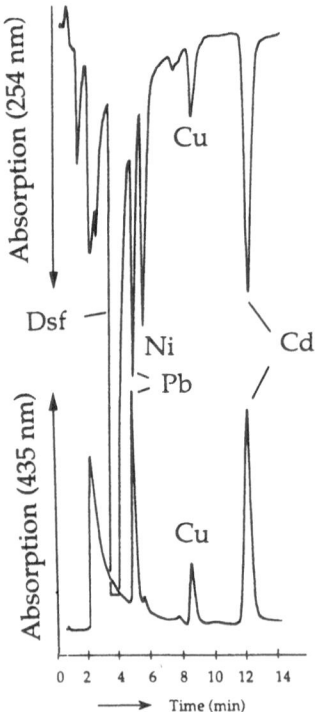

Figure 8. Chromatogram of Ni(DEtTC)$_2$, Cd(DEtTC)$_2$, Pb(DEtTC)$_2$ and Cu(DEtTC)$_2$ with detection at (a) 254 nm and (b) 435 using the copper(II) phosphate reactor.

## POST-COLUMN REACTION DETECTION OF ANIONS

Ligand exchange-based post-column reaction detection can also be used for the trace-level determination of anions. As one example, we recently studied the determination of the potential drug D-myo-1,2,6-inositol trisphosphate (IP$_3$) by ion-pair reversed-phase HPLC combined with a ligand-exchange reaction and fluorescence detection [19], working along lines similar to those used previously for the determination of organosulphur compounds with Pd(II)-calcein [20]. The strong fluorescence of calcein is completely quenched by divalent palladium. If, now, an organosulphur compound elutes from the HPLC column and the column effluent is mixed with a Pd(II)-calcein reagent stream, the high binding strength of Pd(II) to the organosulphur analyte causes the release of an equivalent amount of calcein, the fluorescence of which is then monitored.

In the case of IP$_3$, trivalent aluminium and iron were selected for further study, because they form strong complexes with the analyte of interest. Aluminium, however, did not sufficiently quench the fluorescence of any of the seven ligands tested - sulphosalicylic acid, chromotropic acid, catechol, calcein, methylcalcein, calcein blue and methylcalcein blue (MCB) - which, in actual practice, would mean that the baseline would always be rather high. Restricting further work to the use of Fe(III) as metal ion, therefore, we found sulphosalicylic acid, chromotropic acid and MCB to be efficiently quenched by Fe(III), while they form weaker complexes with this metal ion than does IP$_3$. Comparison of optimum pH ranges for efficient ligand exchange with IP$_3$ as well as best fluorescence intensity of the free ligands showed MCB to be the preferred reagent. It also yields the lowest limit of detection, viz. 10 ng in flow-injection experiments ($\lambda_{exc}$ = 370 nm; $\lambda_{em}$ = 440 nm).

Optimization studies not unexpectedly showed that a Fe(III)/MCB = 3:1 ratio gives best results as regards the fluorescence background and limit of detection. The pH value of the HPLC eluent was set at 7, which is in the range of maximum fluorescence intensity of MCB (pH 4-8), provides fast ligand exchange and is sufficiently high to prevent decomposition of IP$_3$, which occurs at pH 4. Because the MCB fluorescence reaches its maximum at high modifier (methanol) contents, while the HPLC separation itself has to be performed at a relatively low modifier percentage, the Fe(III)/MCB reagent was added post-column dissolved in methanol-water (95:5). Using the above conditions at a Fe(III)/MCB concentration of 6.6 x $10^{-6}$ M, optimum results were obtained at about 10 sec residence time and a temperature of 50-55°C. The detection limit for IP$_3$ was 0.5 ng with excellent linearity up to 1000 ng.

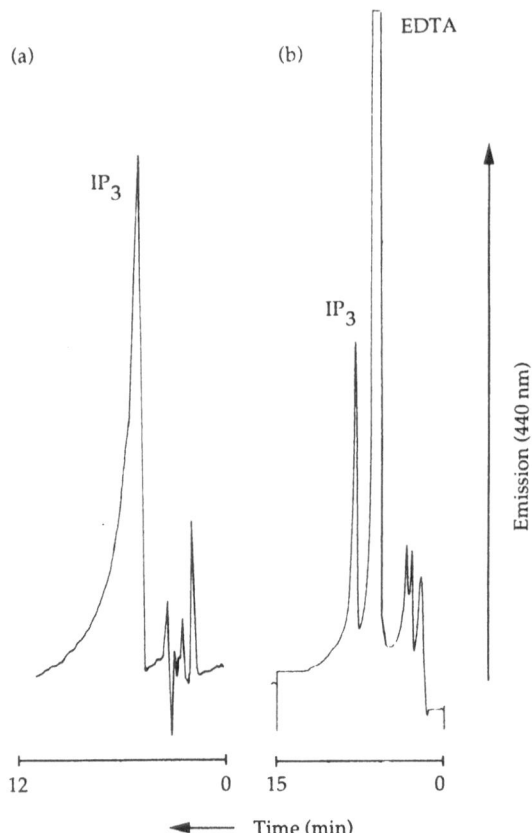

Figure 9. Preconcentration of IP$_3$ (a) without and (b) with addition of EDTA to the sample. Conditions: (a) analytical column, 200 x 4.6 mm I.D. packed with 5 µm RoSil ODS; eluent, acetonitrile/10 mM aqueous Tris buffer (pH 7.0) (50/50, v/v) containing 1 mM TPABr; pre-column, 10 x 2.0 mm I.D. packed with 5 µm RoSil ODS; preconcentration of 1 ml standard solution containing 500 ppb IP$_3$. (b) Analytical column, as (a); eluent, acetonitrile/10 mM aqueous Tris buffer (pH 7.0) (40/60, v/v) containing 1 mM TPABr; pre-column, as (a); preconcentration of 1 ml standard solution containing 40 ppb IP$_3$ and 10 µM EDTA.

In order to achieve detection limits at the low-ppb level a preconcentration technique was developed which is based on ion-pair formation. Addition of alkylammonium-type counter-ions to the sample allowed the enrichment of IP$_3$

on C18-bonded silica from aqueous solutions with breakthrough volumes higher than 25 ml. A detection limit of 0.5 ppb (corresponding to $10^{-9}$ M) was obtained for the preconcentration of 1 ml. At this concentration level it was important to add EDTA (5 $\mu$M) to the sample in order to prevent complexation reactions of IP$_3$ with metal ions present in the mobile phase. If no masking agent such as EDTA is present the IP$_3$ peak exhibits strong tailing and quantitation becomes difficult. A typical chromatogram is shown in Fig. 9 for the preconcentration of IP$_3$ with and without addition of EDTA.

## CONCLUSIONS

The 'traditional' chromatographic determination of ions leaves the ionic species unchanged during all stages of the analysis. Preconcentration and separation are carried out on ion-exchangers and detection is generally performed by means of conductivity detection. This article intends to show an alternative concept which involves complexation of the analyte(s) before, during or after the chromatographic separation. The most important features of this concept are (a) the increase of sensitivity, (b) the increase of selectivity and (c) the compatibility with reversed-phase LC systems.

It is worth mentioning that interferences due to - undesired - complexation reactions can strongly influence the chemical properties of analytes and, therefore, their behaviour in all stages of the HPLC analysis. Ion-pair chromatography of IP$_3$ is an example where metal ions, which derive, e.g., from the stainless-steel parts of the HPLC system or are present in the sample, drastically alter the chromatographic behaviour of the analyte(s). The same holds for the determination of metal ions in samples which also contain complexing agents. The addition of masking agents and the purposive conversion of analytes into metal complexes prior to their introduction into the analytical system - either by the addition of suitable ligands or metal ions - are possibilities to handle this problem.

Preconcentration techniques have been designed for both metal ions and ligands. In the first case, a suitable ligand (diethyldithiocarbamate or -phosphate) has been bound reversibly as an ion-pair on C18-bonded silica. In the second case - which is not described in this paper, since only neutral analytes were determined - metals are bound to chelating phases, their free coordination sites being used for the enrichment of complexing analytes [21,22].

The use of complexation principles in the separation of analytes is illustrated by the determination of metal ions as dithiocarbamate or -phosphate complexes [6,17] and, also, by the determination of diethyldithiocarbamate as its lead(II) complex [12]. In both cases the conversion of the analytes into metal complexes has the function to allow preconcentration and separation on C18-bonded silica, but also to increase the detectability by conventional UV/vis detectors.

Regarding detection, complexation principles can be applied in two different ways: (1) a metal complex with improved detection properties is formed in a pre- or post-column reaction or (2) a fluorescent ligand is released during a ligand-exchange reaction between the analyte and a non-fluorescent metal complex. Both reactions proceed fast (complete conversion in less than 30 sec) and allow the selective and sensitive determination of non-fluorescent and weakly UV-absorbing analytes with detection limits around 1 ng.

## REFERENCES

1.  Drasch, G. , Meyer, L.V. and Kauert, G., Detection of lead and cadmium by HPLC after chelate extraction with Na-Diethyldithiocarbamate. Z. Anal. Chem., 1982, **311**, 695.

2.  Häring, N. and Ballschmitter, K., Chromatography of metal chelates. IX. Adsorptive preconcentration for determination of copper, cobalt, and nickel in microgram/liter region by reverse phase chromatography of diethyldithiocarbamates. Talanta, 1980, **27**, 873.

3.  Smith, R.M. and Yankey, L.E., Determination of metal ions by LC incorporating dithiocarbamates in the eluent. Analyst (London), 1984, **107**, 744.

4.  Ichinoki, S. and Yamazaki, M., Simultaneous determination of nickel, lead, zinc and copper in citrus leaves and rice flour by LC with hexa-methylenedithiocarbamate extraction. Anal. Chem., 1985, **57**, 2219.

5.  Bond, A.M. and Wallace G.G., LC with electrochemical and/or spectro-photometric detection for automated determination of lead, cadmium, mercury, cobalt, nickel, and copper. Anal. Chem., 1984, **56**, 2085.

6.  Irth, H., de Jong, G.J., Brinkman, U.A.Th. and Frei, R.W., Trace enrichment and separation of metal ions as dithiocarbamate complexes by liquid chromatography. Anal. Chem., 1986, **59**, 98.

7.  Kirkbright, G.F. and Mullins, F.G.P., Separation of dithiocarbamates by HPLC using a micellar mobile phase. Analyst (London), 1984, **109**, 493.

8.  Brandsteterova, E., Lehotay, J., Liska, O. and Garaj, J.J., Application of high performance liquid chromatography in the trace analysis of some fungicides. J. Chromatogr., 1984, **286**, 339.

9. Karchmer, J.H., The analytical chemistry of sulfur and its compounds part II. Wiley-Interscience, New York, 1971.

10. Irth, H., de Jong, G.J., Brinkman, U.A.Th. and Frei, R.W., Metallic copper containing post-column reactor for the detection of thiram and disulfiram in liquid chromatography. J. Chromatogr., 1986, 370, 439.

11. Irth, H., de Jong, G.J., Frei, R.W., Brinkman, U.A.Th., Residue analysis of dithiocarbamates by liquid chromatography with selective pre-column or reaction detection systems. Int. J. Environ. Anal. Chem., in press.

12. Irth, H., de Jong, G.J., Brinkman, U.A.Th. and Frei, R.W., Determination of disulfiram and two of its metabolites in urine by reversed-phase LC with colorimetric detection after post-column complexation. J. Chromatogr., 1988, 424, 95.

13. Busev, A.I. and Ivaniutiu, M.I., Dialkyl- and diaryldithiophosphoric acids as analytical reagents. 1. Potentiometric determination of copper by means of nickel diethyldithiophosphate. Zh. Anal. Khim., 1956, 11, 523.

14. Handley, T.H. and Dean, J.A., O, O'-dialkyl phosphorodithioic acids as extractants for metals. Anal. Chem., 1962, 34, 1312.

15. Cardwell, T.J., Caridi, D. and Loo, M.S., HPLC of the metal chelates of dialkyldithiophosphoric acids. II. Lability of cobalt(III) chelates. J. Chromatogr., 1986, 351, 331.

16. Bode, H. and Arnswald, W., Untersuchung über substituierte Dithiophosphate. I. Mitteilung. Die Diäthyldithiophosphorsäure und ihr Natriumsalz. Z. Anal. Chem., 1962, 185, 99.

17. Irth, H., Brouwer, E., de Jong, G.J., Brinkman, U.A.Th. and Frei, R.W., Trace enrichment and separation of As(III), Sb(III), and Bi(III) as diethyldithiophosphate complexes by reversed-phase liquid chromatography. J. Chromatogr., 1988, 439, 63.

18. Irth, H., de Jong, G.J., Brinkman, U.A.Th. and Frei, R.W., Determination of Pb(II), Cd(II) and Bi(III) by reversed-phase LC of their diethyldithiocarbamate complexes with post-column ligand exchange and selective spectrophotometric detection. J. Chromatogr., in press.

19. Irth, H., Lamoré, M., de Jong, G.J., Brinkman, U.A.Th., Frei, R.W., Kornfeldt, R.A. and Persson, L., Determination of D-myo-1,2,6-inositol trisphosphate by ion-pair reversed-phase liquid chromatography with post-column ligand exchange and fluorescence detection. J. Chromatogr., in press.

20. Werkhoven-Goewie, C.E. , Niessen, W.M.A., Brinkman, U.A.Th. and Frei, R.W., LC detector for organosulphur compounds based on ligand-exchange reactions. J. Chromatogr., 1981, 203, 165.

21. Lipschitz, C. , Irth, H., de Jong, G.J., Brinkman, U.A.Th. and Frei, R.W., Trace enrichment of pyrimidine nucleobases, 5-fluorouracil and bromacil on a silver-loaded stationary phase with on-line reversed-phase HPLC. J. Chromatogr., 1989, 471, 321.

22. Irth, H., Tocklu, R., Welten, K., de Jong, G.J., Frei, R.W. and Brinkman, U.A.Th., Further investigations on the (de)sorption processes on metal-loaded phases: behaviour of adenine, adenosine and barbiturates. J. Pharm. Biomed. Anal., in press.

## ELECTROCHEMICAL ION EXCHANGE

PAULINE M. ALLEN, NEVILL J. BRIDGER, CHRISTOPHER P. JONES,
MARK D. NEVILLE AND ANDREW D. TURNER
AEA Technology, Building 429, Harwell Laboratory,
Oxfordshire, OX11 ORA, England.

### ABSTRACT

Electrochemical ion exchange (EIX) is a novel separation process, which has
been developed at Harwell over a number of years. Absorption of ions into
an EIX electrode is controlled by an externally applied potential. Elution
is achieved by simple polarity reversal - no eluant chemicals are required.
This enables multiple use of ion exchange capacity. EIX has been
demonstrated to absorb cations of IA, IIA, transition and post-transition
metals as well as anions. To date, EIX has mainly been applied to nuclear
waste streams. For example, using an inorganic ion exchanger, caesium and
cobalt have been selectively removed from sodium- and lithium-bearing
feeds respectively. With appropriate ion exchangers, anions such as
nitrate, borate, chloride, and sulphate can also be removed by EIX. Other
potential applications include removal of heavy metals (e.g. Hg and Cd)
from industrial effluents and conventional water softening. This paper
gives a summary of the scientific principles involved, as well as details
of operating experience.

### INTRODUCTION

Electrochemical ion exchange (EIX) was first investigated as a process for
brackish water desalination. This novel separation process has since been
developed at Harwell over a number of years. Absorption of ions into an
EIX electrode is controlled by an externally applied potential. Elution is
achieved by simple polarity reversal - no eluant chemicals are required.
This enables multiple use of ion exchange capacity by repeated absorption/
elution cycles (> 2000 cycles over 2 years in one case). As the process is
controlled electrically, EIX has many new and desirable features compared
to conventional ion exchange.

## OBJECTIVES

The primary objective has been to remove (radioactive) cations and anions
from nuclear waste.  Potential non-nuclear applications including
water-softening (Ca, Mg removal), industrial effluent processing (Pb, Cd,
Hg removal), and precious metal recovery (e.g. Ag, Au).

## APPROACH

### Principles of EIX

In EIX, a weak acidic or basic ion exchanger is bonded to a mesh electrode
by means of an elastomeric binder, for the removal of cations or anions
respectively.  During the absorption cycle for a cation, the EIX electrode
is made cathodic.  Electrolysis of water generates a local alkaline
environment within the electrode structure:

$$2H_2O + 2e^- \rightarrow H_2 + 2OH^-$$

The hydroxyl ions deprotonate the exchanger, generating active sites:

$$RCOOH + OH^- \rightarrow RCOO^- + H_2O$$

which then undergo exchange:

$$RCOO^- + M^+ \rightarrow RCOOM$$

The applied potential also induces migration into the ion exchanger,
enhancing the kinetics, and enabling high utilization of capacity.

Elution of cations from the EIX electrode is achieved by reversing the
polarity.  This produces a local acidic environment within the electrode:

$$2H_2O \rightarrow O_2 + 4H^+ + 4e^-$$

Electrogenerated protons displace metal ions from the weak cation exchanger:

$$RCOOM + H^+ \rightarrow RCOOH + M^+$$

The presence of the electric field gradient enables virtually complete
elution.

An equivalent set of reactions is invoked when an anion is absorbed
and desorbed within a weak base anion exchanger.

### Electrode Manufacture

A simple, reproducible method has been developed for the fabrication of EIX
electrodes.  A slurry of ion exchanger/binder/solvent is poured onto a
metal mesh, contained in a Perspex mould.  Solvent evaporation gives a very
strong, non-friable electrode (see Figure 1).

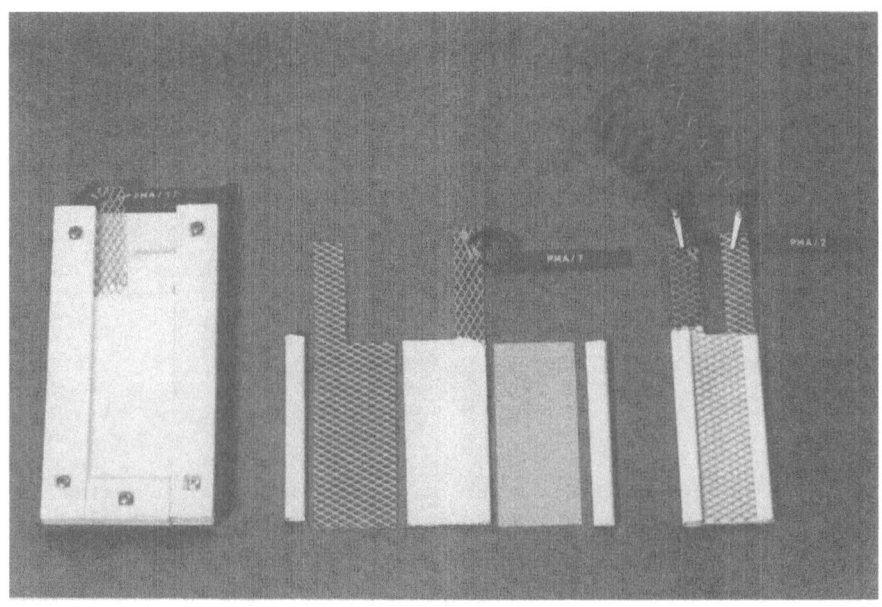

Figure 1.   Fabrication of an EIX module.

Figure 2.   EIX rig used in batch experiments.

Uncoated mesh is used for the counter electrode, which is clamped to the working electrode to form the EIX module. Two experimental configurations have been used. In the first, there is a counter electrode on either side of the EIX electrode, whereas in the second, there is a counter electrode on the front side of the EIX electrode, and an insulating sheet of polythene on the back. The resulting module can be immersed in stirred electrolyte (batch mode) or contained in a flow cell, through which a stream of electrolyte can pass. The rig used for batch experiments is illustrated in Figure 2.

## RESULTS

### Cation EIX

Early work on cation EIX concentrated on organic ion exchange resins. However, these suffered from radiation damage. An alternative inorganic cation exchanger was therefore sought. In 1987, amorphous zirconium phosphate was identified as a candidate absorber. This material is used in portable renal dialysis units, for removal of ammonium ions. Consequently it is commercially available at reasonable cost (from Magnesium Elektron Ltd.).

(a) Selective absorption of caesium cations. Exceptional caesium decontamination results have been obtained with EIX electrodes constructed from amorphous zirconium phosphate. For example, $Cs^+$ has been selectively removed from a 100 ppm $Na^+$ stream, with a decontamination factor (DF) > 5000, to beyond 8% cation loading. Elution on polarity reversal gave rise to a volume reduction factor (VRF) >> 100. Even after $Na^+$ breakthrough, the concentration of $Cs^+$ in the effluent was below the level of detection (ppb).

(b) Absorption/desorption of cobalt cations. Treatment of high concentrations of Co(II) (100 ppm) by conventional EIX resulted in $Co(OH)_2$ precipitation on the electrode, which resulted in poor kinetics. However, it was found that trace levels of $^{60}Co$ could be removed effectively using EIX. Subsequently, modules incorporating amorphous zirconium phosphate have been applied to Co(II) solutions at concentrations below the solubility limit of the hydroxide. Cobalt has been successfully removed from a solution containing 100 ppb Co. The concentration of Co(II) in the effluent stream was below the level of detection (ppt). DF's of $\sim$ 1000 were measured in some experiments.

Elution of the EIX module was found to be pH dependent. Co(II) did not elute into distilled water (pH 7). However, when the pH was adjusted to $\sim$ 2 by addition of nitric acid, quantitative recovery of Co(II) was achieved.

In PWR streams, $Li^+$ is present as well as Co(II). Both are removed by EIX with amorphous zirconium phosphate. However, selective desorption is possible. Lithium hydroxide is freely soluble in water, whereas the solubility of cobalt hydroxide is only 0.0032 g $\ell^{-1}$. By maintaining a high pH (11-12) during desorption, only $Li^+$ was eluted. The pH was then reduced (1.5-2.5) to allow regeneration of cobalt. A VRF > 200 was achieved for a 100 ppb Co initial feed.

## Zirconium Phosphate Hydrolysis

Amorphous zirconium phosphate possesses excellent cation exchange properties for both caesium and cobalt. Unfortunately, ion exchange is accompanied by a small degree of phosphate elution, which indicates hydrolysis of the exchanger:

$$Zr(HPO_4)_2 + 6OH^- \rightarrow Zr(OH)_4 + 2PO_4^{3-} + 2H_2O$$

For this material, hydrolysis results in slower absorption kinetics and reduced capacity, limiting electrode lifetime to $\sim$ 6 years. Work is in progress to minimise this problem.

## Anion EIX

In 1987, experimental work was expanded to include anions as well as cations. Again, a degree of selectivity has been achieved.

(a) Selective absorption of chloride and sulphate anions. EIX electrodes have been constructed using IRA60 and IRA94S anion exchange resins. These electrodes selectively removed chloride and sulphate anions, in the presence of high concentrations of boron (borate/boric acid). Selectivity at pH 5 was an order of magnitude better than at pH 7. At the lower pH, boric acid is almost undissociated, and consequently little borate ion exchange can take place.

(b) Protective coatings for electrodes. During the absorption cycle for an anion, the EIX electrode is made anodic. Electrolysis of water produces an acidic environment within the electrode:

$$2H_2O \rightarrow O_2 + 4H^+ + 4e^-$$

If the anion being absorbed is chloride, a further electrode reaction is also possible:

$$2Cl^- \rightarrow Cl_2 + 2e^-$$

Chlorine, and the oxy-anions derived from it, have a deleterious effect on the anion exchange resin. One way to prevent this is to coat the electrode with a <u>cation</u> exchanger. Electrostatic repulsion prevents $Cl^-$ from entering the negatively charged matrix (Donnan exclusion). Neutral $H_2O$ molecules can pass through the cation exchanger, so the <u>desired</u> electrode reaction can still take place.

During elution of the EIX electrode, the same two reactions ($O_2$, $Cl_2$ evolution) will occur at the counter electrode. As this electrode is in intimate contact with the EIX electrode, it may be necessary to coat the counter electrode as well.

A number of cation exchange materials have been screened (Dowex 50, CG120, ZIE96, IRC84, and amorphous zirconium phosphate). A divided cell was used, in order to monitor the anode reaction alone. Oxidised chlorine species in the anode compartment were determined by means of iodine/ thiosulphate titrations. Dowex 50 and CG120 were identified as suitable cation exchangers to inhibit chloride oxidation.

A bilayer electrode has now been fabricated with a thin protective coating of Dowex 50 (cation exchanger) on the mesh, covered by a thick coating of IRA94S (anion exchanger). Preliminary experiments suggest enhanced performance.


## CONCLUSIONS

EIX is a powerful technique for removing a wide range of cations and anions from aqueous streams. Selectivity may be achieved by careful choice of experimental conditions. Although the technique has been applied initially to the treatment of low level radioactive waste liquors, it is believed to have a much wider potential applicability in water treatment, industrial effluent processing, and metal recovery operations.

Important features of EIX are high utilization of the IX capacity, the electrical elution of ions (thereby avoiding the addition of acid or salt), and multiple use of the exchange through repeated absorption/elution cycling.

This work was undertaken as part of the UKAEA Underlying Research Programme.

ION EXCHANGE SEPARATIONS IN CONJUNCTION WITH ELECTROCHEMICAL
DETECTION

RALPH COCHRANE
PRODUCT SPECIALIST
DIONEX (UK) LTD
ALBANY COURT, CAMBERLEY, SURREY, GU15 2PL

## ABSTRACT

This paper will describe the use of the new Dionex Pulsed
Electrochemical Detector, following ion exchange separations,
for the detection of a wide range of both organic and
inorganic ions.

This detector contains both conductivity and amperometry for
the detection of ionic and oxidisable compounds and is
particularly well suited to compounds with little or no UV
absorbance.

## INTRODUCTION

Pulsed Electrochemical Detector

The Dionex Pulsed Electrochemical Detector combines the
following in one module:

        Conductivity
        D.C. Amperometry
        Pulsed Amperometry
        Integrated Amperometry
        Cyclic Voltammetry
        pH
        Temperature

        This is possible because both conductivity and
amperometry rely on the application of potential waveforms to
generate a resulting current.  It is the type of waveform
applied and the manner in which the current is measured which
determines the detection mode.  Each mode is discussed in
detail below.

220

## DISCUSSION

### Conductivity

Ion Chromatography, through the use of suppressed conductivity detection, is now a well documented and accepted technique. Discovered and patented by Dow Chemicals(1) and manufactured by Dionex, chemical suppression has evolved and allowed conductivity detection to be developed into a powerful HPLC detector for inorganic and organic ions, figures 1 and 2.

## Gradient Separation of Anions

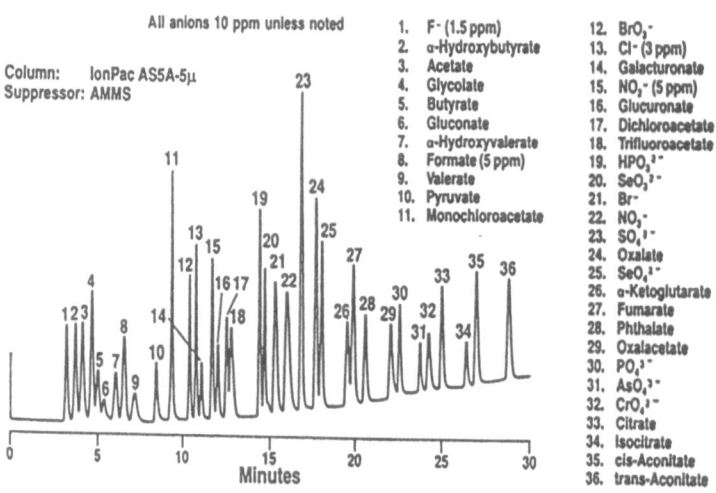

Column: IonPac AS5A-5μ
Suppressor: AMMS

All anions 10 ppm unless noted

1. F- (1.5 ppm)
2. α-Hydroxybutyrate
3. Acetate
4. Glycolate
5. Butyrate
6. Gluconate
7. α-Hydroxyvalerate
8. Formate (5 ppm)
9. Valerate
10. Pyruvate
11. Monochloroacetate
12. BrO₃⁻
13. Cl⁻ (3 ppm)
14. Galacturonate
15. NO₂⁻ (5 ppm)
16. Glucuronate
17. Dichloroacetate
18. Trifluoroacetate
19. HPO₃²⁻
20. SeO₃²⁻
21. Br⁻
22. NO₃⁻
23. SO₄²⁻
24. Oxalate
25. SeO₄²⁻
26. α-Ketoglutarate
27. Fumarate
28. Phthalate
29. Oxalacetate
30. PO₄³⁻
31. AsO₄³⁻
32. CrO₄²⁻
33. Citrate
34. Isocitrate
35. cis-Aconitate
36. trans-Aconitate

## Gradient Separation of Inorganic and Organic Cations

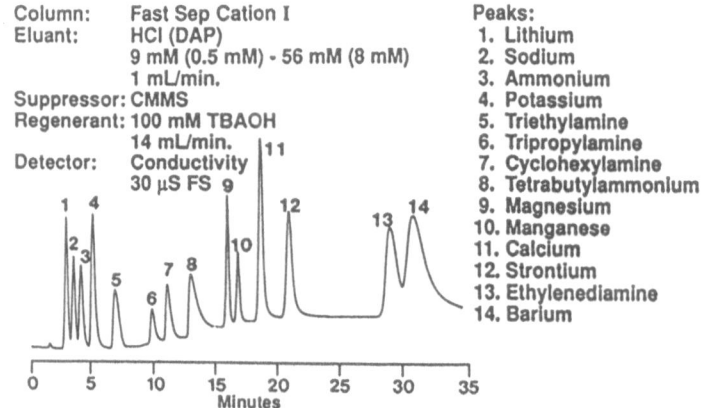

Column: Fast Sep Cation I
Eluant: HCl (DAP) 9 mM (0.5 mM) - 56 mM (8 mM) 1 mL/min.
Suppressor: CMMS
Regenerant: 100 mM TBAOH 14 mL/min.
Detector: Conductivity 30 μS FS

Peaks:
1. Lithium
2. Sodium
3. Ammonium
4. Potassium
5. Triethylamine
6. Tripropylamine
7. Cyclohexylamine
8. Tetrabutylammonium
9. Magnesium
10. Manganese
11. Calcium
12. Strontium
13. Ethylenediamine
14. Barium

Figures 1 & 2

Columns: During this time, although probably less well known, is the evolution of Dionex column technology. The use of pellicular latex resins gives rise to fast and efficient ion exchange separations and the ability to produce a selection of both anionic and cationic columns with different capacities and selectivities. This has recently culminated in a range of new and unique columns with both ion exchange and reverse-phase separation characteristics built into the same column. (2)


## D.C. Amperometry

Electro chemical detection can also be applied for some oxidisable ions and polar organic molecules. Amperometry is a term used to describe oxidation or reduction of part of a component (<10%) at an electrode surface. Direct Current or DC amperometry is the most common form and occurs when a single potential is applied to an electrode and subsequent oxidation or reduction of the analyte results in a current or charge which is measured and displayed as the chromatographic peak. Examples of DC amperometry are shown in figures 3 and 4.


## Sulfide and Cyanide at ppb Levels

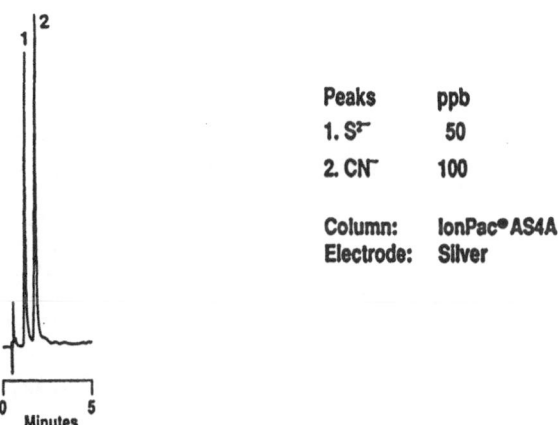

| Peaks | ppb |
|-------|-----|
| 1. $S^{2-}$ | 50 |
| 2. $CN^-$ | 100 |

Column: IonPac® AS4A
Electrode: Silver

Figure 3.

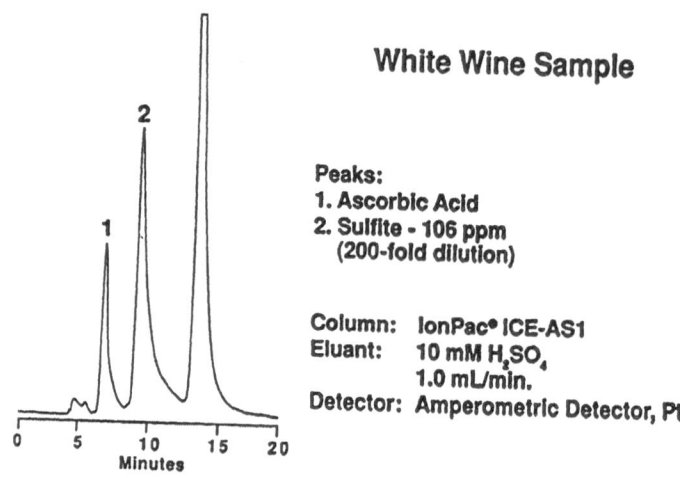

**White Wine Sample**

Peaks:
1. Ascorbic Acid
2. Sulfite - 106 ppm
   (200-fold dilution)

Column:   IonPac® ICE-AS1
Eluant:    10 mM $H_2SO_4$
           1.0 mL/min.
Detector: Amperometric Detector, Pt

Figure 4.

However there are classes of compounds such as carbohydrates, alcohols and amino acids which in the DC mode produces reaction products which poison the surface of the working electrode. This means repeated injection of such compounds resulting in a loss of response with time, figure 5.

Repeated **Injections of an Amino Acid and Carbohydrate Using DC Amperometric Detection**

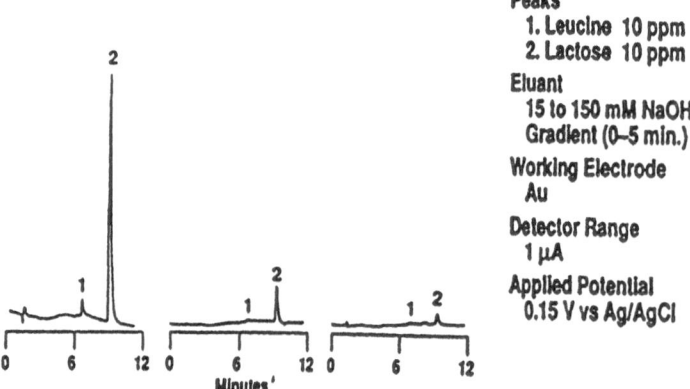

Peaks
1. Leucine 10 ppm
2. Lactose 10 ppm

Eluant
15 to 150 mM NaOH
Gradient (0–5 min.)

Working Electrode
Au

Detector Range
1 µA

Applied Potential
0.15 V vs Ag/AgCl

Figure 5.

Electrochemical detection would be an ideal system for these compounds and others, particularly since alternative detection schemes such as low wavelength UV or RI are both non-specific and insensitive. To get round this problem Dionex have pioneered the use of Pulsed and Integrated Amperometry.

## Pulsed Amperometry

In its original form, a sequence of three pulses was used to allow detection of carbohydrates. The sequence of pulses is shown in figure 6.

# Triple Potential Sequence used in Pulsed Amperometry

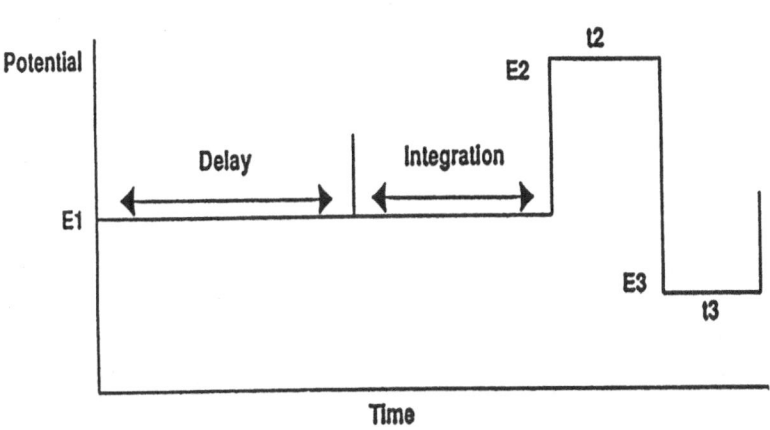

Figure 6.

There are four distinct sections. E1 is the sampling potential during which oxidation occurs. Integration of the oxidative current occurs only after a delay period to allow charging current following the pulse from E3 to decay. E1 is followed by positive and negative cleaning potentials. E2 applied for t2 is used to oxidise off the electrode surface the degradation products produced from the E1 pulse(and thus causing fouling of the electrode)is now itself oxidised and is reduced back to its original form by the pulse E3 applied over time t3. This allows for sensitive and specific detection of all carbohydrate species (3), examples are given in figures 7 and 8.

## Carbohydrates by Gradient Elution

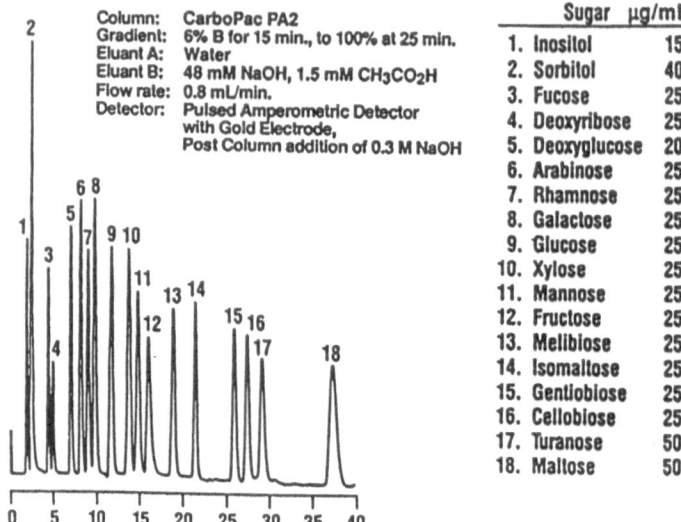

| Column: | CarboPac PA2 |
| Gradient: | 6% B for 15 min., to 100% at 25 min. |
| Eluant A: | Water |
| Eluant B: | 48 mM NaOH, 1.5 mM CH$_3$CO$_2$H |
| Flow rate: | 0.8 mL/min. |
| Detector: | Pulsed Amperometric Detector with Gold Electrode, Post Column addition of 0.3 M NaOH |

| Sugar | μg/mL |
|---|---|
| 1. Inositol | 15 |
| 2. Sorbitol | 40 |
| 3. Fucose | 25 |
| 4. Deoxyribose | 25 |
| 5. Deoxyglucose | 20 |
| 6. Arabinose | 25 |
| 7. Rhamnose | 25 |
| 8. Galactose | 25 |
| 9. Glucose | 25 |
| 10. Xylose | 25 |
| 11. Mannose | 25 |
| 12. Fructose | 25 |
| 13. Melibiose | 25 |
| 14. Isomaltose | 25 |
| 15. Gentiobiose | 25 |
| 16. Cellobiose | 25 |
| 17. Turanose | 50 |
| 18. Maltose | 50 |

Figure 7.

## Dextrin 7 Glucose Polymers

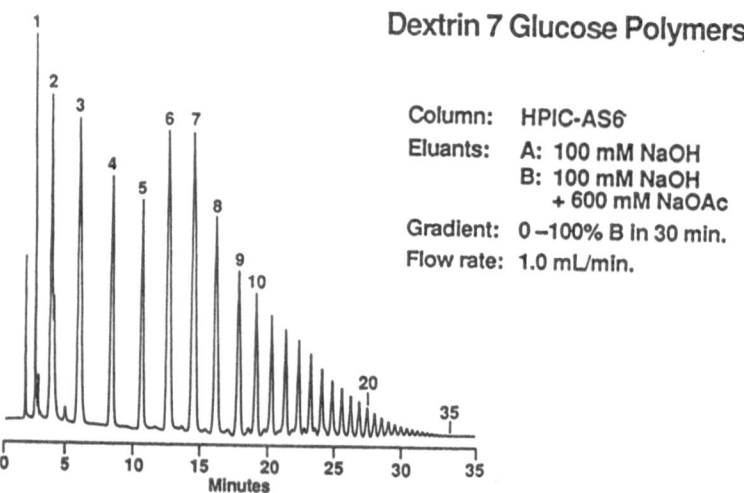

| Column: | HPIC-AS6 |
| Eluants: | A: 100 mM NaOH |
| | B: 100 mM NaOH + 600 mM NaOAc |
| Gradient: | 0–100% B in 30 min. |
| Flow rate: | 1.0 mL/min. |

Figure 8.

This detector is 10 to 100 times more sensitive than either UV or RI for carbohydrates. There are other compounds which can be detected using pulse amperometry, examples are given in figures 9 and 10.

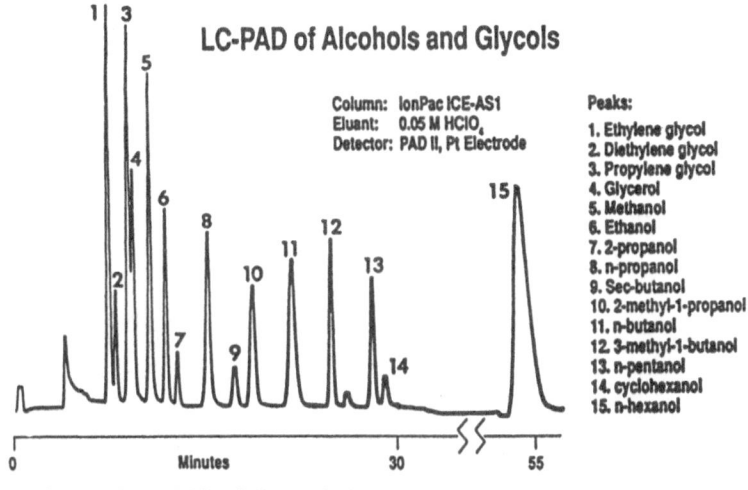

Figure 9.

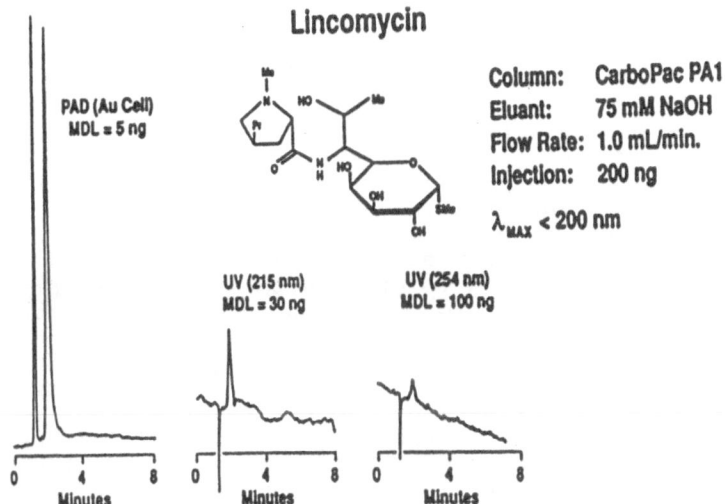

Figure 10.

The latter shows Lincomycin, an aminoglycoside antibiotic which is considerably more sensitive to PAD than UV.

### Cyclic Voltammetry

Cyclic Voltammetry (CV) is commonly used to determine the appropriate potentials for Pulsed Amperometric Detection (PAD). An example of a CV for glucose is given in figure 11 and this is typical for most carbohydrates.

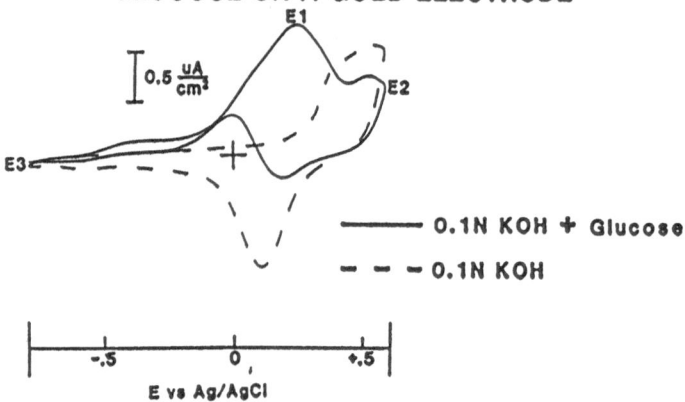

The dashed line is the current from the supporting
electrolyte; i.e. the background current.  Beginning at -0.8V
and sweeping in a positive direction, the background current
is flat until approxiamtely 0.25V, where oxidation of the
surface of the gold electrode to gold oxide begins.  Following
reversal of the potential sweep direction at 0.6V the gold
oxide is reduced back to gold, with the current producing a
trough at 0.1V.  With glucose added to the solution, the
current from glucose oxidation increases as the potential is
swept in a positive direction, peaking at 0.26V.  The current
then decreases because the formation of gold oxide inhibits
further glucose oxidation.  On the reverse scan, the current
actually reverses from reducing to oxidising at the onset of
the gold oxide reduction.  As soon as the reduction of gold
begins, oxidation of glucose also begins.  The optimum
potential to use for E1 is between 0.1 to 0.2 Volts, however
0.05V is often used because there is proportionally less
background noise so there is overall more sensitivity.  Cyclic
Voltammetry  can be performed directly in the PED's amperometry
flow cell, either statically or in the flowing stream.

Figure 11.

## Integrated Amperometry

Some compounds such as amino acids are still not very
sensitive using Pulsed Amperometric Detection.  The reason can
be seen from the CV in figure 12, which shows that the maximum
response for leucine at 0.3V is only just above the
background.  The result is that at 0.3V the oxidation of the
analyte is catalysed by metal oxide formation on the electrode
surface but the presence  of the analyte also serves to
inhibit metal oxide formation.  This gives rise to negative
dips as the background current increases, figure 13.

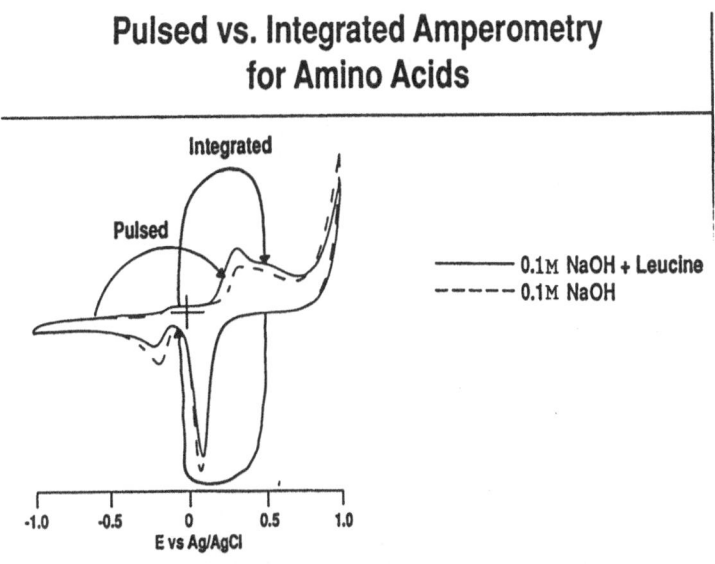

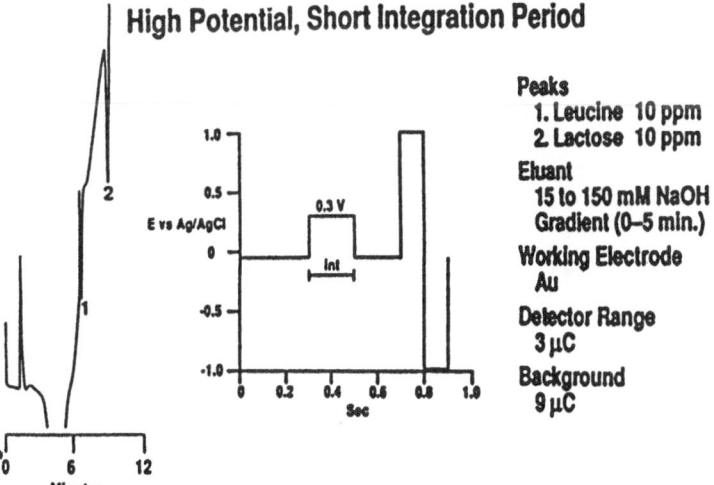

Figures 12 & 13.

This is overcome by the use of Integrated Amperometry (IA) which, instead of measuring the current at E1 only, the current is measured continuously during a single cycle where the electrode is oxidised and then reduced back to its original state, figure 14.

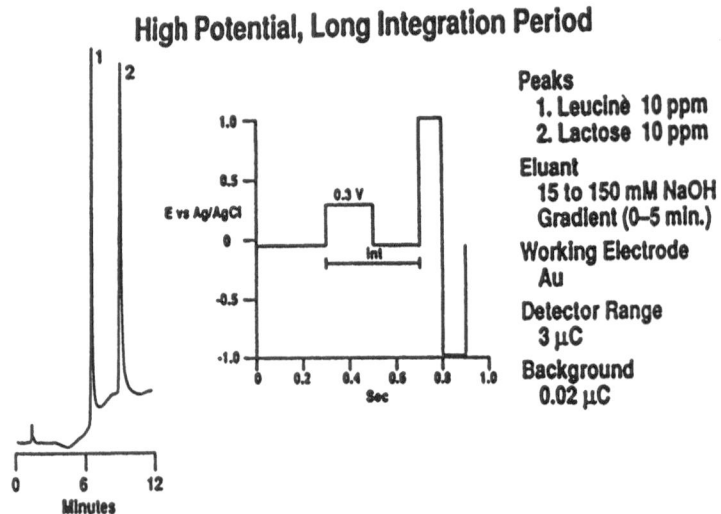

Figure 14.

Thus, integrating for both the forward and reversed pulses cancels changes in the metal oxide formation current by reducing the electrode back to its starting point. The result is a far more stable baseline and better sensitivity.

pH

A reference electrode in the amperometric cell continuously measures pH and is used to minimise baseline drift during pH gradients.

Temperature

A cell thermistor reading corrects conductivity measurements to eliminate drift due to small temperature changes. The cell temperature can also be plotted with the chromatogram.

## CONCLUSION

Both conductivity and amperometry are well established
techniques for the detection of ionic and oxidisable compounds,
both organic and inorganic. Many compounds can be detected by
both methods as shown below.

| CONDUCTIVITY | D.C. AMPEROMETRY | INTEGRATED AMP |
|---|---|---|
| Inorganic Anions | Catecholamines | Carbohydrates |
| Inorganic Cations | Phenols | Aliphatic Amines |
| Carboxylic Acids | Aromatic Amines | Amino Acids |
| Sulphonic Acids | Thiols | Alcohols |
| Phosphonic Acids | Cyanide | Aldehydes |
| Amines | Sulphide | Sulphur Species |
| | Iodide | |
| | Sulphite | |

Compounds with little or no UV absorbance are frequently
detected by low wavelength UV or refractive index (RI), simply
because these detectors are easy to use and are already in
most laboratories. Electrochemical detection can often
provide a more sensitive and specific form of detection as
well as being gradient compatible. Therefore a detector which
combines such a variety of electrochemical modes can provide a
nearly universal methods                                    with
unparalleld capability.

## REFERENCES

1.    Small, H., Stevens, T.S., Bauman, W.C., in Analytical
      Chemistry No. 47 Page 1801, 1975.


2.    Pohl, C., Ionex 90


3.    Dionex Technical Note 20, March 1989.

# REVERSIBLE EXTRACTION OF COBALT(II) FROM AQUEOUS MEDIA USING INORGANIC ION EXCHANGERS IN ELECTROCHEMICAL ION EXCHANGE.

ROBERT J.W. ADAMS AND MICHAEL J. HUDSON,
University of Reading, Chemistry Department, Whiteknights,
P.O. Box 224, Reading, Berkshire, RG6 2AD, U.K.

## ABSTRACT.

Cobalt(II) may be reversibly extracted using electrochemical ion exchange (EIX). In EIX, a current is passed so as to assist the extraction of the metal ions. Elution is achieved by reversal of the current. Electrodes are composed of a platinised titanium expanded metal mesh surrounded by inorganic ion exchanger supported in an unreactive polymer matrix. There is no observable plating out on the metal part of the electrodes and the process of extraction occurs entirely within the inorganic ion exchanger. The rate of extraction is significantly greater and final concentrations substantially lower than is achieved by normal ion exchange using identically supported exchangers.

## INTRODUCTION.

The use of water in all areas of the nuclear power and reprocessing industries inevitably gives rise to large volumes of low specific activity liquid waste streams. Before further treatment it is necessary to increase the concentration of active material held in solution. This requires a selective and energy efficient method to concentrate the active species present, and to allow the inactive bulk to be safely discharged to the environment. The smaller the final volume of active waste for disposal, the more economic the process becomes.

### Electrochemical Ion Exchange.

There is widespread familiarity with ion exchange for waste treatment in the nuclear industry. Electrical processes are less well known, but they do

have their own benefits in that they may be controlled by the extra
variable of the applied potential in addition to the normal parameters. One
such process is EIX, developed over a number of years by the Applied
Electrochemistry Group at Harwell. This is essentially a novel form of an
advanced ion exchange process in which the ion exchange material has been
incorporated into a binder matrix which surrounds an electrode structure
(figure 1). The ion exchange material is fully supported throughout the
process and consequently it is possible to use materials not normally
considered suitable for conventional ion exchange column work, such as the
inorganic layered phosphate cation exchangers.

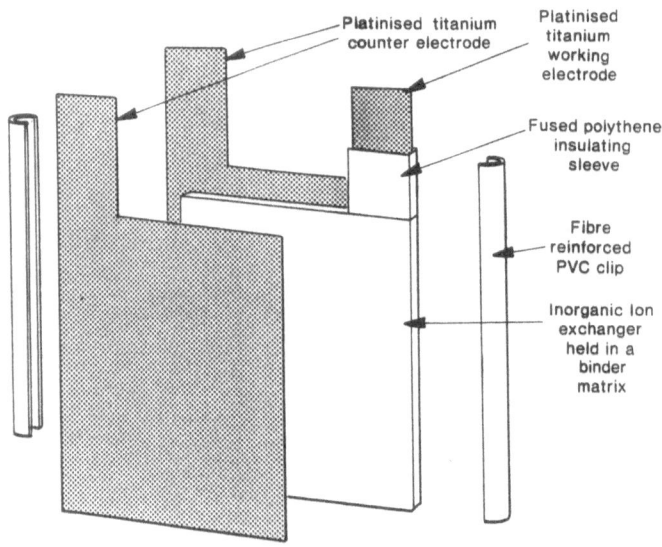

**Figure 1.** Exploded diagram of an EIX module showing the inorganic ion
exchanger coated working electrode and counter electrodes (courtesy of
A.E.A. Technology, Harwell).

For cation exchange under a cathodic applied potential (with respect to
the counter electrodes), cations are first attracted towards the electrode
due to the electrical potential gradient, and then rapidly absorbed into
the ion exchange matrix. This process is enhanced by the alkaline
environment generated in situ within the electrode by the electrochemical
water splitting reaction [1] which also

$$H_2O_{(1)} + e^- = 1/2H_{2(g)} + OH^-_{(aq)} \tag{1}$$

activates weakly acidic cation exchangers (ROH) to cation absorption [2].

$$ROH_{(s)} + OH^-_{(aq)} = RO^-_{(s)} + H_2O_{(l)} \qquad [2]$$

There is also enhanced ion migration between adjacent ion exchanger groups, owing to the high local electric field gradient. The use of an applied electric field enhances ion exchange kinetics; increases the breadth of the pH range available for operation; and allows a significant proportion of the theoretically available capacity to be used.

**Elution.**

By reversing the applied potential in EIX systems (making the ion exchange electrode anodic) exchanged cations may be eluted into a limited volume of solution, thus generating a concentrated product suitable for disposal, perhaps as a highly active waste.

This work has been carried out in conjunction with A.E.A. Technology, Harwell for the decontamination of aqueous waste solutions. Alpha zirconium hydrogen phosphate (ZrP) was chosen as the inorganic ion exchanger for incorporation into the EIX electrodes and cobalt(II) was the cation studied. The reason for using an inorganic ion exchanger is because it shows excellent resistance to radiolytic damage; inorganic ion exchangers are less affected by oxidising media than organic ion exchangers, and show greater potential temperature stability [1,2,3]. Both amorphous and fully crystalline ZrP were used in order to compare the ion exchange extraction and elution properties of the two exchangers.

**EXPERIMENTAL.**

**Reagents:-** All chemicals were of Analar grade, except for zirconium oxychloride octahydrate, which was of reagent grade. Amorphous zirconium bis(monohydrogen orthophosphate)(ZrP) was obtained commercially ( Magnesium Elektron, Manchester). Crystalline alpha ZrP (ZrPx) was produced by a scaled up version of the direct precipitation method of Alberti [4,5].

**Equipment:-** pH measurements were made using an EDT Research model ECM101 pH/Ion meter with temperature compensation probe. Sodium and cobalt ion concentrations were determined by atomic absorption spectroscopy on a Perkin Elmer 1100B Atomic Absorption Spectrophotometer. The power supply used for EIX was a Thurlby dual 30 volt, 1 amp D.C. power supply used in constant current mode. Applied voltage and solution pH were monitored using a JJ Lloyd CR452 dual channel chart recorder. During the elution phase of

EIX pH was held constant using an autotitrator.

### Construction and Use of an EIX Working Electrode Assembly.

The EIX working electrode assembly (figure 1) consisted of 3 main components, as shown. All the current carrying electrodes consisted of 100% platinised expanded titanium mesh of approximate dimensions 14 x 6cm., with a total surface area of approximately 100 $cm^2$. The central "working electrode" was composed of one of these meshes, onto which had been cast a coating of powdered ZrP (amorphous or crystalline, <0.15mm, 45g total) held in an inert hydrophobic organic binder material. Before use, excess material was trimmed off and any bare current feeder covered by an inert non-conducting sealant.

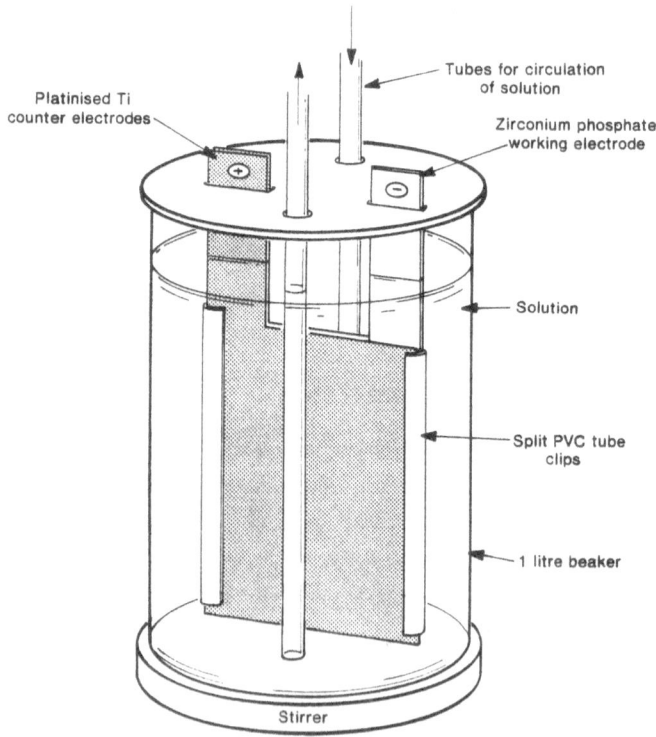

**Figure 2.** Diagram showing the assembled EIX module in the EIX cell with the solution recirculation tubes shown.

**Electrode Pretreatment.**

The completed electrode assembly was immersed in a stirred sodium hydrogen
carbonate solution (1 litre, 0.01M) as shown in figure 2. The working
electrode was then made cathodic (current density 4mA cm$^{-2}$ nominal). The
sodium was quickly reduced to low levels. This pretreatment increased the
conductivity of the working electrode matrix [6,7].

**Cobalt Extraction.**

The pretreated electrode assembly was then immersed in a stirred cobalt(II)
nitrate solution (50mg dm$^{-3}$, 0.85mmol.) and the working electrode again
made cathodic (as above).  In subsequent extraction cycles it has not been
found necessary to re-load the electrode with sodium ions since the
increase in conductivity appears to be a permanent effect.

**Cobalt Elution.**

Once an EIX electrode had been loaded with cobalt(II) the ions were eluted
by immersing the assembly in a dilute nitric acid solution and reversing
the polarity. Elution into solutions at pHs 2 and 3 are reported below.

## RESULTS AND DISCUSSION.

**EIX versus IX.**

Figure 3 shows the extraction of cobalt(II) from aqueous solution
using a sodium pretreated EIX electrode containing amorphous ZrP as the ion
exchanger. The electrode was, however, used in a passive or normal ion
exchange mode (IX). It can be seen that the rate of extraction was slow,
with a time to half concentration ($t_{1/2}$) of 2500 minutes. This was a
particularly long time due to the low effective surface area of the ion
exchanger in the electrode which limited the rate at which cobalt was
absorbed. Figure 4 (upper curve) shows the performance of an identical
electrode being used in the active EIX mode. The $t_{1/2}$ was approximately 25
minutes, corresponding to a rate 100 times faster than under IX conditions.
For crystalline ZrP (lower curve), the $t_{1/2}$ is about 12 minutes. This
demonstrates the enhancement in performance that may be achieved by EIX.
The applied potential gradient electrostatically enhanced the attraction of
ions into the exchanger matrix where they were absorbed by the
electrochemically activated groups.

**Crystalline versus Amorphous ZrP.**

The crystalline and amorphous forms gave approximately equivalent decontamination factors of 500 and 250 respectively, but the crystalline material reached the lowest ambient concentration more quickly. Both of the materials produced solutions of lower cobalt concentration than IX under similar conditions.

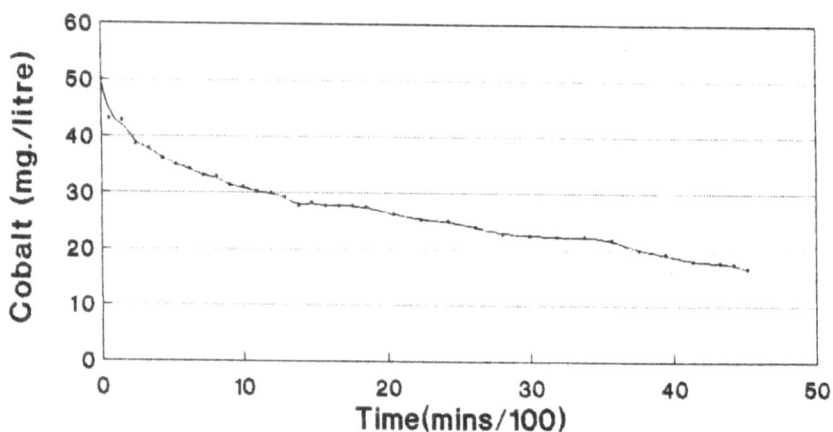

**Figure 3.** Use of a sodium preloaded EIX electrode to measure ion exchange kinetics for cobalt absorption.

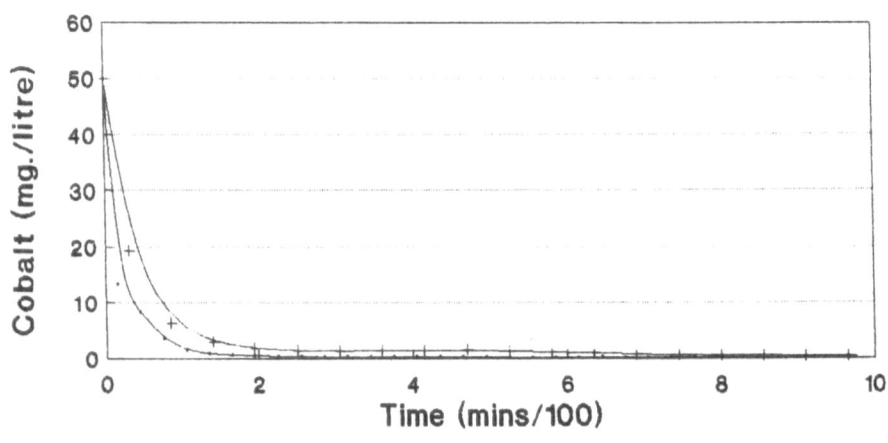

**Figure 4.** Comparison between the performance of crystalline (lower curve) and amorphous (upper curve) ZrP exchangers in EIX.

**Elution.**

Figure 5 shows the different characteristics observed for the elution of identical EIX electrodes composed of crystalline and amorphous ZrP. For the crystalline material, when elution was studied at an ambient pH of 3, a small proportion of the absorbed cobalt was eluted, with a maximum of 33% being released. This was followed by a rapid decrease in the concentration of cobalt in solution (point A). This result has been ascribed to reabsorption of the cobalt by ion exchange sites present in the surface layers of the EIX working electrode. These sites were activated to cation exchange by the intimate contact of the hydroxyl-producing counter electrode. This effect was promoted because the solution pH was above the equilibrium pH of cobalt over crystalline ZrP, which has been previously determined as being 2.30 units [8]. The cobalt ions, once absorbed, were not released from these surface sites.

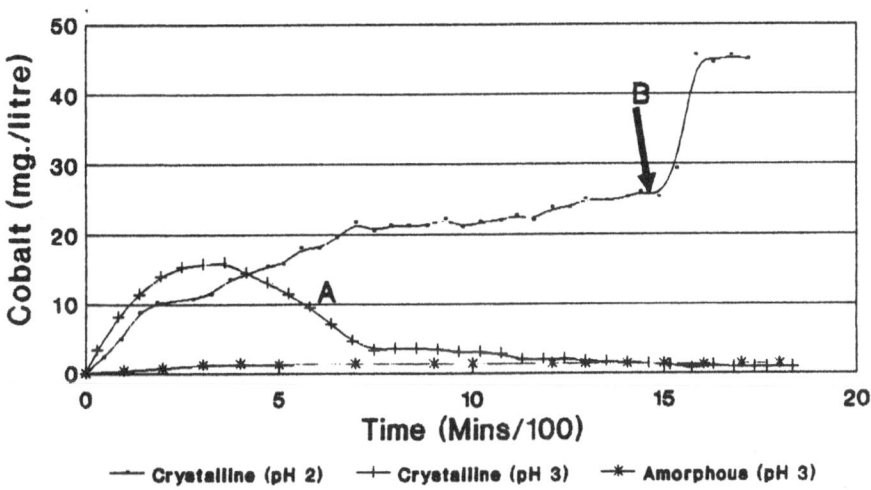

**Figure 5.** Elution performance of amorphous and crystalline ZrP at pHs 2 and 3 under EIX conditions.

Elution of crystalline material at pH 2 gave a gradual increase in cobalt concentration whilst the eluting current was maintained. Just under 50% of the absorbed cobalt was retained in the working electrode. However, on removal of the current (after point B) a rapid increase in solution cobalt concentration to over 90% elution was observed indicating that surface absorption was still being encountered in the surface layers. On

removal of the applied potential the previously activated surface ion
exchange groups came to equilibrium with the solution pH. This resulted in
the ion exchange equilibrium being altered more in favour of the free
cobalt species, thus explaining the rapid rise in solution cobalt
concentration on removal of the applied voltage.

The lowest line shows the elution of cobalt from amorphous ZrP with a
solution pH of 3. The amount of cobalt eluted was negligible (less than
4%). Thus elution was less satisfactory from amorphous than crystalline
ZrP. This may be ascribed to the difference in the hydration environments
found in amorphous and crystalline ZrP, which mean that more energy was
required to remove an ion from an exchange site in the amorphous form [9].

Figure 6 shows the gradual decrease in the rate of cobalt extraction
as the EIX electrode is taken through a number of absorption and elution
cycles. The loss of performance of the electrode when used for EIX has been
attributed to the gradual alkaline hydrolysis of the inorganic ion
exchanger caused by the high pHs generated in the cation loading phase of
EIX when hydroxyl ions are generated "in situ" within the electrode. This
leads to a slow loss of the monohydrogenphosphate groups present on the
surface of the ion exchanger. Studies at Harwell have indicated that
suitable materials are available which minimise this problem.

Figure 7 shows the change in elution performance of an EIX working
electrode with the number of absorption and elution cycles performed.

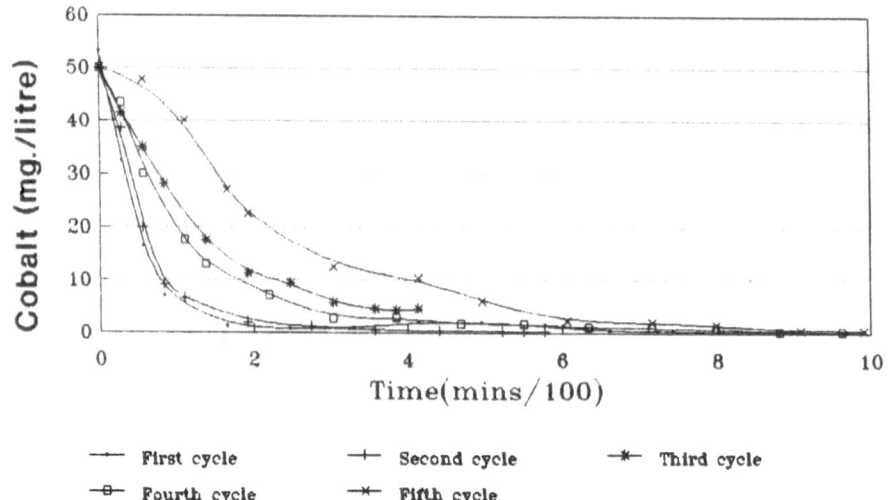

**Figure 6.** Graph showing the reduction in rate of cobalt extraction with
number of cycles of EIX carried out with the electrode.

It can be seen that the rate of cobalt elution increases with the number of cycles. This effect is again a result of hydrolysis of the ion exchanger matrix due to the aforementioned processes. Since the high pH associated with the cation loading phase is localised near the central current feeder it is to be expected that the core of the working electrode will suffer the greatest levels of alkaline hydrolysis, with the peripheries being relatively untouched. The sections of the curves after the arrows (marked) show that cobalt desorption continues when the applied potential is removed.

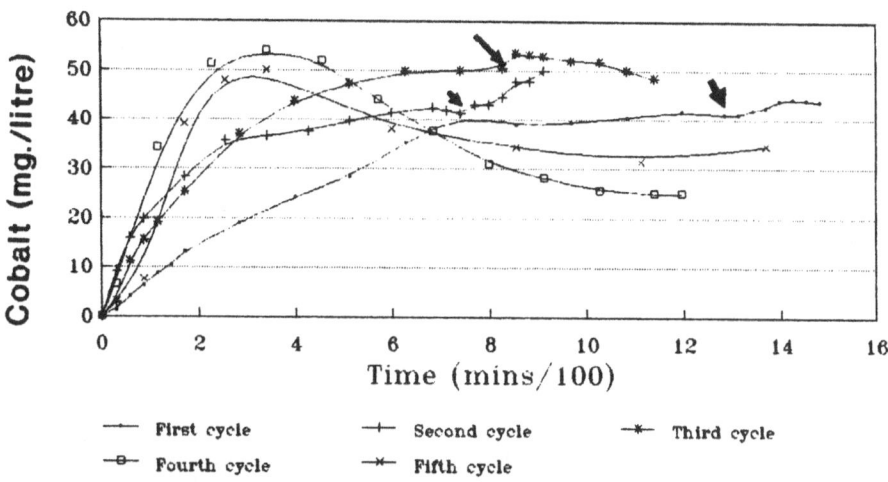

**Figure 7.** Graph showing the increase in the rate of elution with the number of cycles of EIX made with the electrode.

### CONCLUSIONS.

Cobalt may be reversibly extracted and eluted using the technique of EIX. It has been found to give enhanced rates of cobalt absorption compared to IX, under similar conditions. The use of crystalline ZrP in EIX leads to a further increase in the rate of cobalt extraction and elution compared to the amorphous ZrP. During the elution stages it was found that there was surface reabsorption of cobalt from solution but this effect may be ameliorated by removal of the applied potential, allowing the electrode to release the surface exchanged cobalt ions.

### ACKNOWLEDGEMENT.

The research was funded by the Underlying Research Programme of the U.K. Atomic Energy Authority (A.E.A. Technology, Harwell).

**REFERENCES.**

1.  G. Alberti, M. Casciola and U. Costantino; Inorganic ion exchange membranes made of acid salts of tetravalent metals. A short review.; J. Memb. Sci., 1983, 16, 137.

2.  G. Alberti and U. Costantino; Recent progress in the field of synthetic inorganic exchangers having a layered or fibrous structure.; J. Chrom, 1974, 102, 5.

3.  V. Vesely, V. Pekarek and A. Ruvarac; Correlation of sorption properties and hydrolysis behaviour for various types of zirconium phosphates.; Bull. Soc. Chim. France, 1968, 18, 32.

4.  G. Alberti and E. Torracca; Crystalline insoluble salts of polybasic metals II. Synthesis of crystalline zirconium or titanium phosphate by direct precipitation.; J. Inorg. Nucl. Chem., 1968, 30, 317-8.

5.  G. Alberti, U. Costantino, and R. Giulietti; Preparation of large crystals of alpha $Zr(HPO_4)_2.H_2O$.; J. Inorg. Nucl. Chem., 1980, 42, 1062-3.

6.  G. Alberti and E. Torracca; Electrical conductance of amorphous zirconium phosphate in various salt forms.; J. Inorg. Nucl. Chem., 1968, 30, 1093.

7.  A. Clearfield, A. Oskarsson and C. Oskarsson; On the mechanism of ion exchange in crystalline zirconium phosphates VI. The effect of crystallinity of the exchanger on $Na^+/H^+$ exchange; Ion Exch. Membr., 1972, 1, 91.

8.  S. Allulli, A. La Ginestra, M.A. Massucci, M. Pelliccioni and N. Tomassini; Uptake of transition metal ions by crystalline zirconium hydrogen phosphate.; Inorg. Nucl. Chem. Letters; 1974, 10, 337.

9.  A. Clearfield and J. M. Kalnins; On the mechanism of ion exchange in zirconium phosphates XIII. Exchange of some divalent transition metal ions on alpha-zirconium phosphate.; J. Inorg. Nucl. Chem., 1976, 38, 849-52.

# Part 5

# NEW MATERIALS

# A NEW SOLVENT RESISTANT ION EXCHANGE MATERIAL FOR ANALYTICAL CHROMATOGRAPHY

Christopher A. Pohl
Research and Development Department
Dionex Corporation
1228 Titan Way
Sunnyvale, California 94088-3603, USA

## ABSTRACT

A pair of new solvent resistant pellicular cation exchangers is described. The new materials make use of a latex-based cation exchange gel coating attached to a highly crosslinked polyvinyl aromatic core. The polymeric core can be either nonporous or porous. In the latter case, the pellicular cation exchanger allows a variety of reversed phase retention modes to occur simultaneous to cation exchange retention. The ability to add solvents to the mobile phase significantly expands the degree of selectivity control achievable with these new materials. A comparison with ion-pair chromatography performed on a conventional bonded phase silica reversed phase column demonstrates the high selectivity of the new materials.

## INTRODUCTION

The literature is replete with examples of the highly selective nature of ion exchange gels(1). The primary difficulty with using such materials stems from their poor physical properties. The combination of poor mass transport and shrink/swell characteristics has greatly limited the utilization of fully functional ion exchange gels in high performance analytical applications. In the early 1970's the first in a series of alternative ion exchange materials was developed(2). These pellicular materials consisted of a glass core upon which a polymeric film was deposited. The film was subsequently functionalized to produce the pellicular ion exchange material. These materials exhibited superior mass transport properties but were limited in their use due to their relatively large size (30-50 microns), low capacity, and somewhat delicate nature.

By the middle of the 1970's the glass core pellicular ion exchanger had largely been supplanted by the bonded phase microparticulate silica ion exchange materials. These ion exchangers had the advantage of ion exchange capacities comparable to those of gel type ion exchangers as well as the higher efficiencies to be expected from these small particle materials. A major disadvantage of these microparticulate ion exchangers is their rather poor chemical stability. Another limitation of these materials stems from the rather two dimensional nature of this class of ion exchanger. Unlike gel type ion exchangers,where selectivity is a function of the three dimensional distribution of  charged and neutral regions in the ion exchanger, microparticulate ion exchange materials have a rather two dimensional ion exchange surface with charged groups distributed at a fixed distance from the surface of the silica to which the charged groups are bonded. As a result, these ion exchange materials have generally lower selectivity than classical gel type ion exchangers. Recently, several groups have partially overcome this deficiency with the development of mixed mode ion exchange materials(3).

By the end of the 1970's the technique known as ion pair chromatography began to supplant ion exchange as the method of choice in the high performance chromatographic separation of ionic species(4,5). The main advantage of the technique stems from its relative versatility. With a single reversed phase column one can analyze anions, cations and neutral compounds. While this technique is often presented as a direct replacement for ion exchange, the selectivity of the two techniques is actually quite different and  highly complementary(6). Ion pair generally provides the best selectivity for ions of low hydration energy and for ions  which contain a hydrophobic moiety, whereas ion exchange generally offers superior selectivity for ions of high hydration energy and for ions which have a valency greater than one. Thus a stationary phase capable of performing both modes simultaneously should provide superior overall selectivity.

During the middle 1970's Hamish Small developed an entirely new type of pellicular ion exchange material(7). Instead of using a glass core, the entire bead including the core was constructed of polymeric materials. In the originally developed material, attachment of the ion exchange pellicle was accomplished via an electrostatic mechanism. In the case of the original anion exchange material the core consisted of a 2% crosslinked  polystyrene bead which had been lightly surface-sulfonated so as to provide a negative surface charge to the core. The ion exchange pellicle was applied by mixing this sulfonated core with a colloidal dispersion of anion exchange particles (generally 0.2-0.3 microns in diameter) . This results in an irreversible electrostatic attachment of the anion exchange particles to the core.

The ion exchange materials produced by Small's method offer several advantages over previous approaches. First, it allows for completely separate manufacturing processes for the core and the pellicle. This greatly simplifies the research and development of new materials as well as improving the manufacturability of these materials. Second, the combination of a polymeric core and electrostatic attachment greatly improves the ruggedness of these materials over the glass core type pellicular materials. Third, the selectivity advantages of gel-type ion exchangers are maintained in as much as the pellicle consists of a coating of gel-type ion exchange microparticles. Fourth, the

thin pellicular layer preserves the mass transport advantages of earlier glass core pellicular ion exchange materials. And finally, Small's method allows for facile reduction of
the pellicular particle diameter thus overcoming the low chromatograghic efficiency
problems of the larger particle glass core pellicular ion exchange materials.

The pellicular material developed by Small is certainly very useful in high performance chromatographic separations. These materials are widely used as chromatographic media in the field of ion chromatography. However, materials of this sort contain
a design flaw which limits their use with organic solvents. The problem rests with the
nature of the polymeric core. The 2% crosslink is too low to prevent excessive swelling
of the core in the presence of organic solvents. In a previous paper (8),we described a
pair of new pellicular anion exchange materials which have been developed to address
the use of solvents. Other authors have described the use of these new materials for use
in pharmaceutical applications (9). In this report we describe two new cation exchange
materials which are designed to allow the use of organic solvents while maintaining all
of the stated advantages of the pellicular materials originally developed by Small.

The pellicular particles used in the PCX-100 column consist of a latex coating on
a nonporous core. The nonporous core consists of an 8.5 micron diameter 60%
crosslink polyvinylaromatic polymer bead. The high crosslink of the core assures virtually no swelling in organic solvents. The core particles are synthesized so as to directly
obtain a spherical particle with negative surface charge. While pellicular anion exchange materials can be made via electrostic attachment of the anion exchange particles to this surface, the preparation cation exchange materials from this same core
particle requires a double layer attachment process. The primary pellicle consists of an
anion exchange latex coating. The latex chosen for this purpose is .04 microns in diameter. It is a 3% crosslink copolymer of vinylbenzyl chloride and divinylbenzene which has
been quaternized with trimethylamine. The size and composition of the latex was
chosen so that the magnitude of anion exchange retention is relatively minor. The
secondary pellicle in the ion exchange particle consists of a cation exchange latex coating. This layer comprises the analytical ion exchange layer. It consists of a 5% crosslink
polyvinylaromatic polymer which has been fully sulfonated.

The particles used in the PCX-500 column are identical to those of the cation
exchange material used in the PCX-100 column with the exception of the core. In the
case of the PCX-500 pellicular material the core is produced in such a way as to create
a high surface area neutral hydrophobic core while still maintaining a negatively
charged exterior surface to which the primary and secondary pellicles are attached. The
core has an surface area of 450 meters$^2$/gram and an average pore diameter of .006
microns. The small size of the pores effectively excludes ion exchange particles from the
interior of the core particle so that all ion exchange sites are confined to the exterior
surface of the core.

## MATERIALS AND METHODS

### Chemicals

(±)-Norephedrine hydrochloride (NE), (±)-ephedrine hydrochloride (E), (-)-methylephedrine (ME), (-)-norpseudoephedrine (NPE), (+)-pseudoephedrine (PE), (-)-N-methylpseudoephedrine (MPE), phenethylamine hydrochloride (PA), N-methylphenethylamine (MPA), (+)-N,N-dimethyl-1-phenethylamine (DPA), and cesium chloride, optical grade were obtained from Aldrich (Milwaukee, WI, U.S.A.); doxepin hydrochloride, propylamine and propranolol (1 mg/ml in methanol) from Sigma (St. Louis, MO, U.S.A.); acetonitrile, distilled-in-glass grade, from Burdick & Jackson Labs. (Muskegon, MI, U.S.A.); sodium chloride, puriss. grade, from Fluka (Ronkonkoma, NY, U.S.A.); sodium heptanesulphonate monohydrate, HPLC grade, Pierce (Rockford, IL, U.S.A.); lithium chloride, potassium chloride, hydrochloric acid (37%), phosphoric acid (85%), reagent grade, from J.T.Baker (Phillipsburg, NJ, U.S.A.).

### Mobile Phases

All mobile phases were prepared by proportioning via the gradient pump using 100% acetonitrile, 100% 18 megohm-cm DI water, and a third ionic solution. The third solution was either 0.0250 M sodium heptanesulphonate with 0.010 M phosphoric acid, 0.210 M hydrochloric acid or 0.200 M alkali metal chloride with 0.010 M hydrochloric acid. While performing reversed phase ion pair chromatography with an amine modifier, a fourth eluant consisting of 0.010 M propylamine and 0.010 M phosphoric was also proportioned. In all cases the eluants were sparged for 10 minutes with helium after which a constant helium pressure of 10 psi was applied to all eluant bottles.

### Equipment

A Dionex gradient pump (model GPM-2), Dionex polymeric microinjection valve (model MIV-1), eluant degas module (model EDM-2), variable wavelength detector (model VDM-2), and AI 400 data system were used. The data system included a Hewlett-Packard Vectra computer and ThinkJet printer. The columns used were either a Nova-Pak $C_{18}$ from Waters (150 mm x 3.9 mm I.D.), an OmniPac PCX-100 (250 mm x 4 mm I.D.), or an OmniPac PCX-500 (250 mm x 4 mm I.D.). All columns were operated at a flow rate of 1 ml/min. and at ambient temperature. A Brookhaven model BI-90 was used to measure the diameter of the ion exchange latex particles.

## RESULTS AND DISCUSSION

While latex-based ion exchange materials have been widely used in ion chromatography, the use of solvents in the mobile phase has been rather limited. This is due in
part to the rather limited effect solvent has on the ion exchange selectivity of inorganic
ions normally analyzed in ion chromatography. For this reason, a series of closely related aromatic amines were chosen with the expectation that solvent effects would be
more significant. They consist of the ephedrine series, the pseudoephedrine series, and
the phenethylamine series. Within each series the variable is the degree of substitution
at the amine. For example, norephedrine is phenylpropanolamine ; ephedrine is N-
methylphenylpropanolamine; and methylephedrine is N,N-dimethylphenylpropanolam-
ine. Because phenylpropanolamine contains two asymmetric centers there are diastereomeric pairs for each of these phenylpropanolamines. The ephedrine series represents
the asymmetrical diastereomers (1S,2R or 1R,2S) while the pseudoephedrine series
represents the symmetrical diastereomers (1S,2S or 1R,2R). Together the two sets of
phenylpropanolamines represent an effective means of evaluating the relative selectivity of various chromatographic systems in as much as diastereomeric pairs are often
difficult to resolve in liquid chromatography. The phenethylamine series was chosen to
allow study of the effect of minor variations in the hydrophobicity of the aromatic
substituent attached to the amine.

Table 1 shows the tabulated retention data for the PCX-100 column and the
PCX-500 column with a cesium chloride eluant system. A number of general trends are
evident from the data listed in Table 1. For example, the effect of increasing levels of
solvent in the mobile phase with either column is to reduce retention. This effect, which
appears to be quite general, is probably due to the disruptive effect solvent molecules
have on the water structure in the mobile phase. This has the overall effect of reducing
the difference in water structure between the mobile and stationary phases thus resulting in reduced retention for most ions. Of course, a secondary cause of reduced retention with increasing solvent levels is the ability of solvent molecules to reduce the extent
of hydrophobic interaction between the solute and the polymer matrix surrounding the
ion exchange site. While this factor may be relevant here, it can not explain the similar
reduction in retention observed with highly hydrated inorganic ions as solvent level is
increased.

With cesium as the eluant cation the elution order within a given series is opposite that which would be expected if solute hydrophobicity were the dominant factor in
retention. What appears to be the determining factor with cesium is the relative size of
the solute, with the smallest ions eluting last. While such an elution order is consistent
with the general observation that hydrated ionic radius indeed plays a major role in

determining elution order in ion exchange, the fact that large, relatively hydrophobic cations elute in order of decreasing size may be taken as an indication that hydrophobic interactions play a relatively minor role in determining elution order with this system.

Indeed, some degree of confirmation of this hypothesis may be obtained by noting the differences between the retention data of the PCX-100 column and the PCX-500 column (Figure 1). When comparing the two columns it is invariably observed that the reversed phase character of the PCX-500 results in reduced resolution within a given series. This is due to the fact that hydrophobicity and hydrated ionic radius exert opposing effects on retention. The reversed phase component of retention in the PCX-500 results in the largest   retention for the most hydrophobic ion. Since the most hydrophobic ion within a given series is also the largest ion which results in the lowest retention by cation exchange, these two factors tend to oppose one another. Even though in this case the overall effect of the reversed phase component of retention is to reduce resolution, this still serves to illustrate the value of two independently controlled retention processes within a single stationary phase. By manipulating the relative mag-nitude of  reversed phase and ion exchange retention exceptional control of selectivity is possible.

TABLE 1

Cesium chloride eluant system retention data*

| Solute | 40% ACN** PCX-100 | 40% ACN PCX-500 | 30% ACN PCX-100 | 30% ACN PCX-500 | 20% ACN PCX-100 | 20% ACN PCX-500 |
|---|---|---|---|---|---|---|
| NE | 2.70 mins. | 3.03 | 4.00 | 4.43 | 6.05 | 6.95 |
| E | 2.13 | 2.58 | 3.20 | 3.83 | 5.02 | 6.40 |
| ME | 1.80 | 2.37 | 2.70 | 3.55 | 4.40 | 6.40 |
| NPE | 2.88 | 3.18 | 4.43 | 4.82 | 7.05 | 7.87 |
| PE | 2.17 | 2.60 | 3.30 | 3.92 | 5.35 | 6.70 |
| MPE | 1.67 | 2.23 | 2.45 | 3.27 | 3.95 | 5.77 |
| PA | 3.57 | 3.93 | 6.13 | 6.48 | 10.77 | 11.48 |
| MPA | 2.67 | 3.15 | 4.60 | 5.20 | 8.25 | 9.55 |
| DPA | 2.20 | 2.73 | 3.62 | 4.32 | 6.45 | 7.98 |

* Eluant contained 0.080 M CsCl and 0.004 M HCl in addition to the indicated level of acetonitrile.
**ACN = acetonitrile

The use of hydronium ion as the eluting cation results in a remarkably different selectivity for the test solutes (Table 2). Within a given series the elution order reverses. This is probably due to the hydration energy differences between hydronium ion and cesium ion. While cesium and the aromatic test solutes have roughly comparable hydration values, hydronium ion is significantly more hydrated. As a result of the large difference between the hydronium ion and the solutes in terms of hydration it appears that hydration energy factors dominate the cation exchange retention process. In such an eluant system ions with the lowest hydration energy will elute last.

Because hydration energy and hydrophobicity are generally inversely related, the effect of cation exchange retention which shows increasing retention with decreasing hydration is cooperative with the effect of reversed phase retention which shows increasing retention with increasing hydrophobicity. The comparison of the PCX-100 and the PCX-500 retention data (Table 2) confirms this. With one exception the addition of a reversed phase retention mode results in an increase in resolution within a given series. The lone exception to this trend is the pseudoephedrine series. The explanation for this discrepancy is unclear but it is possibly related to the effects of stereochemistry on hydration and hydophobicity.

TABLE 2

Hydrochloric acid eluant system retention data*

| Solute | 40% ACN** PCX-100 | 40% ACN PCX-500 | 30% ACN PCX-100 | 30% ACN PCX-500 |
|---|---|---|---|---|
| NE | 4.98 mins. | 4.98 | 6.58 | 6.70 |
| E | 5.28 | 5.33 | 7.43 | 7.63 |
| ME | 5.52 | 5.65 | 8.23 | 8.53 |
| NPE | 5.37 | 5.35 | 7.33 | 7.33 |
| PE | 5.43 | 5.52 | 7.82 | 7.92 |
| MPE | 5.15 | 5.35 | 7.62 | 7.92 |
| PA | 9.07 | 8.93 | 13.67 | 13.27 |
| MPA | 9.93 | 9.88 | 15.75 | 15.78 |
| DPA | 12.15 | 12.15 | 19.37 | 18.95 |

* Eluant contained 0.084 M HCl in addition to the indicated level of acetonitrile.

**ACN = acetonitrile

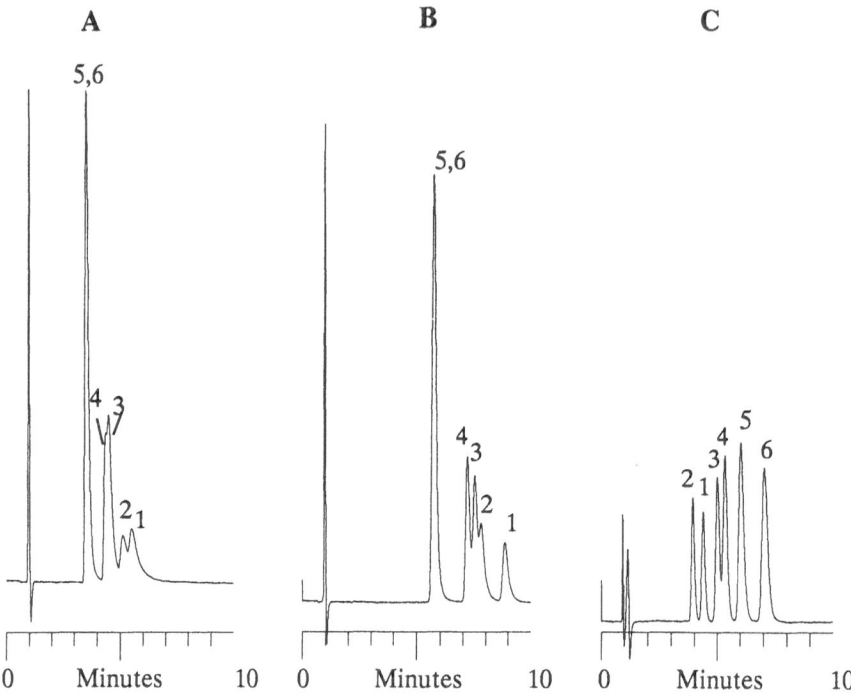

Figure 1. Comparison of cation exchange selectivity to cation ion pair selectivity. Solute identities: 1 = ME; 2 = MPE; 3 = E; 4 = PE; 5 = NE; 6 = NPE. Chromatographic conditions for Fig. 1A: column, Waters $C_{18}$ Nova Pak (150 x 3.9 mm I.D.); mobile phase was 0.005 M sodium heptanesulphonate, 0.002 M phosphoric acid and 20% acetonitrile. Chromatographic conditions for Fig. 1B: column was same as 3A; mobile phase was 0.005 M sodium heptanesulphonate, 0.006 M phosphoric acid, 0.002 M propylamine and 18% acetonitrile. Chromatographic conditions for Fig. 1C: column was Dionex OmniPac PCX-100 (250 x 4 mm I.D.); mobile phase was 0.080 M cesium chloride, 0.004 M hydrochloric acid and 20% acetonitrile. Flow rate for all three chromatograms was 1 ml/min. and detection wavelength was 215 nm.

A negative consequence of the hydrochloric acid eluant system is that selectivities are generally inferior to those obtained with the cesium chloride eluant system. Apparently, selectivity based on hydrated ionic radius is a more powerful tool in performing difficult separations. In studies with other alkali metal eluant cations it was found that a progressive trend exists. Hydronium ion exhibits hydration energy dominated retention. Cesium ion exhibits hydrated ionic size dominated retention and the intervening cations exhibit selectivity intermediate between them. Thus, the most selective of the eluant systems are those with hydronium and cesium since these ions are on the extreme ends of column 1A of the Periodic Table (except of course for francium which is unsuitable as an eluant ion for safety and cost reasons). Potassium results in only slightly less selectivity as an eluant cation and thus may be prefered in light of the relatively high cost of cesium salts.

Figures 1 and 2 demonstrate the relative selectivity of the PCX-100 column and an ion pair based separation on a reversed phase column. The poor peak shape for the test solutes is a result of their interaction with silanols on the silica surface. While an amine modifier can be added to the mobile phase to suppress these interactions, the selectivity of the PCX-100 column is still significantly better. Even in cases where the selectivity is adequate by ion pair chromatography the PCX-100 still exhibits significantly superior performance (Figure 3).

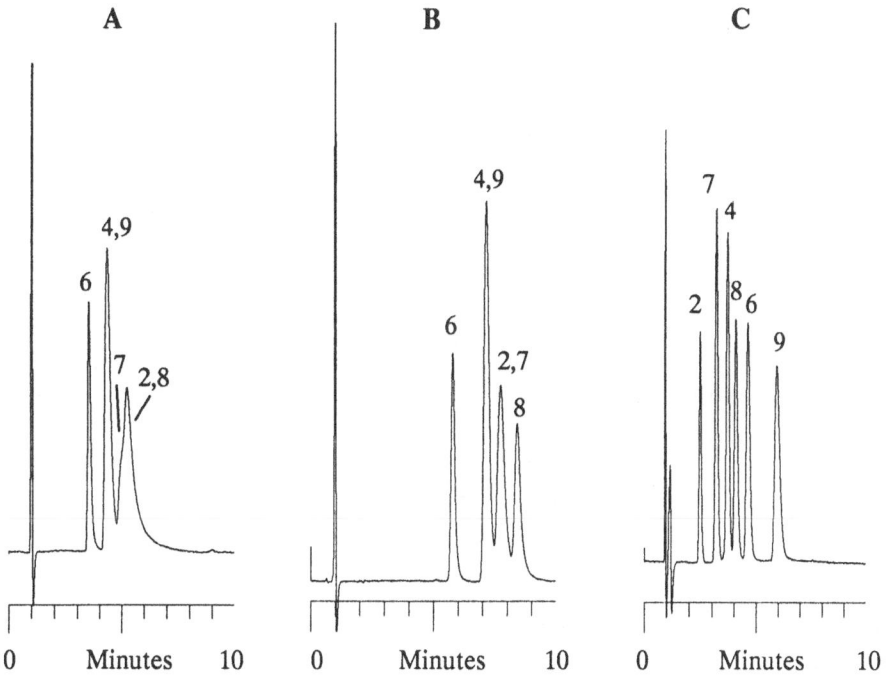

Figure 2. Comparison of cation exchange selectivity to cation ion pair selectivity. Solute identities: 2 = MPE; 4 = PE; 6 = NPE; 7 = DPA; 8 = MPA; 9 = PA. Chromatographic conditions for Fig. 2A were the same as Fig. 1A. Chromatographic conditions for Fig. 2B were the same as Fig. 1B. Chromatographic conditions for Fig. 2C were the same as Fig. 1C except that the concentration of cesium chloride and hydrochloric acid were doubled relative to Fig. 1C.

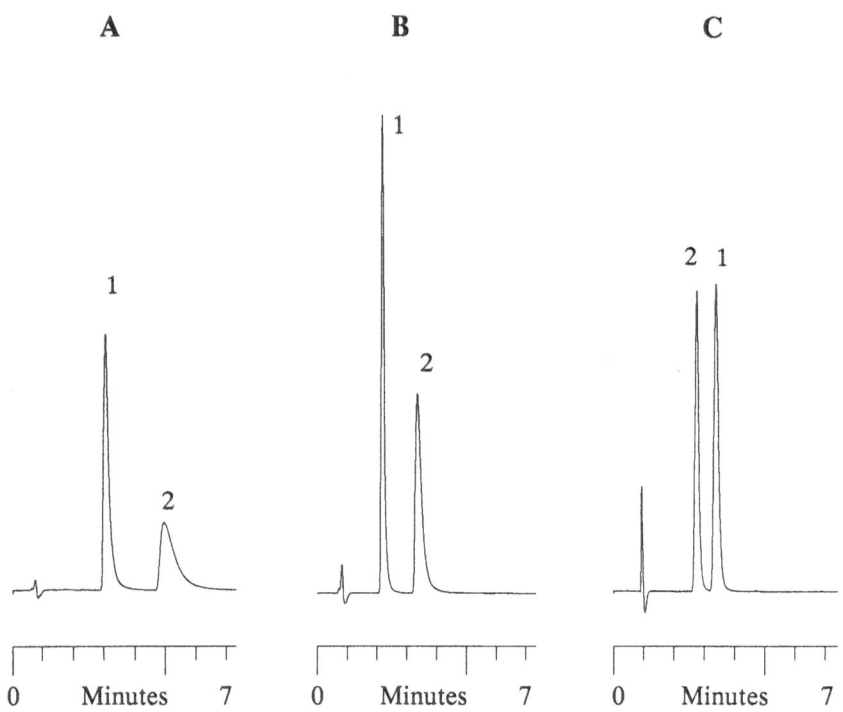

Figure 3. Comparison of cation exchange selectivity to cation ion pair selectivity. Solute identities: 1 = propranolol; 2 = doxepin. Chromatographic conditions for Fig. 3A were the same as for Fig. 1A except that the acetonitrile level was increased to 50%. Chromatographic conditions for Fig. 3B were the same as for Fig. 1B except that the level of acetonitrile was increased to 40%. Chromatographic conditions for Fig. 3C were the same as for Fig. 1C except that the level of acetonitrile was increased to 40%.

## CONCLUSIONS

Two new solvent compatible pellicular cation exchange materials have been evaluated with a series of test solutes. The data indicates that these columns are highly selective. The most selective eluant systems make use of either potassium or cesium as eluant cations. The addition of a reversed phase retention mode to the PCX-500 provides an additional degree of selectivity control not present in the cation exchange retention mode. Future investigation will study the effects of counter anions and ion pair reagents on the selectivity of the PCX-500.

## REFERENCES

1. Samuelson, O., <u>Ion Exchange Separations in Analytical Chemistry</u>, John Wiley and Sons, New York, N.Y., 1963, Ch. 15 and 16.

2. Horvath, C., Pellicular ion exchange resins in chromatography. In <u>Ion Exchange and Solvent Extraction</u>, ed. J. A. Marinsky and Y. Marcus, Marcel Dekker, New York, N.Y., 1973, Vol.5, Ch.3.

3. Issaq, H. J. and Gutierrez, J., Mixed packings in high performance liquid chromatography: II. Mixed packings vs. mixed ligands. <u>J. Liq. Chrom.</u>, 1981, 11 (14), 2851-2861.

4. Cantwell, F. F. and Puon, S., Mechanism of chromatographic retention of organic ions on a nonionic adsorbent. <u>Anal. Chem.</u>, 1979, 51 (6), 623-632.

5. Bidlingmeyer, B. A., Deming, S. N., Price, Jr., W. P., Sachok, B. and Petrusek, M., Retention mechanism for reversed-phase ion-pair liquid chromatography. <u>J. Chromatogr.</u>, 1979, 186, 419-434.

6. Pohl, C. A., Potts, M. E., Heberling, S. S., Stillian, J. R., Summerfelt, V. E., and Barreto, V. B., "Novel, pellicular, covalent latex bonded, polymer ion exchangers for high performance chromatographic separations", paper presented at the 40th Pittsburgh Conference and Exposition on Analytical Chemistry and Applied Spectroscopy (Atlanta, Georgia, 1989), abstract #753.

7. Small, H., Stevens, T. S. and Bauman, W. C., Novel ion exchange chromatographic method using conductivity detection. <u>Anal. Chem.</u>, 1975, 47 (11), 1801-1809.

8. Stillian, J. R. and Pohl, C. A., New latex bonded pellicular anion exchangers with multi-phase selectivity for high performance chromatographic separations. Accepted for publication in <u>J.Chromatogr.</u>

9. Slingsby, R. W. and Rey, M., Determination of pharmaceuticals by multiphase chromatography : combined reversed phase and ion exchange in one column. Accepted for publication in <u>J. Liq. Chromatogr.</u>

# ION EXCHANGE ADSORPTION OF METAL IONS ON AMINE AND PYRIDINE TYPES OF CHELATING RESINS

KATSUTOSHI INOUE, KAZUHARU YOSHIZUKA and YOSHINARI BABA
Department of Applied Chemistry, Saga University
Honjo-machi, Saga 840, Japan

## ABSTRACT

Distribution equilibria were investigated for the adsorption of metal ions and acids on Sumichelate MC-10 and CR-2 resins, polyethylene-polyamine and-pyridine types of novel chelating resins, respectively. The total exchange capacities of MC-10 and CR-2 resins for hydrochloric acid were 5.6 and 4.2 mol/kg dry resin respectively while the adsorption equilibrium constants of the acids on MC-10 resin were much greater than those on CR-2 resin. Adsorption of various divalent transition metals was investigated from hydrochloric acid on the chloride form of the resins. The adsorption took place in the order Cd>Zn>Cu>Co>Mn for each resin. Adsorption of silver(I), zinc (II) and copper(II) was investigated from 1 mol/dm$^3$ aqueous ammonium nitrate solution on the free base form of the resins. It took place in the order Cu>Ag>Zn in the adsorption on MC-10 resin, while the sequence was Ag>Cu>Zn in the adsorption on CR-2 resin.

## INTRODUCTION

Like the solvent extraction technique, chelating ion exchange is an effective and powerful unit operation for the separation/puri-fication/concentration of metal ions from aqueous solutions, especially from very dilute solutions. A number of synthetic chelating resins containing various functionalities such as iminodiacetic acid, aminophosphonic acid, dithiocarbamic acid and so on have been developed to date [1], and some of them are now in every day extensive use on a commercial scale. However, the majority of these chelating resins have not shown satis-factorily high selectivity toward the specific metal ion to be separated, thus requiring the use of very large amounts of resin per unit volume of process solution.
Recently, Sumitomo Chemical Co. Ltd. has developed a novel

macroporous chelating resin containing polyethylene-polyamine functionality, Sumichelate MC-10. In the present paper, we conducted fundamental investigations on the adsorption of metal ions on this resin and on Sumichelate CR-2 resin, a similar type of macroporous chelating resin containing pyridine functionality for comparison. The officially reported physical properties of these resins are shown in Table 1.

TABLE 1
Physical properties of MC-10 and CR-2 resins

| resin | MC-10 | CR-2 |
|---|---|---|
| appearance | white or pale yellow spherical beads | white spherical beads |
| specific gravity | 1.2 - 1.3 | 1.1 |
| moisture content | 60 - 70 % | 35 - 45 % |
| particle size | 20 - 48 mesh | 10 - 50 mesh |

## EXPERIMENTAL

The Sumichelate MC-10 and CR-2 resins, kindly supplied from Sumitomo Chemical Co. Ltd., were conditioned according to the conventional method and dried in vacuum at room temperature to a constant weight before use. The chloride forms of the resins were used for the adsorption of metals from hydrochloric acid solutions while free base forms were used for that from aqueous ammonium nitrate solutions and for the adsorption of hydrochloric and nitric acids.
Reagent grades of $CdCl_2 \cdot 2.5H_2O$, $ZnCl_2$, $CuCl_2 \cdot 2H_2O$, $CoCl_2 \cdot 6H_2O$ and $MnCl_2 \cdot 4H_2O$ were used as chlorides of these metals; also, reagent grades of $Cu(NO_3)_2 \cdot 3H_2O$, $AgNO_3$ and $Zn(NO_3)_2 \cdot 6H_2O$ as nitrates of these metals. The metal chlorides were used by dissolving in different concentrations of hydrochloric acid solutions while the metal nitrates were used by dissolving in 1 $mol/dm^3$ ammonium nitrate solution, the pH of which was adjusted by adding small amounts of nitric acid or aqueous ammonia solution. For the adsorption of metals from hydrochloric acid solutions, the initial metal concentrations were around 5 m $mol/dm^3$ while, for the adsorption from ammonium nitrate solutions, the initial concentrations of copper and zinc were around 10 m $mol/dm^3$ and that of silver was around 1 m $mol/dm^3$. All experiments were carried out batchwise. A weighed amount of the resins was shaken overnight with a constant volume of metal-containing aqueous solution in a glass-stoppered flask immersed in a water-bath incubator maintained at 303 K. In a preliminary experiment , all metals were confirmed to attain equilibrium within a few hours. The amount of metal adsorbed on a unit gram of the dry resin was calculated from the metal concentrations before and after equilibrium and the weight of the dry resin used. The initial and equilibrium concentrations of metals except for silver were determined by means of titration with EDTA; 1-(1-hydroxy-2-naphthylazo)-6-nitro-2-naphthol-4-sulfonic acid, sodium salt (BT) was used as an

indicator for zinc, cadmium and manganese, 1-(2-pyridylazo)-2-
naphthol (PAN) and purpuric acid, ammonium salt (Murexide) were
used as those for copper and cobalt respectively.   Those of
silver were determined by means of    atomic absorption
spectrochemical analysis using a Nippon Jarrell-Ash model AA-
782 spectrophotometer.   The distribution of metal ions between
resin and aqueous phases was represented in terms of the
distribution ratio, D, the ratio of the amount of metal
adsorbed on unit kg of the resin to the equilibrium metal
concentration in the aqueous phase.

## RESULTS AND DISCUSSION

### Adsorption of Hydrochloric and Nitric Acids
Prior to the investigation of the adsorption of metals, the
adsorption of hydrochloric and nitric acids were investigated
in order to evaluate the total exchange capacities of the
resins.   Figure 1 illustrates the relationship between the
amount of adsorption of hydrochloric and nitric acids on free
base forms of the resins and the equilibrium acid concentration.   The adsorption of nitric acid is slightly greater
than that of hydrochloric acid for each resin.   In the low
concentration region, the plots are lying on straight lines
with the slope of unity for each acid and for each resin and
they tend to approach constant values corresponding to each
acid and to each resin with increasing acid concentration in
the high concentration region.

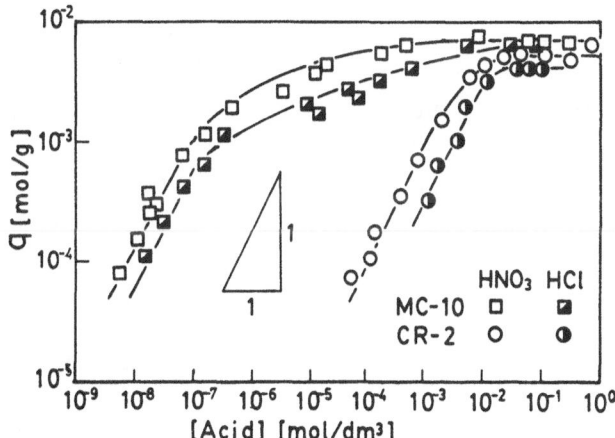

FIGURE 1. Adsorption of hydrochloric and nitric acids on the
MC-10 and CR-2 resins.

On the assumption that the acids (HA) are adsorbed on the
resins according to the reaction described by Eq.(1), the
amount of acid adsorbed is expressed by Eq.(2).

$$\overline{RNH_2} + HA = \overline{RNH_3^+A^-} \quad ; \quad K_a \qquad (1)$$

$$q = [\overline{RNH_3^+A^-}] = K_a[\overline{RNH_2}][HA] \qquad (2)$$

The total exchange capacity of the resins, Q, is expressed by

$$Q = [\overline{RNH_3^+A^-}] + [\overline{RNH_2}] \qquad (3)$$

The combination of Eqs.(2) and (3) gives Eq.(4) as the ultimate expression for the equilibrium amount of adsorption of the acids.

$$q = K_a[HA]Q/(1 + K_a[HA]) \qquad (4)$$

In the low concentration region of the acids, Eq.(4) is approximated by Eq.(5) while it tends to a constant value, Q, in the high concentration region.

$$q = K_a[HA]Q \qquad (5)$$

According to Eq.(5), the apparent adsorption equilibrium constants, $K_a$, for each acid is evaluated from the intercepts of the straight lines with the slope of unity with the ordinate at [HA] = 1 and values of Q as listed in Table 2 along with Q.

TABLE 2
Total exchange capacity (Q) and apparent adsorption equilibrium constant ($K_a$) of MC-10 and CR-2 resins

|  |  | MC-10 | CR-2 |
|---|---|---|---|
| Q [mol/kg] | HCl | 5.6 | 4.2 |
|  | $HNO_3$ | 6.3 | 5.0 |
| $K_e$ [$dm^3$/mol] | HCl | $1.1 \times 10^6$ | $7.5 \times 10^1$ |
|  | $HNO_3$ | $2.9 \times 10^6$ | $2.0 \times 10^2$ |

Adsorption of Metals from Hydrochloric Acid Solutions

Figures 2 and 3 illustrate the relationship between the distribution coefficient of various divalent transition metals, cadmium(II), zinc(II), copper(II), cobalt(II) and manganese(II) and the mean activity of hydrochloric acid for the adsorption from different concentrations of hydrochloric acid solutions on the chloride forms of the MC-10 and CR-2 resins respectively. Here, the mean activity of hydrochloric acid was calculated from its concentration determined by titration and literature values of mean activity coefficients[2]. The adsorption takes place in the order, Cd>Zn>Cu>Co>Mn, either on the MC-10 resin or on the CR-2 resin though the adsorption of manganese was not observed on the latter resin. The sequence is in accordance with that observed in the solvent extraction with Primene JMT, a commercial primary amine [3], or in accordance with that of the stability of chloro-complexes of these metals.

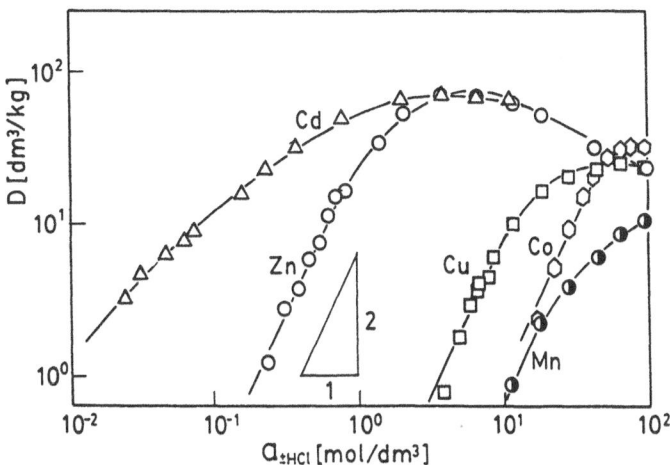

FIGURE 2. Adsorption of divalent transition metals on the chloride form of the MC-10 resin from different concentrations of hydrochloric acid.

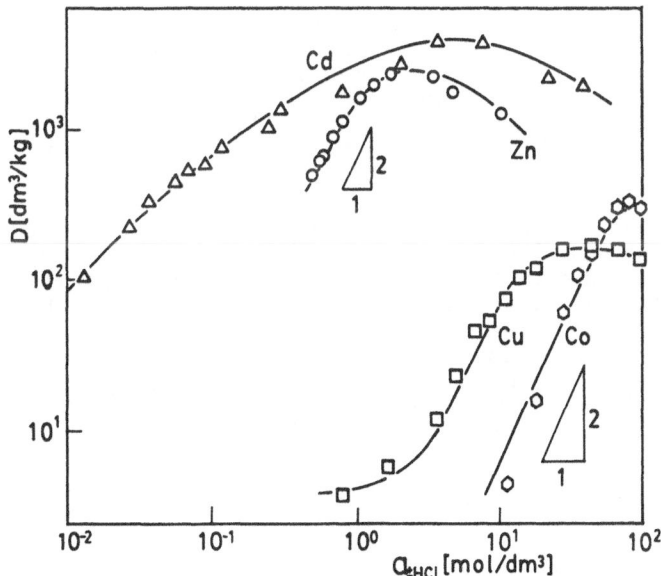

FIGURE 3. Adsorption of divalent transition metals on the chloride form of the CR-2 resin from different concentrations of hydrochloric acid.

The adsorption increases with increasing hydrochloric acid concentration in its low concentration region and, after passing through the maximum, it decreases in the high concentration region for the adsorption of cadmium and zinc in particular. In the region of low loading, the plots are straight lines with the slope of 2 for each metal and for each resin except for cadmium. These results appear to be analogous to those observed in the solvent extraction of these metals from hydrochloric acid with high molecular weight amines and suggest that the anionic species of the tetrachloro complexes of these divalent metals are adsorbed on the chloride form of the resins according to the anion exchange reaction described by Eq.(6) to give rise to an ion-pair complex as shown below.

$$\overline{(R\text{-}NH_3^+Cl^-)_2} + MCl_4^{2-} = \overline{(RNH_3^+)_2MCl_4^{2-}} + 2\ Cl^-;\ K_{e1} \qquad (6)$$

According to the reaction of Eq.(6), the amount of adsorption of metal is described by

$$q = K_{e1}[\overline{(RNH_3^+Cl^-)_2}][MCl_4^{2-}]/a_{\pm HCl}^2 \qquad (7)$$

where the concentration of the tetrachloro complex in the aqueous phase, $[MCl_4^{2-}]$, is expressed in terms of total metal concentration, $C_M$, and the stability constants of each chloro complexes, $\beta_i$, as follows;

$$[MCl_4^{2-}] = \beta_4 a_{\pm HCl}^4 C_M / (\ 1 + \sum_{i=1}^{4} \beta_i a_{\pm HCl}^i\ ) \qquad (8)$$

Since, under the present experimental conditions, the initial metal concentration is low enough for the amount of metal adsorbed on the resins to be neglected in comparison with the total exchange capacities of the resins, the amount of the free quarternary ammonium chloride form of the functional groups of the resins, i.e. those free from the interaction with the metal-tetrachloro complex anion, are nearly equal to the total exchange capacity, Q. Ultimately, the distribution coefficient is expressed as follows:

$$D = q/C_M = K_{e1}(Q/2)\beta_4 a_{\pm HCl}^2 / (\ 1 + \sum_{i=1}^{4} \beta_i a_{\pm HCl}^i\ ) \qquad (9)$$

It is further approximated by Eq.(10) in the low concentration region of hydrochloric acid.

$$D = K_{e1}(Q/2)\beta_4 a_{\pm HCl}^2 \qquad (10)$$

Equations (9) and (10) can qualitatively interprete the experimental results shown in Figs. 2 and 3.

## Adsorption of Metals from Aqueous Ammonium Nitrate Solutions

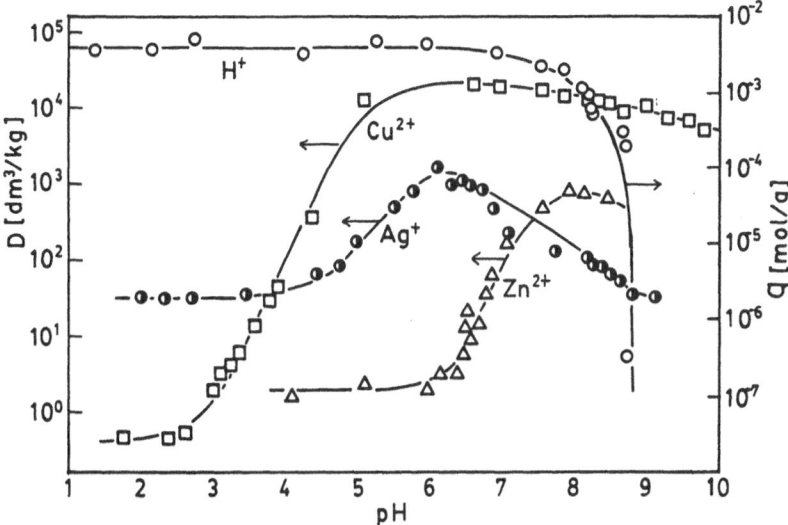

FIGURE 4. Adsorption of copper(II), zinc(II) and silver(I) on the free base form of the MC-10 resin from 1 mol/dm$^3$ aqueous ammonium nitrate solutions.

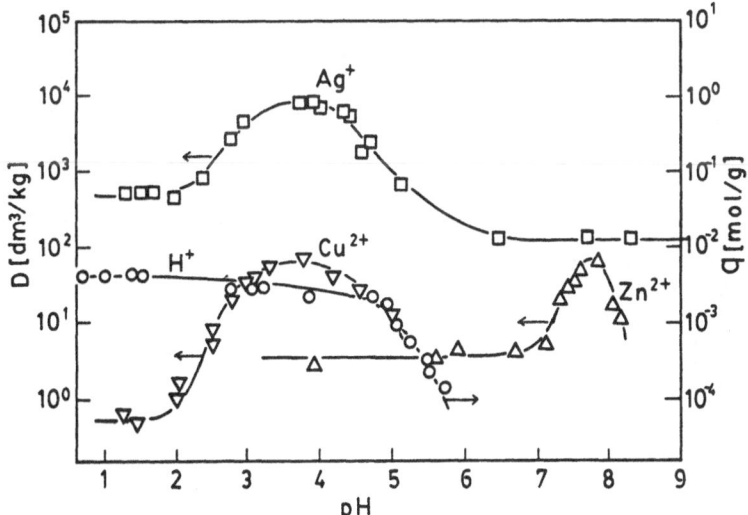

FIGURE 5. Adsorption of copper(II), zinc(II) and silver(I) on the free base form of the CR-2 resin from 1 mol/dm$^3$ aqueous ammonium nitrate solutions.

Figures 4 and 5 illustrate the plots of the distribution
coefficients of silver(I), copper(II) and zinc(II) together
with the amount of adsorption of hydrogen ion against pH in the
adsorption from 1 mol/dm$^3$ aqueous ammonium nitrate solution on
the free base form of the MC-10 and CR-2 resins respectively.
With increasing pH, the amount of adsorption of hydrogen ion
decreases and, conversely    that of metal adsorption increases
though it decreases with further increase of pH after passing
through the maximum at high pH in both figures.   The adsorption
of metals takes place in the sequence, Cu>Ag>Zn, on the MC-10
resin as is shown in Fig. 4 whilst it takes place in the
sequence, Ag>Cu>Zn, on the CR-2 resin as is shown in Fig. 5.
The sequence on the former resin is in accordance with that of
the stability of the amine   complexes of these metals.[4]
These results suggest that, as pH increases, hydrogen ions
desorb from the resins leaving the active sites of free amine
or pyridine functional groups opened to metal ions in the
aqueous phase.   The adsorption of metal ions, for example, on
the MC-10 resin gives rise to polymer-like amine   complexes as
shown below according to the reaction described by Eq.(11).

$$\overline{(RNH_2)_m} + M^{m+} + m\ NO_3^- = \overline{(RNH_2)_m M \cdot mNO_3}; \qquad K_{e2} \qquad (11)$$

( m = 1 for silver and = 2 for copper and zinc )

The decrease of the distribution coefficient at high pH is
considered to be due to the increase of the concentrations of
the higher metal- amine complexes difficult to be adsorbed on
the resin.
On the other hand, at very low pH, the distribution coefficient
for each metal is independent of pH and that of silver is much
greater than those of copper and zinc at very low pH.   Taking
into consideration the fact that silver(I) is apt to give rise
to an ion-pair complex with nitrate anion [5], this might be
considered to suggest that these metals are adsorbed on the
nitrate form of the resin by weakly interacting with the
nitrate anion on the resin to give rise to a kind of ion-pair
complex    the example of which is shown below for silver.

The adsorption on the CR-2 resin is also considered to take place by a similar mechanism.

## ACKNOWLEDGEMENT

The authors are greatly indebted to the kind supply of the samples of Sumichelate MC-10 and CR-2 resins for Sumitomo Chemical Co. Ltd..

## REFERENCES

1. Sahni, S.K. and Reedijk, J., Coordination chemistry of chelating resins and ion exchangers. Coord.Chem.Rev., 1984, 59, 1-139.
2. Harned, H.S. and Owen, B.B., Physical Chemistry of Electrolytic Solutions", Reinhold, New York, 1958, pp. 716
3. Shmidt, V.S., Amine Extraction (English edition), Israel Program for Scientific Translations, Jerusalem, 1971, pp. 134
4. Freiser, H. and Fernando, Q., Ionic Equilibria in Analytical Chemistry (Japanese edition), Kagaku Dojin, Kyoto, 1967, pp. 254
5. Hogfeldt, E., Stability Constants of Metal-Ion Complexes: Part A, Pergamon Press, Oxford, 1982, pp. 125

# SPECIFIC RESINS FOR METAL ION SEPARATION.
## THE Cr(III), Fe(III), Al(III) SYSTEM.

D.PETRUZZELLI, L.LIBERTI, R.PASSINO, G.TIRAVANTI.

Istituto Ricerca sulle Acque/Consiglio Nazionale delle Ricerche
5, Via De Blasio, 70123 Bari, Italy.
Istituto Chimica Applicata, Università di Bari.
200, Via Re David, 70125 Bari, Italy.

### ABSTRACT

Among established technologies for metal control in the environment it is worth mentioning selective ion exchange, membrane separation, and electrolysis.
Ion exchange, in particular, seems to offer additional advantages over environmental operation in terms of separation and reuse of valuable metallic by-products.
In reference to specific ion exchange applications, however, insufficient literature data are available and this is particularly true for some multivalent ionic systems, thus making the theoretical prediction of the general behaviour of resins towards reference systems difficult.

In this paper data are presented and discussed relative to the separation of Cr(III), Fe(III) & Al(III) by means of a process based on the use of selective reactive polymers. Starting from a complex mixture of the three metals and organic chelating agents, simulating the sulphuric acid extract of a typical tannery sludge, metal separation and recovery is achieved by means of a simple set-up. A weak base resin (Duolite A-7), as "specific" chelating agent for the Fe(III) complex, and a strong cation resin (Purolite C 160), for "selective" separation of Al(III) from Cr(III) complex, are efficiently used to the purpose.
Details for separation of ionic species based on their speciation in the liquid-phase are given.

## INTRODUCTION

Control technologies for inorganic waste management, in general, and heavy metals, in particular, are essentially based on "destructive" technologies such as precipitation-coprecipitation [1] coupled to pre- or post-oxidation [2], reduction [2,3] and concentration [4,5,6] of the species of interest. These operations convert the metals into inert forms with a minimal environmental impact, before final disposal in landfill [7,8].

Generally speaking it is predictable that environmental technologies would be based on "source-control" of pollutants (clean technologies) [1,9] rather than on "end-of-line" treatments (prevention is better than cure); in those cases in which there is no option, however, preferred end-of-line technologies should eventually be conservative toward pollutants allowing for recovery and reuse for both environmental and economic advantage of the operation.

It is known that tanneries are among those activities mostly blamed for their environmental impact: ≈50% of the incoming chemicals (essentially Cr(III) salts) are disposed of in the form of sludges from wastewater treatment operations. About 120,000 tons/yr (dry weight) of sludges (2-5% Chromium content) are generated in Italy, whose final destination is landfilling.

On the other hand the disposal operation referred to, turns to be increasingly impracticable with site saturation and a possible alternative would be represented by agricultural land application of sludges after detoxification.

Ion exchange as an established unit operation appears to be reliable for this purpose [4,10] allowing at the same time recovery of metal byproducts, to be eventually reused in the productive streams, and sludge detoxification for safe land disposal.

In this connection the paper discusses a process for chromium ion separation and recovery from aluminum and ferric ions resulting from a sulphuric acid extract of tannery sludges by ion exchange. Together with a process description, details on separation of ionic species based on their speciation in the liquid-phase are given.

## PROCESS DESCRIPTION

The general layout of the process includes the following steps (see Fig.1):
a) Sulphuric acid extraction of metals from sludges at pH 1;
b) Washing of extracted sludges;
c) Filtration for solid-liquid separation;

d) Chemical oxidation of ferrous ions;
e) Selective ion exchange for Cr(III), Al(III), Fe(III) separation;
f) Elution of chromium complex from ion exchange sections;
g) Recovery of Al and Fe from spent regeneration eluates;
h) Agricultural land application of detoxified sludges.

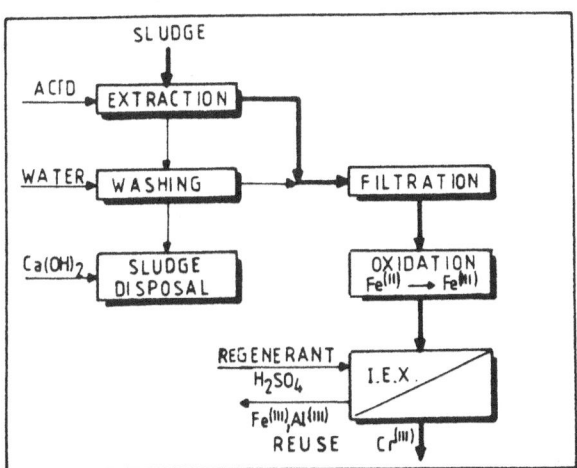

Fig.1. General flow-sheet of the process.

Fig.2 presents a conceptual scheme of the ion exchange section. The flowsheet is based on a sulphate form aminophenol weak-base resin (Duolite A-7), capable of selective chelation of ferric ions, and a sulphonic strong cation resin (Purolite C 160) for separation of Al(III) from Cr(III) ionic complexes. Trivalent chromium is quantitatively eluted from both ion exchange sections as a non-exchangeable sulphate complex.

Fig.2. Conceptual scheme of the ion exchange section for Fe(III), Al(III), Cr(III) separation and recovery from sulphuric acid extract of tannery sludge.

## MATERIALS AND METHODS

Tab. 1 reports the principal physico-chemical characteristics of the resins investigated. Before any experimental determination resins were sieved in the range 16-40 mesh and conditioned by three subsequent acid-base conversions ($H_2SO_4$/NaOH 1M, 5BV).

TABLE 1

Physico-Chemical characteristics of resins investigated.

| | RESIN | FUNCTIONAL GROUPS | MATRIX | CROSSLINKING DEGREE (% DVB) | POROSITY | PHYSICAL FORM | EXCHANGE CAPACITY ($eq/l_r$) |
|---|---|---|---|---|---|---|---|
| strong cationic | Amberlite IR 120 | $-SO_3H$ | Styrenic | 8 | gel | spherular | 1.9 |
| | Amberlite 252 | $-SO_3H$ | Styrenic | 15 | macro | spherular | 1.8 |
| | Amberlite 200 | $-SO_3H$ | Styrenic | 20 | macro | spherular | 1.7 |
| | Purolite C 100 | $-SO_3H$ | Styrenic | 10 | gel | spherular | 2.2 |
| | Purolite C 160 | $-SO_3H$ | Styrenic | 20 | macro | spherular | 2.3 |
| | Duolite C 3 | $-SO_3H$ | Phenolic | n/a | macro | granular | 1.2 |
| weak anionic | Duolite A 7 | $-NHR, -NH_2$ | Phenolic | n/a | macro | granular | 2.5 |

5 g ($\approx$20 cm$^3$) samples of conditioned resins in regenerated form ($H^+$ and $SO_4^-$ respectively for cationic and anionic) were loaded into a set of thermostatted (20+-3°C) laboratory columns (30 cm bed height) and submitted to a cycle including the following steps:
- complete exhaustion;
- displacement of the interstitial solution;
- full regeneration.

Preliminary experiments showed that the exchange kinetics were relatively slow. For this reason exhaustion was carried-out at flow-rates of 4BV/h using a synthetic solution reproducing the average composition of the sulphuric acid extract from tannery sludges (see Table 2 for average composition).

TABLE 2

Average composition of the synthetic solution simulating the acid extract from tannery sludge.

| | | |
|---|---|---|
| Cr(III) | 450 | mg/l |
| Fe(III) | 430 | " |
| Al(III) | 440 | " |
| Ca(II) | 220 | " |
| pH | 1 | |

Regeneration steps were carried-out at a flowrate of 2BV/h using 2N sulphuric acid solutions. After regeneration resin Duolite A-7 was partially hydrolysed by washing at pH 2 for best performances.
The influence of main variables in the liquid-phase such as solution pH, ionic strength, and anionic background were evaluated in order to understand resin performances in terms of ionic speciation. Analytical determinations on total metals were made by flame atomic absorption spectroscopy using a Mod.951 system from Instrumentation Laboratory, Wilmington, MA, USA. An ion chromatograph Mod.4000i from Dionex, Sunnyvale, CA,USA was used for the theoretical metal ion speciation in the liquid-phase.
Computer program MINEQL from Massachusetts Institute of Technology [11] was used for theoretical determination of metal ionic species in the liquid-phase.
Basic chromic sulphate from Carlo Erba, Milan, Italy, was used to simulate the tanning bath. Freshly prepared solutions using reagent grade chemicals (sulphate salts) were used throughout the experiments.

## RESULTS AND DISCUSSION

a) Selective removal of Fe(III) on aminophenol anion resin.
It is reported [12,13] that aminophenol anion resins in sulphate form were effective in ferric and cupric ion selective removal from aqueous solutions. Based on reference experiences commercial weak base aminophenol resin Duolite A-7, regenerated in $SO_4$-form, (see Table 1) was tested for this purpose. In Fig.3 are reported breakthrough curves for ferric ion removal for experiments carried-out at pH 1 and 2.

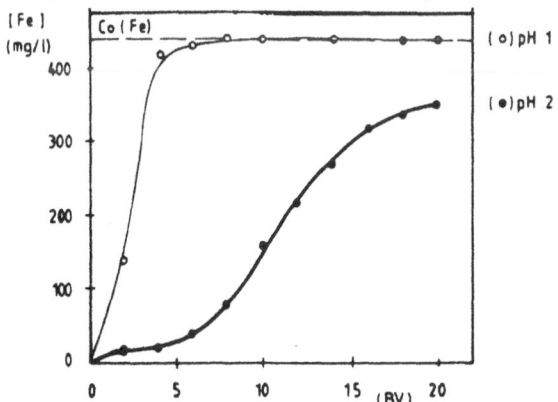

Fig.3. Breakthrough curves of Fe(III) complex on resin Duolite A-7 at different pH.

As reported [12], and confirmed by our experiments, the pH of
the influent solution has a critically controlling effect on
the behaviour of the resin, with best performances in the range
1.8-2.0.
These findings are interpreted in terms of the following two
factors:
- ionic speciation of ferric species in the liquid-phase;
- ligand availability for metal chelation.
As shown in Fig.4 at pH values substantially lower than 1.5 the
predominant ferric species in sulphate solutions is represented
by the cationic complex [FeSO$_4$]$^+$ which is definitely excluded
by the Donnan effect from the resin-phase [14], being the tertiary
amino functional groups of the resin essentially in the
protonated salt form.

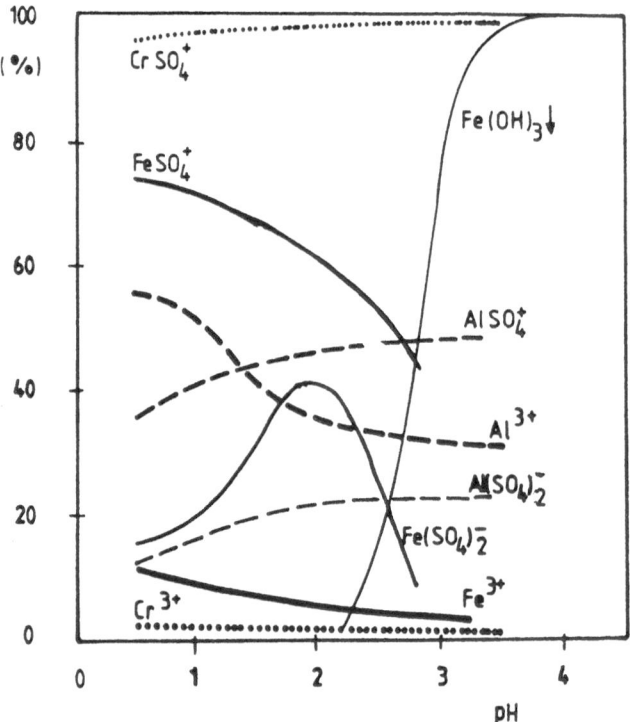

Fig.4. Theoretical distribution diagram of ionic species
for the system in consideration.

Protonated quaternary ammonium nitrogen groups, on the other hand
have no electron pairs available for donation to metals for
chelation.    Both these factors contribute to the poor resin
performance of the resin at low pH conditions.
At pH 2 the steady decrease  of the cationic complex [FeSO$_4$]$^+$

(excluded by the resin) is accompanied by a maximum peak for the anion complex $[Fe(SO_4)_2]^-$, which freely diffuses into the solid-phase, allowing for best performances of the resin toward ferric ion complex retention. Beginning at pH values higher than 2.3 the precipitation of $Fe(OH)_3$ occurs, as well as that of the other metal hydroxides.

Fig. 5 reports general performances (exhaustion-regeneration) of sulfate-form resin Duolite A-7 from the solution simulating the acid extract reported in Table 2.

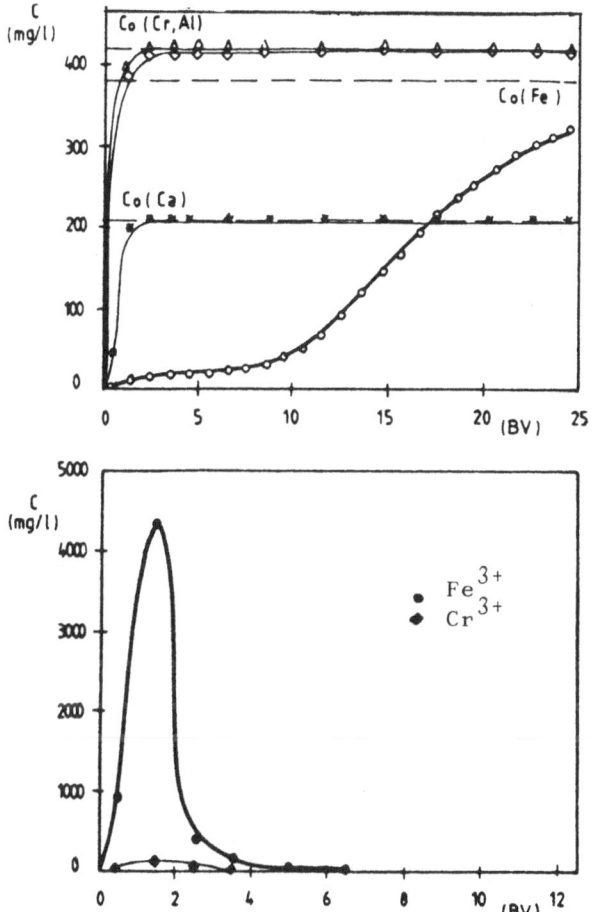

Fig.5. Exhaustion-regeneration performances
of resin Duolite A-7 ($SO_4$-form).

Chromium, aluminium and calcium ionic species are readily eluted from the resin whereas ferric complex is retained. It was confirmed experimentally that, as expected [12], ferrous ions are not retained by the resin.

Once again breakthrough behaviour of all other metals can be interpreted in terms of ionic speciation in the liquid-phase. From Fig. 4 it is clearly evident that both chromium and aluminium species are present in the form of cationic complexes ( $[CrSO_4]^+$ ≥97% and $[AlSO_4]^+ + Al^{3+}$ ≥80%) and are thus excluded from the solid phase by the Donnan effect.

The (uncomplexed) cation of calcium is similarly excluded.

b) Selective separation of Al(III) and Cr(III) ionic complexes by strong cation sulphonic resin.

Experiments have been carried out on the set of sulphonic cation resins reported in Table 1. The resins differ in total exchange capacity, degree of crosslinking' porosity of the resin matrix, and the nature of polymeric matrix itself. General behaviour of all resins toward the system under discussion was essentially equivalent, thus confirming the slight influence of resin characteristics on the performances. Therefore, the investigation was directed toward evaluation of the influence of the liquid-phase characteristics (essentially ionic speciation). Accordingly experiments at different values of solution pH, relative concentration of metals, and ionic strength, (all capable of influencing the stability of ionic complexes in the liquid-phase) were carried out. Results relative to resin Purolite C 160 are reported here, since they were representative of those obtained on the set of resins tested.

In Fig.6 are reported breakthrough curves for chromium and aluminium ions for tests at different solution pH, e.g., 0.5; 1.0; 2.0; 2.5. Higher pH values were ruled out because of metal hydroxide precipitation.

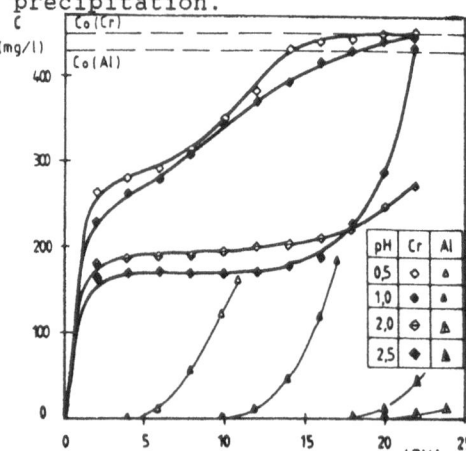

Fig.6. Breakthrough curves for the Cr(III)/Al(III) system on resin Purolite C 160 at different pH.

The strong influence of ionic speciation on resin performances
is clearly shown by the separation of the mixture of the metal
ions.

Based on strict thermodynamic considerations, for pH values
lower than 1.0, where predominant cationic species in solution
are in the order (see Fig.4) :

$[CrSO_4]^+$ ($\approx$98%); $[FeSO_4]^+$ ($\approx$75%); $[AlSO_4]^+$ ($\approx$40%); $Al^{3+}$ ($\approx$50%)
in spite of the lower relative fraction of $Al^{3+}$ ions with respect to
other cationic species, the high charge-density trivalent
aluminium cation should be preferred by the resin because of
the electroselectivity effect [15]. This was found to be the case as
confirmed by the shape of breakthrough curves reported in Fig.7:

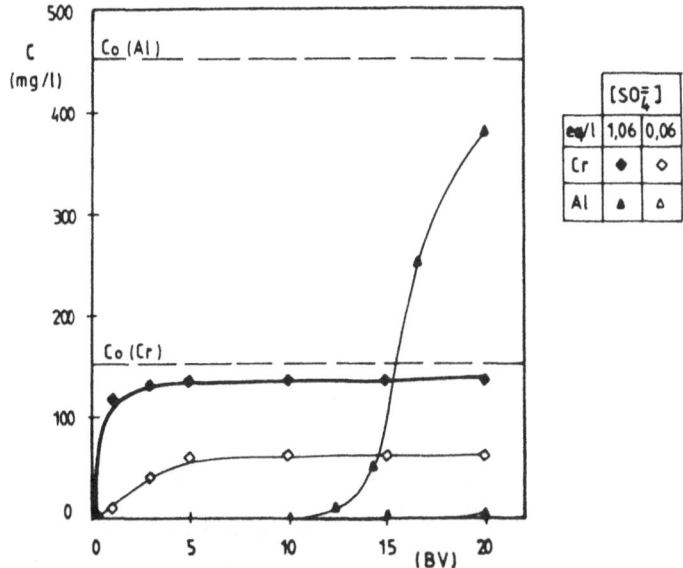

Fig.7. Breakthrough curves for the Cr(III)/Al(III)
system on resin Purolite C 160 at different
sulphate concentrations.

Trivalent aluminium ion is, in fact, selectively retained,
chromium ionic complexes being excluded.
For pH values higher than 1.0 (see Fig.4) a significant
reduction of the $Al^{3+}$ fraction in favor of $[AlSO_4]^+$ complex
occurs; this leads to a corresponding reduction of the
electroselectivity effect favoring aluminium retention: the end
result is (see Fig.7) an increasing selectivity for chromic
species over aluminium on the resin.
The overall aluminium retention, in terms of resin capacity,
although less selective, is increased, however, due to the fact
that now two aluminium cationic species contribute to the
exchange process.

To substantiate the influence of accompanying anion concentration on metal speciation, results of experiments carried-out at different sulphate concentration in the liquid-phase keeping constant the pH and concentrations of all other species are also shown in Fig. 7. At low sulphate concentration, e.g., 0.06 N, a significantly higher chromium retention is observed as compared to experiments carried out at higher (≈1.0N) sulphate concentration. Speciation calculations relating to both high and low sulphate concentration in fact indicate a higher fraction of free $Cr^{3+}$ (40% vs. ≤ 5%).

Regeneration of the resin, using 2N sulphuric acid solutions, led to quite pure aluminium solutions with less than 5% of chromium contamination. Accordingly, chromium species pass through both resin sections, allowing quantitative recovery in the effluent solution.

## CONCLUSIONS

Ion exchange technology for metal removal and recovery from wastewaters of different origin is nowadays sufficiently reliable. Nevertheless in spite of the recognized potential ion exchange has still not reached a stage of complete maturity compared with alternative control technologies.

Results reported above are an example of potential use. A set-up made of a commercial $SO_4$-form aminophenol resin, and a conventional strong cation sulphonic resin can be efficiently used for specific ferric ion retention and selective chromium-aluminium separation respectively.

The process is now undergoing experimentation on a bigger scale using a laboratory pilot plant. In this context particular attention will be devoted to the influence of organic substances present in the acid extract as potential fouling agents for the resin, to different kinetic aspects of metal complex formation in the liquid-phase, as well as to resin retention kinetics themselves. Based on the considerations reported here, the process might well be amended; this will be the object of further investigation.

## Ackowledgements

Authors are indebted to the Italian Tannery Association, Milan, for their cooperation.

John R. Millar, M.A., consultant of Istituto di Ricerca sulle Acque, is gratefully acknowledged for his helpful advice throughout the experimental work.

Messers N.Limoni, R.Ciannarella, G.Bagnuolo are acknowledged for their fruitful technical assistance.

## REFERENCES

1. Patterson, J.W., Metals separation and recovery. In Metal Speciation Separation and Recovery, ed. J.W.Patterson, R.Passino, Lewis Publishers, Chelsea, MI, 1987.
2. Nemerow, N.L., Theories and practices of industrial waste management, Addison Wesley Publishing Co.,Reading, MA, 1963.
3. Eckenfelder, W.W.,Jr, Industrial Waste Pollution Control, McGraw-Hill Publishing Co., New York, NY, 1966.
4. Dillman, T.R., Metal ion removal and recovery. In Ion exchange for pollution control, ed.C.Calmon, C.R.C. Press, Boca Raton, FL, 1979.
5. Gold, H., Metal Finishing wastes, in Ion exchange for pollution control, ed. C.Calmon, C.R.C. Press, Boca Raton, FL, 1979.
6. Clifford, D., Subramanian, S., Song, T, Recovery of dissolved inorganic contaminants from wastes, Environ.Sci.Technol. 1986, 20,11,1072.
7. Griffin, F.A., Frost R.R., Au, A.K., Robinson, G.D., Shimp, N.F., Attenuation of pollutants in municipal landfill leachate by clay minerals. Part II, Heavy metal adsorption. Environmental Geology Notes, April 1977, n°79.
8. Hoeks, J., Pollution of soil and groundwater from land disposal of solid-wastes, Inst.Land Water Management Research Wageningen, The Netherlands, Tech.Bull.n°96, 1976.
9. Ling, J.T., Low- or non-pollution technology through pollution prevention. An Overview. 3M Company Report to the United Nations Environmental Programme, Office of Industry and Environment, 1982.
10. Anderson, R.E., Toxic metal decontamination: Does ion exchange fit?, Amber Hi Lites n°181, Rohm&Haas Co 1987.
11. Westall, J.C., Zachary, J.L., Morel, F.M.M., MINEQL Computer program, Tech.Note no.18, Dept.Civil Eng. Massachussets Institute of Technology.
12. Hodgkin, J.H., Eibl, R., Ferric ion chelation by aminophenol resins, React.Polym. 1986,4,285.
13. Chanda, M., Driscoll, K.F., Rempel, G.L., Polybenzilimidazole based resins new chelating agents. Cupric ion selectivity of resins. React.Polym. 1988,9,277.
14. Helfferich, F.G., Ion Exchange, McGraw-Hill Pub.Co. New York, NY, 1962, p.135.
15. Helfferich, F.G., Ion Exchange, McGraw-Hill Pub.Co. New York, NY, 1962 p.156.

# IMPROVEMENTS IN MACRORETICULAR PARTICLES FOR ION-EXCHANGE AND GEL-PERMEATION CHROMATOGRAPHY

J. S. WATSON
Chemical Technology Division
Oak Ridge National Laboratory
P.O. Box 2008
Oak Ridge, Tennessee 37831-6046, USA

## ABSTRACT

The effectiveness of chromatographic processes is often limited by the resistance to mass transfer into the exchanger particles. The likelihood of the exchange process being limited by particle diffusion resistance is especially high when one or both of the exchanging ions has a high molecular weight, and thus large molecular size and low mobility within the particles. Separation of ionic species from bioreactors is likely to involve such high molecular weight species and slow diffusion rates. Macroreticular particles are more effective in these applications because the larger species can diffuse more freely into the inner regions of the particles through large pores. The effects of small viscous flow rates through the larger pores of the macroreticular particle structure are examined in this paper to show how such flow can increase exchange rates. The importance of such flow is evaluated in terms of an effective diffusivity within the particle. The performance of an ion-exchange material can be affected strongly by the dimensions of its internal structure. The approach used to describe and evaluate the performance of ion-exchange materials with high-molecular-weight ions can be applied to macroreticular particles for separation of biological species and polymers by gel permeation.

## INTRODUCTION

Macroreticular ion-exchange resins and adsorbents that contain two or more principal pore sizes have received attention in recent years because of their improved diffusion kinetics relative to conventional homogeneous resins. The larger pore structure in macroreticular resins provides relatively open structures through which the exchanging ions (or adsorbing molecules) can diffuse into the inner regions of the particles. This becomes especially important when large molecules are involved, such as those important in biotechnology. Small pore sizes can seriously hinder the diffusion of larger molecules that are not much smaller than the pore dimensions. A macroreticular resin provides more rapid diffusion of the ions into the particles, and the path lengths for diffusion into the smaller pores can be much less than the particle radius. This is the conventional view of macroreticular resins and adsorbents [1-4], and it has worked well in describing many ion-exchange systems.

This paper, however, takes a slightly different view of the performance and design of macroreticular resins or adsorbents for liquid systems, such as ion exchange. The larger void regions in macroreticular materials may also be able to transmit a small, but significant, flow of the liquid through the particles. This viscous flow through the particles, even if small, could have significant effects on the performance of the resin and even provide an additional mechanism for transport of exchanging ions (or solutes) into the resin particles.

The combined effects of viscous flow and diffusion through homogeneous material, such as very porous gel particles, have been estimated in earlier work by McKenzie [5] and by McKenzie and Watson [6]. Flow through some gel materials appears to play a significant role in gel-permeation chromatography.

This paper addresses the slightly different problem found in macroreticular particles in which viscous flow through the particles occurs in only one of the two (or more) sizes of pores. A very recent paper [7] has shown that macroreticular materials constructed as clusters of smaller particles demonstrate remarkable adsorption rates for large biologically important molecules. Those results could be particularly important and affect the development of numerous ion-exchange resins and adsorbents in the future.

## DESCRIPTION OF THE MODEL

The physical model for considering viscous flow through macroreticular particles is illustrated in Figure 1(a). The macroreticular particles are considered to consist of a cluster of smaller particles that are attached. Although this is only one possible way to build a macroreticular material with two very different pore sizes, it will be sufficient to illustrate the phenomena to be discussed. For instance, the resin could consist of large tubular-shaped pores surrounded by a media with much smaller pores. In a randomly packed bed, the flow pattern through each particle will be different and often complex. However, the average flow through the particles will be simple vertical (axial) flow because the average pressure gradient in the column is always in the vertical (axial) direction for downflow. Clearly, there will be local pressure gradients that cause complex flow patterns in local particles, but the average flow pattern will be simple, as shown in Figure 1(a). The flow rate within the particle can be, and probably will be, much smaller than that outside the particle, but even small flow rates can be important contributions to the rate of mass transfer into and out of the larger particle.

The case where radial diffusion and vertical viscous flow are both contributing to the rate of mass transfer into homogeneous particles has been discussed by McKenzie [5] and by McKenzie and Watson [6]. This paper addresses the case where ion exchange occurs within the inner small subparticles, and vertical viscous flow through the macropores is much more important than diffusion. In this case, the mass transfer rate into the particle can be viewed as the average of several small vertical columns of different lengths that are assembled to approximate the spherical shape of the macroreticular resin particle, illustrated in Figure 1(b). The loading ions will appear first along the upper surface of the particle (for downflow), that is, in the upper portion of each of the imaginary columns that make up the overall particle. The loading front will then move down each column or each portion of the overall particle at the same rate. Breakthrough will occur first on the outer regions of the particle and last in the lower regions along the center line. By summing the rate at which ion exchange occurs in each of these small imaginary columns, the rate of ion exchange within such a particle under a sharp loading front can be calculated. This exchange rate can then be compared with the rate expected under radial diffusion, and the result can be expressed in terms of an effective (or apparent) diffusion coefficient for the overall particle. It is convenient to express the results predicted by the model in terms of an effective diffusion coefficient in the particle because that is the parameter most often used to describe mass transfer rates within ion-exchange media.

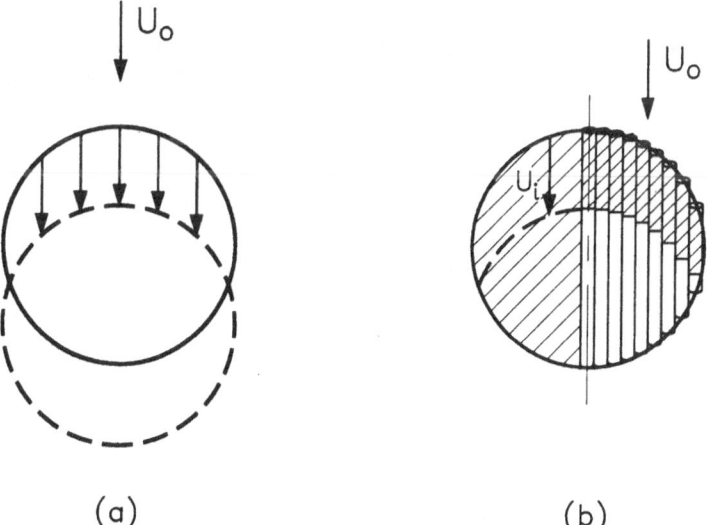

(a)                              (b)

Figure 1.  The model for movement of a loading front
through a porous resin particle.

**Rapid Kinetics within the Subparticles**

This is the simplest case discussed in this paper, but it best illustrates the possible role of viscous flow in macroreticular particles. With a "favorable" isotherm (negative curvature), the loading front will be sharp and will move down the particle at a constant rate:

$$U_L = U_p/K ,\qquad (1)$$

where

$U_L$ = the rate at which the front moves through the particle,

$U_p$ = the superficial velocity of fluid in the pores,

$K$ = equilibrium loading of the ion of interest in a unit volume of the larger particle, including the ions in the solid subparticles and the liquid, divided by the concentration in the liquid.

This definition of K is not used universally, but is a convenient form. Equation (1) could be written in terms of the conventional distribution coefficient, $K_d$, based on unit volumes of resin and solution. Note that the regions (i.e., the imaginary columns) near the edge (the equator) of the particle will become saturated first, because the length of the band path is shortest, while the last part of the particle to become loaded will be that region near the bottom of the particle, along with the axial center line, where the length that the front must travel is the longest.

The loading rate in such cases can be easily predicted from geometrical considerations. Consider the loading front to be simply a spherical shell that progresses down through the particle at the front velocity, $U_L$, and the region of the particle above the shell is the region loaded. Unlike the case where the particle loads by diffusion controlled processes, the particle will become completely loaded in a finite time; that is, when

$$t_{final} = 2R/U_L ,\qquad (2)$$

with R being the radius of the particle.

The fraction of the particle that is loaded at an intermediate time, t, is

$$Q/Q_{final} = (3/4)(U/K)[(t/R) - (1/12)(U/K)^2(t/R)^3]$$

$$= (3/4)U_L[(t/R) - (1/12)U_L^2(t/R)^3] .\qquad (3)$$

The equation can be normalized by defining a dimensionless time

$$T = (U_L/R)t .\qquad (4)$$

Then

$$Q/Q_{final} = (3/4)[T - (1/12)T^3] .$$ (5)

The particle will be one-half loaded when $Q/Q_{final} = 0.5$.

$$0.5 = (3/4)[T_{1/2} - (1/12)T_{1/2}^3]$$

or

$$T_{1/2} = 0.695 .$$ (6)

The loading curve for resin loading controlled by viscous flow with rapid kinetics (mass transfer within the subparticles) and a favorable isotherm is shown in Figure 2.

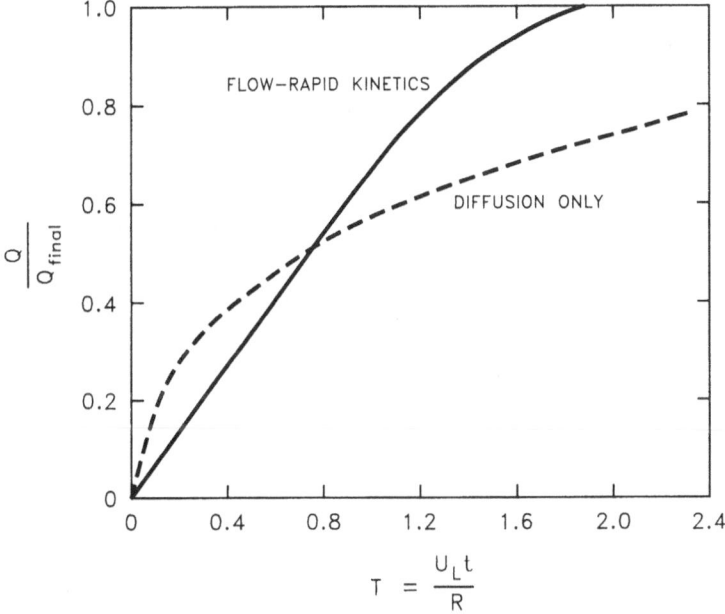

Figure 2. Loading rates with flow through a porous resin and comparison with diffusion loading rates.

### Comparison with Diffusion-Controlled Loading

Available and commonly used analyses of ion-exchange or adsorption processes are based on more conventional kinetic processes, usually on diffusion rates in the particle and/or through the liquid film surrounding the particle. To use such analyses, it is necessary to relate the behavior from this model for macroreticular resins to the terms used in conventional column analyses. Of course, the comparison is most appropriately made with diffusion within the particle, the most similar mechanism usually suggested by conventional analyses. Cases in which diffusion resistance within the resin particles is the controlling or the major resistance are most interesting because these are the cases where flow through the particle can be more helpful. Film resistance would be affected if the flux of fluid into the particle is sufficient to affect the film thickness. In most cases where the proposed mechanism is likely to apply, the liquid film on the leading edge of the particle is expected to be reduced by the flow into the particle and become less important.

The loading curve for flow through the particles shown in Figure 2 differs from that expected with diffusion-controlled loading, particularly at high loadings, or long times. Since the viscous flow model predicts complete loading after a finite time and the diffusion model predicts an infinite time for complete loading, the results will be most different at high loadings. Near zero loading, the diffusion model (without film resistance) predicts infinite loading rates, and the relative difference between the models also will be large. However, over a considerable portion of the curve, an approximate comparison can be made between the diffusion and viscous flow models. For this comparison, the two curves are "fit" at the 50% loading point, and the "effective" diffusion coefficient required to match the viscous flow model at that loading is evaluated. At 50% loading for the viscous flow model,

$$(U_p/K)(t_{1/2}/R) = 0.695 \ . \tag{7}$$

For the diffusion model,

$$D_{eff}t_{1/2}/R^2 = 0.0305 \ . \tag{8}$$

Equating the times required for 50% loading and solving for $D_{eff}$,

$$D_{eff} = 0.0439 \quad U_p R/K = 0.0439 \ U_L R \ . \tag{9}$$

### Flow Rate through the Particle

The structure of macroreticular resins can differ considerably, but we will consider the case in which the particle is assumed to be composed of a cluster of smaller subparticles packed in the same manner as a randomly packed bed. In this case, well-known equations for flow through packed beds can be used. Perhaps the most frequently used equation is the Ergun relation:

$$dP/dz = [150\mu U_p/(2r)^2](1 - e)^2/e^3 \, , \tag{10}$$

where r is the radius of the small subparticles, $\mu$ is the viscosity of the liquid, and e is the void fraction. This equation can be used for essentially any macroreticular ion-exchange particle because flow through the particles will be very slow and thus in the viscous region. The equation was developed originally by approximating the flow through the bed as a series of cylindrical channels, and it is likely to be reasonably accurate for most geometries of uniform-size macropores. The smaller pores of the macroreticular resin are not considered large enough for significant flow.

If the flow down the packed bed of macroreticular particles is also in the viscous region, the same equation can be used to predict the pressure drop and flow rate relationship. Substituting $e_o$ for the void fraction in the packed bed of larger particles, the equation would then read

$$dP/dz = [150\mu U_o/(2R)^2](1 - e_o)^2/e_o^3 \, , \tag{11}$$

where $U_o$ is the superficial velocity through the overall packed bed. The pressure drop through the bed and, thus, through the macroreticular particles is the same. If the void fractions are the same,

$$U/U_o = (r/R)^2 = (d/d_o)^2 \, , \tag{12}$$

where d and $d_o$ are diameters of the subparticles and larger resin particles, respectively. In general, the void fractions in the packed bed and within the macroreticular resin will not be the same. In those cases, an additional term must be used:

$$U/U_o = (r/R)^2 \, C_r \, , \tag{13}$$

where

$$C_r = [(1 - e_o)^2/e_o^3]/[(1 - e)^2/e^3] \, . \tag{14}$$

Then, for viscous flow in the packed bed as well as through the particles,

$$D_{eff} = 0.0439 \, C_r U_o r^2/(KR) \, . \tag{15}$$

## EFFECTS OF FLOW THROUGH PARTICLES
## WITH VERY SLOW KINETICS

The preceding discussion indicates that the performance of a macroreticular resin constructed as a cluster of smaller particles improves as the size of the smaller particles is increased. This occurs because the factor limiting the loading rate is flow through the particle, not the kinetics or diffusion through the smaller subparticles. Those assumptions will not be valid if the size of the subparticles is increased too much; eventually the kinetics (probably diffusion into the subparticles) will control the loading process. Optimization of the performance of macroreticular particles can be complex in the most general case; next consider the opposite case where the kinetics (usually diffusion into the smaller subparticles) is very slow. In this case, the loading front is extremely diffuse and all subparticles will see essentially the same solution concentration at all times; that is, the loading rate in all subparticles is essentially the same. The loading curve will then have a shape exactly like one would expect for diffusion into the larger overall sphere, but the rate will be much faster:

$$D_{eff}/D_s = (R/r)^2 , \qquad (16)$$

where $D_s$ is the real diffusion coefficient within the smaller subparticles.

Note that under conditions controlled by diffusion into the small subparticles, the effective diffusivity is independent of flow rate and <u>increases</u> with <u>decreasing</u> size of the subparticles. The independence of the effective diffusion coefficient of liquid flow rate offers an experimental test to determine when this limited case applies.

Of course, the lack of dependence on flow rate can also occur for particles where there is no flow through the particles and the effective diffusion coefficient measured is simply the diffusion coefficient within the larger particle.

## COMMENTS ON THE INTERMEDIATE CASES
## WITH FINITE KINETICS

It is difficult to make generalizations about the intermediate cases where diffusion into the small subparticles is slow enough to affect the loading rate but not slow enough to control the loading/exchange rate. Reliable predictions of the behavior in those cases require information about the equilibria isotherms as well as about diffusion into the smaller subparticles. Detailed discussions of such cases will be left to later papers that can address specific systems. The discussion for this paper will be limited to general comments.

By continuing to use the model illustrated in Figure 1(b), in which the particles are approximated by a series of small imaginary columns with the lengths corresponding to lengths of the different flow paths through the particle, one can begin with conventional transit analyses of columns to predict breakthrough curves and sum the breakthrough curves from the imaginary columns to estimate the performance of the particle. Such analyses require that the equilibration loading/exchange isotherm and the kinetics of diffusion within the small subparticle be specified. The analyses most frequently quoted (and probably most frequently used) are for linear isotherms that will not apply to many ion-exchange applications. The most serious difficulty with the use of conventional column analyses is the usual assumption that the column is "long," and the loading front has developed a constant pattern. This is not likely to be the case for many of the flow paths within a macroreticular resin particle, which makes the mathematics more involved since numerous "short" paths (imaginary columns shown in Figure 2) must be evaluated to describe the behavior of a macroreticular particle with internal viscous flow. The numerical calculations required are significant and will be more easily justified for narrow ranges of conditions of specific experiments than for a generalized discussion in this paper.

The short paths were less serious conditions for the two limiting cases discussed quantitatively. In the first case, of rapid kinetics, the loading front entered the particle as a sharp front, and rapid kinetics (with a favorable isotherm) would not let the front spread. Actually, strongly favorable isotherms will help keep the front sharp so that this limiting case can apply over a somewhat wider range of kinetic parameters (diffusion coefficients in the subparticles). In the second case, very slow kinetics, the breakthrough occurred almost instantly as the breakthrough front was very diffuse (almost flat). The shorter path length simply makes this behavior more likely.

## APPLICATIONS TO GASEOUS SYSTEMS

This discussion of the role of viscous flow through ion-exchange or adsorbent particles was limited to liquid systems, and the considerations presented should be limited to those cases. These considerations will not apply to many gas-solid adsorption cases, even if significant "flow" occurs through the particles. If the macropores through the particles are sufficiently small, any flow through those pores will be by molecular (Knudsen) flow, which is driven by the gradient in the partial pressure of each component of the gas. This is essentially the same as diffusion of each gas through the pores, and conventional analyses based on diffusion processes alone should be sufficient.

## SUMMARY AND CONCLUSIONS

Flow can occur through macroporous ion-exchange resins and can have a significant effect on mass transfer rates (effective diffusion coefficients). For the flow rate to be important, it will usually be necessary for the pores through the resin to be relatively large. If the pores are created by forming the macroreticular resin particle from clusters of smaller subparticles, the flow rate through the particles will be enhanced by increasing the size of the smaller subparticles used to form the larger overall particle. Viscous flow through the macropores will be more important than diffusion when

$$U_p/KR >> D_p/R^2 \,, \tag{17}$$

where $D_p$ is the diffusion coefficient in the macropores. For ion-exchange processes, $D_p$ can generally be assumed to be significantly lower than the diffusion coefficient in water. Optimization of a macroreticular resin particle depends on the particular chemical system to be used. Simple cases are presented for predicting the performance for systems under limiting conditions. Generalized predictions of resin performance and optimization can be made for specific systems.

## ACKNOWLEDGMENTS

The Oak Ridge National Laboratory is operated by Martin Marietta Energy Systems, Inc., under contract DE-AC05-84OR21400 with the U.S. Department of Energy.

## REFERENCES

1.  Ruckenstein, E., Vaidyanathan, A.S. and Yongquist, G.R., Sorption by solids with bidisperse pore structures. Chem. Engr. Sci., 1971, 26, 1305-1318.

2.  Ma, Y.H. and Lee, T.Y., Transient diffusion in solids with a bipore distribution. AIChE J., 1976, 22, 147.

3.  Lee, L.-K., The kinetics of sorption in a biporous adsorbent particle. AIChE J., 1978, 24, 531.

4.  Niederjaufner, G. and Pontoglio, A., Study of a separation process through adsorption of molecular sieves: application to a chlorotoluene isomers mixture. Chem. Engr. Sci., 1984, 39, 383-393.

5.  McKenzie, P.E., M.S. thesis (in preparation).

6.  McKenzie, P.E. and Watson, J.S. (paper in preparation).

7.  Regnier, F.E. and Afeyan, N.B., High throughput chromatographic packing for protein purification. Paper presented at the 1989 Annual Meeting of the American Institute of Chemical Engineers, San Francisco, California, Nov. 5-10, 1989.

# SOME PROPERTIES OF NEW ION EXCHANGERS WITH ORGANOPOLYSILOXANE STRUCTURE

FENGMING CHEN, GERARD COTE and DENISE BAUER
Laboratoire de Chimie Analytique (Unité associée au CNRS n° 437)
E.S.P.C.I., 10, rue Vauquelin, 75005 Paris, France

## ABSTRACT

New ion exchangers have been recently synthesized at DEGUSSA by the polycondensation of suitable bifunctional silane monomers. Such ion exchangers can be obtained as spherical or egg-shaped particles with a grain size of about 0.1 to 1.5 mm and surface areas up to 700 m2 / g. They are composed of a siliceous core and contain a high concentration of functional groups so that their exchange capacity is comparable to that of conventional organic systems. The functional groups are bound to the matrix via carbon-silicon bonds which are fairly stable towards hydrolysis (up to pH = 11 or 12). The main advantages of the new exchangers lie in the absence of swelling or shrinking during ion exchange operations, in their high thermal stability (up to 200 °C) and in the great variety of functional groups now available. The ion exchange properties of these exchangers have been investigated by studying the fixation of Ag(I) and Cu(I), As(III) and Sb(III), Bi(III), Cu(II) and Ni(II), Ga(III), Pd(II) and Pt(IV).

## INTRODUCTION

Silica surfaces modified by organofunctional silanes find numerous applications as stationary phases for chromatography [1,2]. Organofunctionalized silanes are a relatively new class of compounds. They were first prepared in the early 1940s, after the discovery of the Mueller-Rochow-process and the hydrosilylation reaction [3]. Silanized inorganic materials have a number of advantages compared to organic polymers. Nevertheless, their properties as supports for immobilized functional groups and active compounds are not optimal. Because of the relatively low concentration of surface hydroxy groups, on average only 1 to 1.5 of three possible bonds are formed between the organofunctionalized silane and the support. As a consequence, the concentration of the functional groups is relatively low and the silane to surface bond is labile, especially in alkaline aqueous media.

In view of this DEGUSSA has developed a new concept which provides an interesting alternative . Polycondensation of suitably bifunctional silane monomers leads to solids with a siliceous core and a high concentration of functional groups, which, moreover, are bound to the matrix via carbon-silicon bonds that are fairly stable towards hydrolysis [3]. In the present paper, we report the properties of a series of 14 ion exchangers which have been prepared according to such a method. The corresponding functional groups or combinations of functional groups are : -SH;    -S-;    -SH + -S-;    -NH- + -S-;    -NH-[CH$_2$]$_2$-NH$_2$; -N=CH-C$_5$H$_4$N;   -N=CH-C$_6$H$_4$OH;   -NH-C[=S]-NH-;   -NH-C[=S]-N-[CH$_2$]$_2$-N[CH$_3$]$_2$; -NH-C[=S]-N-C[=O]-C$_6$H$_5$;  -NH-C[=S]-S-;  -P[C$_6$H$_5$]-;  P[C$_6$H$_5$]$_2$-  and -SO$_3$H. It is of interest to note that some functional groups such as the phosphines       are rather unusual in ion exchange. The ion exchange properties of these exchangers have been investigated by studying the fixation of a series of metal species representing the main families of metal ions, namely Ag(I) and Cu(I), As(III) and Sb(III), Bi(III), Cu(II) and Ni(II), Ga(III), Pd(II) and Pt(IV).

## MATERIALS AND METHODS

### Reagents
The ion exchangers were kindly supplied by DEGUSSA Company and used as received. Aqueous gallium(III) solutions were prepared from gallium oxide (Rhône Poulenc). The other reagents from various suppliers were all analytical grade.

### Analytical
Microanalyses were performed by the Service Central d'Analyse (CNRS, Vernaison, France). Metal species titrations in aqueous phases were carried out either by atomic absorption spectrophotometry with a Video 11 Instrument Laboratory spectrophotometer or by ICP with an ICP 1500 Plasma-Therm instrument.

### Sorption experiments
Measured amounts (typically 0.2 or 0.5 g of wet exch.) of each exchanger were vigorously shaken with definite volumes (10 or 20 mL) of metal solutions of known concentrations for 24 hours at 20 ± 2 °C. The equilibrium sorption was calculated from the residual metal concentration of the sorbate in the equilibrated solution.

## RESULTS AND DISCUSSION

### General properties
The process developed by DEGUSSA Company allows the preparation of spherical or ovoidal particles with grain sizes of about 0.1 - 1.5 mm (mean value close to 0.6 mm) and surface areas up to 700 m2 / g [3]. The main  advantages of the new exchangers lie in the absence of swelling or shrinking during ion exchange operations, in their high thermal       stability (up to 200 °C) and in the great variety of functional groups now available. Their density is always greater than one.

The elemental composition and water regain of the various ion exchangers studied in the present work are reported in table 1. Examination of this table shows that the number of functional groups varies  between  1.2 x 10$^{-3}$ and  4.0 x 10$^{-3}$ mol / g of dry exchanger or between 0.5 x 10$^{-3}$ and 1.9 x 10$^{-3}$ mol / g of wet exchanger. It can also be noticed that all exchangers the functional groups ·  which contain both sulphur and nitrogen atoms (i.e. exchangers 8 to 11)   have   a deficiency of sulphur.

In figure 1, the percentage of dissolution of several exchangers has been plotted as a function of pH. Examination of this figure shows that the ion exchangers developed by DEGUSSA Company are particularly  resistant to alkaline hydrolysis in spite of their siliceous core. Moreover, it has been observed that most of these exchangers are fairly stable in highly acidic media (e.g. 6 mol / L HCl) so that they can be used over a large range of acidities       i.e. from [HCl] > 6 mol / L up to about pH 11 or 12.

TABLE 1

Nature and elemental composition of the various ion exchangers

| Exchanger no. | Nature of the functional groups | Elements (% wt, dry exch.)* (mole/g dry exch.)* | | Number of functional groups (mole/g dry exch.)* (mole/g wet exch.) | Water regain (% water in wet exch.) |
|---|---|---|---|---|---|
| 1 | -SH | **S** 6.57 % 2.05 x 10$^{-3}$ | | 2.0 x 10$^{-3}$ 0.8 x 10$^{-3}$ | 61 |
| 2 | -S- | **S** 10.4 % 3.24 x 10$^{-3}$ | | 3.2 x 10$^{-3}$ 1.3 x 10$^{-3}$ | 61 |
| 3 | -SH + -S- | **S** 11.9 % 3.71 x 10$^{-3}$ | | 3.7 x 10$^{-3}$ 1.3 x 10$^{-3}$ | 65 |
| 4 | -NH- + -S- | **N** 3.46 % 2.47 x 10$^{-3}$ | **S** 4.96 % 1.55 x 10$^{-3}$ | 4.0 x 10$^{-3}$ 1.2 x 10$^{-3}$ | 71 |
| 5 | -NH-[CH$_2$]$_2$-NH$_2$ | **N** 7.01 % 5.00 x 10$^{-3}$ | | 2.5 x 10$^{-3}$ 0.9 x 10$^{-3}$ | 65 |
| 6 | -N=CH-C$_5$H$_4$N | **N** 6.04 % 4.31 x 10$^{-3}$ | | 2.2 x 10-3 1.4 x 10-3 | 34 |
| 7 | -N=CH-C$_6$H$_4$OH | **N** 4.64 % 3.31 x 10$^{-3}$ | | 3.3 x 10$^{-3}$ 1.9 x 10$^{-3}$ | 42 |
| 8 | -NH-C[=S]-NH- | **N** 8.30 % 5.92 x 10$^{-3}$ | **S** 6.76 % 1.88 x 10$^{-3}$ | 1.9 x 10$^{-3}$ 4.5 x 10$^{-4}$ | 76 |
| 9 | -NH-C[=S]-N-[CH$_2$]$_2$-N[CH$_3$]$_2$ | **N** 8.75 % 6.24 x 10$^{-3}$ | **S** 5.21 % 1.62 x 10$^{-3}$ | 1.6 x 10$^{-3}$ 1.1 x 10$^{-3}$ | 34 |
| 10 | -NH-C[=S]-N-C[=O]-C$_6$H$_5$ | **N** 9.52 % 6.79 x 10$^{-3}$ | **S** 7.61 % 2.37 x 10$^{-3}$ | 2.4 x 10$^{-3}$ 1.5 x 10$^{-3}$ | 35 |
| 11 | -NH-C[=S]-S- | **N** 5.27 % 3.76 x 10$^{-3}$ | **S** 11.07 % 3.45 x 10$^{-3}$ | 1.7 x 10$^{-3}$ 5.7 x 10$^{-4}$ | 67 |
| 12 | -P[C$_6$H$_5$]- | **P** 4.70 % 1.52 x 10$^{-3}$ | | 1.5 x 10$^{-3}$ 5.9 x 10$^{-4}$ | 61 |
| 13 | -P[C$_6$H$_5$]$_2$ | **P** 3.70 % 1.19 x 10$^{-3}$ | | 1.2 x 10$^{-3}$ 4.7 x 10$^{-4}$ | 60 |
| 14 | -SO$_3$H | - | | - | - |

(*) after the exchanger has been dried at 75 °C for 24 hours.

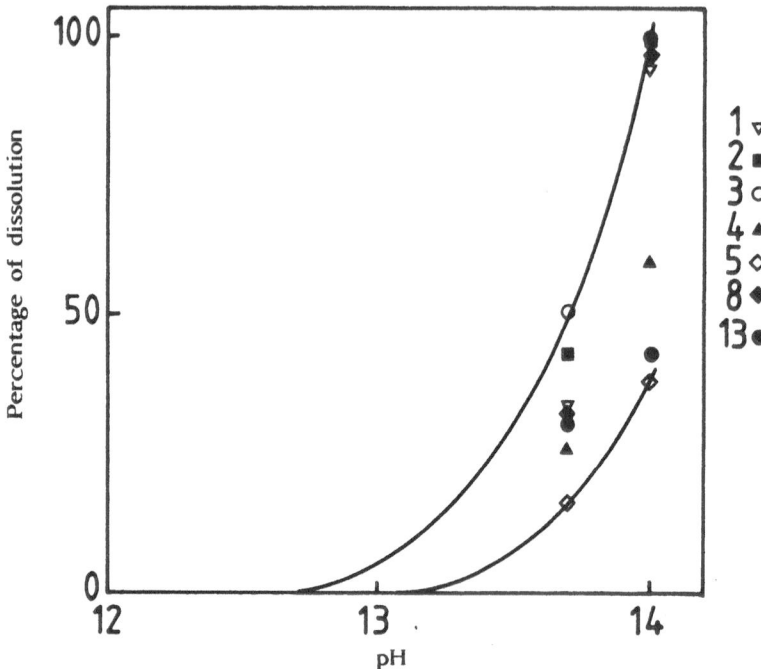

Figure 1. Percentage    dissolution of various ion exchangers as a function of pH at 20 °C.
(Contact time = 72 hrs,  initial exch. charge 50 g [wet] / L).The numbers refer to the
exchanger numbers (see table 1).

## Metal sorption properties

The metal sorption properties of the ion exchangers have been investigated by dividing the
whole family of metal ions into   eight groups [4] and by studying      each of these groups,
except the alkali metal  and alkaline earth metal groups,  the sorption of one or more metal
species on the various exchangers. So six groups of metal ions have been  considered  in the
present paper :  (1) Ag(I), Cu(I) and Tl(I);  (2) As(III), As(V), Sb(III), Sb(V), Ge(IV),
Mo(VI), W(VI), V(V); (3) Pb(II), Pb(IV), Bi(III), Hg(II); (4) Mn(II), Ni(II), Co(II), Zn(II),
Cd(II), Cu(II); (5)Be(II), Al(III), Ga(III), Fe(III), Cr(III), etc., (6) Pd(II), Pt(IV), etc. The
experimental sorption study has been performed with the underlined species.

*Ag(I), Cu(I)* and Tl(I) : The equilibrium data for sorption isotherms of silver(I) from
0.1 mol / L $KNO_3$ are given in figure 2. Examination of these data shows that all sulphur -
containing exchangers have a good affinity for silver(I). At saturation , the mol ratio of Ag /
mol of S in the exchanger remains lower than one (i.e. 0.83, 0.52, 0.71, 0.80 and 0.51 for
exchangers 1,2,3, 9 and 10, respectively), except for exchangers 8 (1.3) and 11 (1.5) which
present a great deficiency of sulphur compared to nitrogen (see table 1). It is of interest to note
that the "diphenylphosphine" exchanger (exchanger 13) has also a good affinity for silver(I).
On the other hand, the "monophenylphosphine" exchanger (exchanger 12) is much less
efficient. At saturation, the mol ratio of Ag / mol of P is equal to 0.36 and 0.93 in exchangers
12 and 13, respectively. In all cases the sorbed Ag(I) is easily stripped by a 2 mol / L
thiosulphate solution.

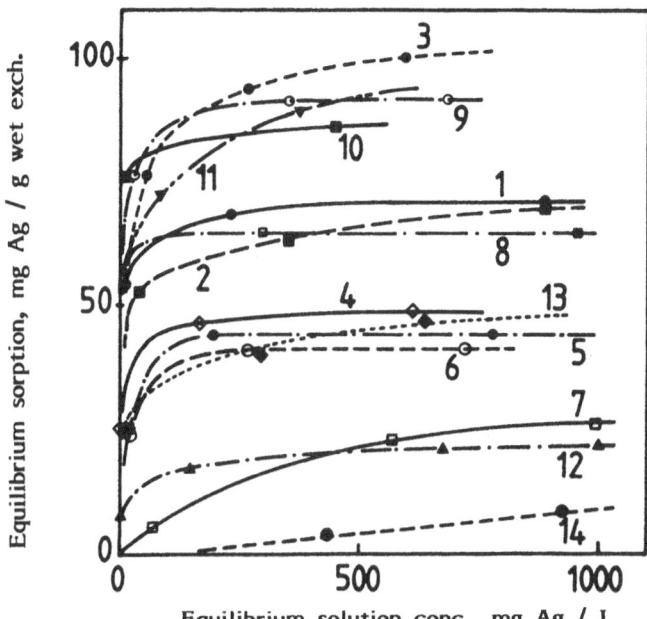

Figure 2 . Equilibrium sorption of Ag(I) from 0.1 mol / L KNO$_3$ at 20 °C. Exchanger charge 20 g (wet) / L. The numbers refer to the exchanger numbers (see table 1).

Copper(I) can exist in concentrated hydrochloric acid media. The results reported in table 2 show that various exchangers can extract copper(I) from such HCl solutions. After sorption, copper(I) is conveniently stripped by a 0.1 mol / L NH$_4^+$ + 0.9 mol / L NH$_3$ solution.

TABLE 2

Copper(I) sorption by various exchangers at 20 ° C [a]

| Exchanger no. | Cu in aq. phase after extraction (g / L) | Equilibrium sorption mg Cu / g dry exch. |
|---|---|---|
| 1 | 1.20 | 100 |
| 2 | 0.92 | 121 |
| 3 | 0.82 | 143 |
| 4 | 1.62 | 90 |
| 5 | 1.95 | 47 |
| 8 | 1.32 | 147 |
| 13 | 1.47 | 76 |

(a) Initial concentration of copper(I) 2.5 g / L in 3 mol / L HCl; exch. charge 33.3 g (wet) / L. The experiments were performed under nitrogen.

**As(III)**, As(V), **Sb(III)**, Sb(V), etc. : The equilibrium data for sorption isotherms of As(III) and Sb(III) from HCl solutions are reported in figures 3 and 4, respectively. Among the 8 exchangers tested, only exchangers 1 (-SH) and 3 (-SH + -S-) which both possess thiol groups significantly retain As(III) (figure 3). Their capacity for As(III) is however relatively low.

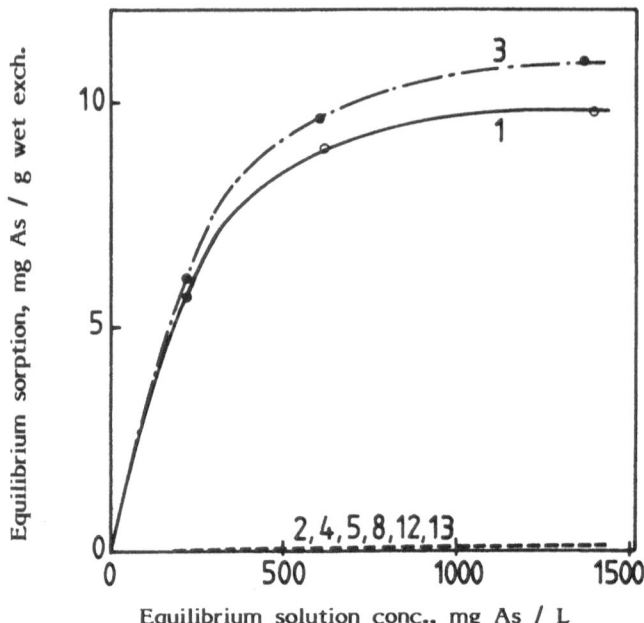

Figure 3. Equilibrium sorption of As(III) from 2 mol / L HCl on various exchangers at 20 °C. Exch. charge 20 g (wet) / L. The numbers refer to the exchanger numbers (see table 1).

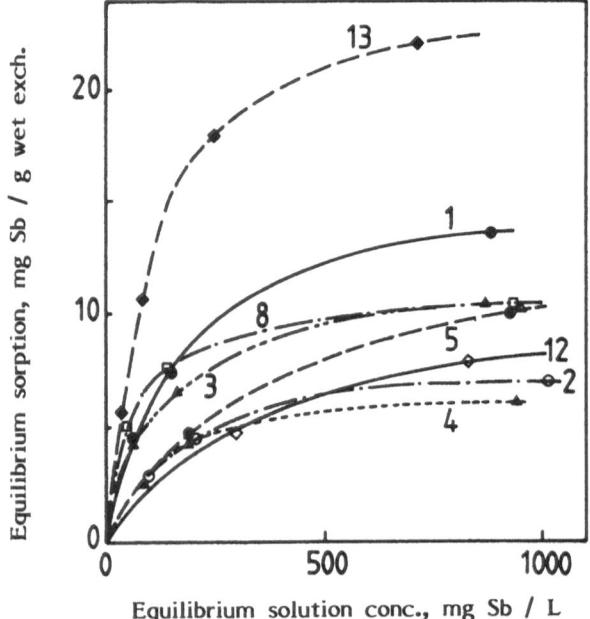

Figure 4. Equilibrium sorption of Sb(III) from 5 mol / L HCl on various exchangers at 20 °C. Exch. charge 20 g (wet) / L. The numbers refer to the exchanger numbers (see table 1).

Examination of figure 4 shows that the "diphenylphosphine" exchanger (exchanger 13) has a good affinity for Sb(III). At saturation, the mol ratio of Sb / mol of P in the solid phase is close to 0.4. Such a value suggests that there is formation of a 1 : 2 metal : ligand complex the formula of which could be $SbCl_3(PPh-)_2$. It is of interest to note that several complexes of this type have already been reported with various metals and triphenylphosphine (PPh3) : $SnCl_4(PPh_3)_2$ [5], $FeCl_3(PPh_3)_2$ [6], $PdCl_2(PPh_3)_2$ [7], etc. However, the possibility that the sorption of Sb(III) may be the result of a reaction of anion exchange cannot be totally excluded . Indeed, Sb(III) exists as anionic chlorocomplexes in 5 mol / L HCl [8] and the $PPh_2$- functional groups of this exchanger may react with hydrogen chloride to form the monocationic phosphonium salt $[-PPh_2H^+]Cl^-$ which can operate as an ion exchanger [9]. As pointed out in the case of the sorption of silver(I), the "monophenylphosphine" exchanger (exchanger 12) has a much lower affinity for Sb(III) than the "diphenylphosphine" one. The other exchangers tested can also retain Sb(III), but to a lesser degree compared to the "diphenylphosphine" exchanger.

Pb(II), Pb(IV), *Bi(III)*, Hg(II) : The equilibrium data for Bi(III) initially present in 5 mol / L HCl are plotted in figure 5. Examination of this figure shows that the "diphenylphosphine" exchanger (exchanger 13) is the best exchanger for the sorption of Bi(III). The value of the mol ratio of Bi / mol of P in the solid phase at saturation cannot be determined accurately from the isotherm given in figure 5, it appears that this ratio is greater than 0.3. Such a value suggests that the sorption of Bi(III) on exchanger 13 may be a result of the formation of a complex the formula of which could be $BiCl_3(PPh_2-)_n$ with n = 3 or more likely 2. However, as in the case of Sb(III), a mechanism of anion exchange cannot be precluded for the sorption of Bi(III) on exchanger 13 since Bi(III) also exists as anionic species in concentrated HCl solutions [10 to 12]. Here again, one can observe that the "monophenylphosphine" exchanger (exchanger 12) is much less efficient than its "diphenylphosphine" analogue. Finally, it should be noticed that exchangers 5 (-NH-[CH2]2-NH2) and 8 (-NH-C[=S]-NH-) have also a good affinity for Bi(III).

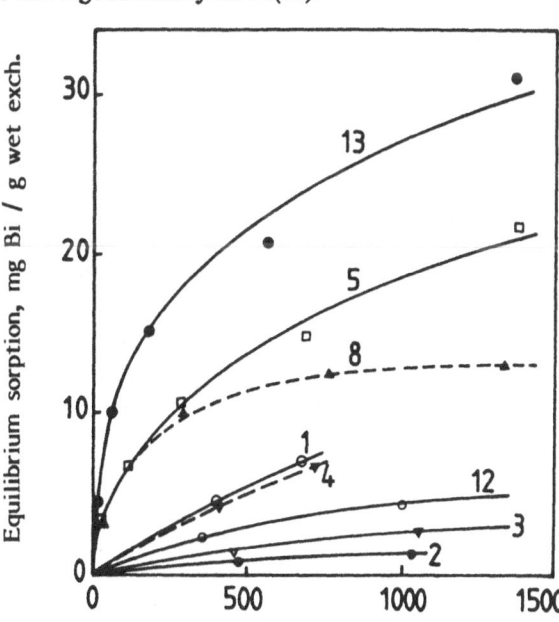

Figure 5

Equilibrium sorption of Bi(III) from 5 mol / L HCl on various exchangers at 20 °C. Exch. charge 20 g (wet) / L. The numbers refer to the exchanger numbers (see table 1).

**Mn(II), _Ni(II)_, Co(II), Zn(II), Cd(II), _Cu(II)_** : The sorption of the elements belonging to this group has not been investigated in detail. Nevertheless, it has been observed that nickel(II) is extracted only at a very small extent (a few mg Ni / g wet exchanger) by exchangers 1, 2, 3, 4, 5, 8 and 13 between pH = 3 and pH = 5, and not extracted at all below pH = 3. On the other hand, copper(II) is moderately extracted by exchangers 7 (21 mg Cu / g wet exchanger) and 11 (24 mg Cu / g wet exchanger), at pH = 4.7.

**Be(II), Al(III), _Ga(III)_, Fe(III), Cr(III), etc.** : The extraction of gallium(III) from acidic solutions has been extensively investigated [13 to 16]. In particular, the recovery of gallium from the acid leach liquors of zinc ores still attracts great attention [15]. In relation to    such a problem, we have chosen to study here the sorption of Ga(III) on exchangers 1, 2, 3, 4, 5, 8 and 13. In fact, none of these exchangers has been found to significantly extract Ga(III) from aqueous sulphuric acid / sulphate solutions between pH = 0 and pH = 3 (1 mol / L $H_2$ / $Na_2SO_4$); [$Ga(OH)_3$ precipitates for pH > 3]. On the other hand, the majority of the exchangers considered in table 1 should extract more or less Ga(III) from concentrated hydrochloric acid solutions according to various mechanisms leading either to the formation of neutral complexes (exchangers 2, 3, 4 and 11) [14] or to the formation of ion-pairs (exchangers 4,5, 6, 7, 8, 9, 10, 12 and 13) [13, 16]. Further investigations on such potential behaviours are now in progress.

**_Pd(II)_, _Pt(IV)_, etc.** : The equilibrium data for sorption isotherms of Pd(II) initially present in 2 mol / L HCl are given in figure 6. Examination of this figure shows that exchangers 1 (-SH), 2 (-S-) and 3 (-SH + -S-) have a good affinity for palladium(II). At saturation, the mol ratio of Pd / mol of S in the solid phase is equal to 0.56 for exchanger 2. Such a value is close to 0.5 and suggests that the sorption of Pd(II) on this exchanger may be a result of the formation of a 1 : 2  metal : ligand complex, probably $PdCl_3(-S-)_2$. We should point out that  1 : 2  metal : ligand complexes of this type are formed during liquid-liquid extraction of Pd(II) with dialkylsulphides [17]. Similarly, the sorption of Pd(II) on exchanger 13 ($PPh_2$-) may result from the formation of a 1 : 2  metal : ligand complex (mol ratio Pd / P = 0.54 at saturation), in agreement with the literature [7]. However the capacity of exchanger 13 is lower than the one of exchanger 2 because of a lower concentration of the functional groups.

Examination of the data given in table 3 reveals that all the exchangers tested can efficiently extract platinum(IV) from 2 mol / L HCl when they are added in excess.  Such a result is surprising in the case of exchanger 2 and shows that the latter does not behave exactly as the dialkylsulphides. Indeed, it is well known that the dialkylsulphides    do not extract platinum(IV) from hydrochloric acid solutions [18].

TABLE 3

Comparison of Pd(II) and Pt(IV) sorption by various exchangers at 20 °C (a)

| Exchanger | Yield of extraction (%) | |
|---|---|---|
| no. | Pt(IV) | Pd(II) |
| 1 | 100 | 100 |
| 2 | 99 | 100 |
| 3 | 100 | 100 |
| 4 | 98 | 100 |
| 5 | 98 | 97 |
| 8 | 100 | 100 |
| 13 | 100 | 100 |

(a) Initial concentration of metal ion 0.5 g / L in 2 mol / L HCl; exch. charge 100 g (wet) / L.

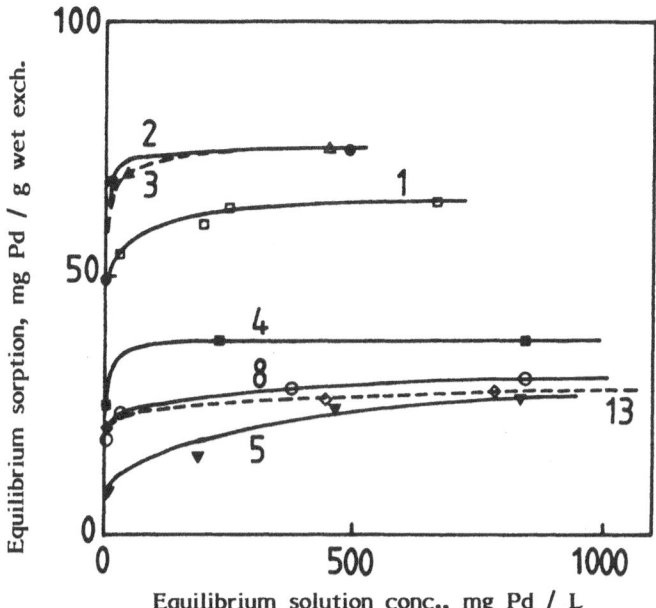

Figure 6. Equilibrium sorption of Pd(II) from 2 mol / L HCl on various exchangers at 20 °C. Exch. charge 20 g (wet) / L. The numbers refer to the exchanger numbers (see table 1).

## POTENTIAL APPLICATIONS AND CONCLUSION

The ion exchangers presented in this work can find many applications, especially in the field of the precious metals. They can also be useful for various unsual separations. For instance, the "diphenylphosphine" exchanger may be advantageously used for removal of traces of antimony and bismuth in effluents.

## ACKNOWLEDGEMENTS

This work has been financed by Degussa Company. The cooperation of Dr. Kleinschmit and Dr. Panster is gratefully acknowledged.

## REFERENCES

1.   Leyden, D.E. and Luttrell, G.H., Preconcentration of trace metals using chelating groups immobilized via silylation. Anal. Chem., 1975, 47, 1612 - 7.

2.   Rosset, R., Caude, M. and Jardy, A., Manuel Pratique de Chromatographie en Phase Liquide , 2nd edition, Masson, Paris, 1982, pp. 141 - 77.

3.   Deschler, U., Kleinschmit, P. and Panster, P., 3-Chloropropyltrialkoxysilanes - key intermediates for the commercial production of organofunctionalized silanes and polysiloxanes, Angew. Chem. Int. Ed. Engl., 1986, 25, 236 - 52.

4.   Charlot, G., Les Réactions Chimiques en Solution Aqueuse et Caractérisation des Ions, 7th edition, Masson, Paris, 1983, pp. 173 - 305.

5. Ohkaku, N. and Nakamoto, K., Metal isotope effect on metal - ligand vibrations. X. Far-infrared spectra of trans adducts of tin(IV) tetrahalide with unidentate ligands. Inorg. Chem., 1973, **12**, 2440 - 6.

6. Walker, J.D. and Poli, R., FeCl₃ - phosphine adducts with trigonal - bipyramidal geometry. Influence of the phosphine on the spin state. Inorg. Chem., 1989, **28**, 1793 - 801.

7. Mojski, M., Extraction of platinum metals from hydrochloric acid medium with triphenylphosphine solution in 1,2-dichloroethane, Talanta, 1980, **27**, 7 - 10.

8. Haight, G.P., Jr. and Yoder Ellis, B., Solubility studies on substituted ammonium salts of halide complexes. V. Tristetramethylammonium enneachlorodiantimonate(III), Inorg. Chem., 1965, **4**, 249 - 50.

9. Yamashoji, Y., Kawaguchi., T., Matsushita, T., Wada, M., Tanaka, M. and Shono, T., Selective solvent extraction of various metal ions with tris(2,6-dimethoxyphenyl)phosphine from hydrochloric acid solutions, Anal. Sci., 1988, **4**, 431 - 2.

10. Newman, L. and Hume, D.N., A spectrophotometric study of the bismuth - chloride complexes, J. Am. Chem. Soc., 1957, **79**, 4576 - 81.

11. Oertel, R.P. and Plane, R.A., Raman study of chloride and bromide complexes of bismuth(III), Inorg. Chem., 1967, **6**, 1960 - 7.

12. Kankare Jouko, J., Computation of equilibrium constants for multicomponent systems from spectrophotometric data, Anal. Chem., 1970, **42**, 1322 - 6.

13. Sato, T., Nakamura, T. and Ishikawa, S., Liquid - liquid extraction of gallium(III) from hydrochloric acid solutions by organophosphorus compounds and high - molecular weight amines. Solvent Extr. Ion Exch., 1984, **2**, 201 - 12.

14. Baba, Y., Nakamura, H. and Inoue, K., Extraction equilibria of gallium(III) and thallium(III) with dihexylsulfide from hydrochloric acid. J. Chem. Eng. Jpn., 1986, **19**, 497 - 502 and references therein.

15. Jacquin, O., Faux-Mallet, S., Cote, G. and Bauer, D., The recovery of gallium(III) from acid leach liquors of zinc ores using selective ion exchange resins. In Recent Developments in Ion Exchange, ed. P.A. Williams and M.J. Hudson, Elsevier Applied Science Publishers, London, 1987, pp. 213-220

16. Yamashoji, Y., Matsushita, T., Tanaka, M., Shono, T. and Wada, M., Ion - pair extraction of the gallium(III) ion from hydrochloric acid with various methoxy - substituted triarylphosphines. Polyhedron, 1989, **8**, 1053 - 9 and references therein.

17. Daamach, S., Cote, G. and Bauer, D., Extraction du palladium(II) par les sulfures de dialkyle et les sulfures de trialkylphosphine : nature des complexes formés en milieu acide chlorhydrique et valeurs des constantes d'extraction. C. R. Acad. Sc. Paris, 1987, t. 304, Série II, n° 15, 889 - 92.

18. Edwards, R.I., Selective solvent extractants for the refining of platinum metals. Proc. ISEC'77, C.I.M. Special volume 21, The Canadian Institute of Mining and Metallurgy, Montreal, 1979, pp. 24 - 31.

# EXTRACTION OF THE PERRHENATE ANION USING GOETHITE SURFACE-MODIFIED WITH HYDROPHOBIC QUATERNARY AMINES

MICHAEL J. HUDSON and DEBRA J. TYLER
Department of Chemistry,
University of Reading,
P.O. Box 224, Whiteknights,
Reading RG6 2AD, Berks, U.K.

## ABSTRACT

Goethite [FeO(OH)], the surface of which we modified by thin coatings of hydrophobic quaternary amines, has been used to extract the perrhenate anion $(ReO_4)^-$ from alkaline aqueous solutions (pH 11.3). The quaternary amines were bound to the surface of the goethite by ionic association between the negatively charged inorganic surface and the positive quaternary ammonium group. However, further free amine groups were retained by hydrophobic interaction with the adsorbed layer leaving other quaternary ammonium groups free on the surface. These were capable of selectively extracting polarisable anions such as perrhenate from non-polarisable anions such as nitrate or hydroxide. The kinetics of extraction were fast ($t_{1/2}$ less than five minutes) and the capacity for perrhenate varied between 10 and 60 m mol/kg for the different modified-goethites.

## INTRODUCTION

It has been established [1] that anions may be extracted from acidic solutions using an inorganic oxide such as zirconium(IV) oxide. Titanium(IV) oxide [2], has also been shown to be capable of extracting anions from solutions of low pH. The amorphous form of titanium(IV) oxide [3] was produced at the rate of one ton per year. Manganese(IV) oxide, when freshly prepared from solution, had a higher surface area than the corresponding metal(IV) oxides [4,5] and also extracted anions from acidic solution. Goethite, [FeO(OH)], has been shown [6] to extract anions at low pH. However, the pH stability of all of the metal(IV) oxides in these low pH solutions remains to be studied. None of the

above inorganic compounds are capable of extracting anions from neutral
or basic solutions. The principal advantages with inorganic ion
exchangers are that they are readily available, inexpensive and are
resistant to radiolysis [7]. However, there are no readily available
inorganic anion exchangers for extraction of anions from alkaline
solutions and they are frequently unselective for anions at low pH. The
adsorption of low molecular weight, oligomeric and polymeric species onto
inorganic surfaces has been extensively studied [8,9] particuarly with
respect to flocculation [10]. It has been shown [11], that quaternary
ammonium compounds are adsorbed onto inorganic substances at pH's above
that corresponding to the isoelectric point. Under these conditions,
the surface has a net negative charge and, for low coverages, there is a
neutralisation of the effective negative charges [12] by the quaternary
ammonium groups which is accompanied by flocculation [13]. At higher
coverages, the charges invert so that the surface, which was previously
negative becomes positive. This effect has not been previously ultilised
to extract anions such as perrhenate. The adsorption of cationic polymers
has been previously studied and may be used to provide surface coatings
[14,15] of a strongly adherent, cationic layer.


## EXPERIMENTAL

The hydrophobic quaternary compound LQA, was supplied as reagent TS 1198
by ABM Chemicals. Goethite [FeO(OH)], was prepared by a standard method
[16] in which the iron(III) nitrate (BDH) was hydrolysed with ammonia
(14 M). The final pH was above 6.5 which is the isoelectric point for
goethite. The surface area of the goethite was 60 $m^2 g^{-1}$ as measured by
the BET method. The particle size was less than 250 micrometres. (Figure 1).

Surface films of the reagent LQA were prepared by refluxing the
goethite (15 g) for twenty minutes with the reagent (10 g) in alcohol
(90 g). The mixture was placed in a rotatory evaporator and the alcohol
removed by suction at ambient temperature. The reaction produced a
modified material the surface area of which was 57 $m^2 g^{-1}$ (BET). With
respect to the other reagents, the names and formulae of the reagents are
listed in Table 2. Analytical figures for all the modified goethites
are displayed in Table 3 and the surface area data can be seen in Table 4.
Fourier Transform infrared (FT-IR) spectra of various modified and
unmodified goethites are shown in Figures 2 and 3. Potassium perrhenate

TABLE 1

$^{13}$C N.M.R. analysis of LQA

| PEAK POSITION / ppm | SPECIES CAUSING PEAK |
|---|---|
| 12 | CH in alkyl chain |
| 14 | $CH_3$ at ends of alkyl chains |
| 20 - 37 | $CH_2$ in alkyl chains |
| 40 | CH next to nitrogen atom |
| 46 | CH next to aryl group |
| 56 - 73 | $CH_2$ in ethylene oxide groups |
| 124 - 134 | CH in aryl ring |

TABLE 2

Reagent names and formulae

| NAME OF REAGENT | FORMULA | ABBREVIATION |
|---|---|---|
| Tetramethylammonium bromide | $[(CH_3)_4N]^+Br^-$ | TMB |
| Octadecyltrimethylammonium bromide | $[C_{13}H_{37}N(CH_3)_3]^+Br^-$ | ODTB |
| Didodecyldimethylammonium bromide | $[(C_{12}H_{25})_2N(CH_3)_2]^+Br^-$ | DDB |
| Tricaprylmethylammonium chloride | $[(C_8H_{17})_3NCH_3]^+Cl^-$ | A336 |
| Tetraoctylammonium bromide | $[C_8H_{17})_4N]^+Br^-$ | TOB |
| Tetradodecylammonium bromide | $[(C_{12}H_{25})_4N]^+Br^-$ | TDB |

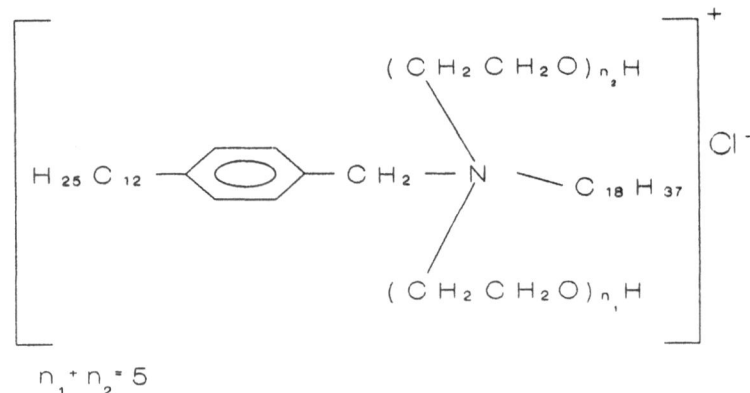

$$n_1 + n_2 = 5$$

Figure 1.   The probable structure of the quaternary ammonium
            compound LQA.

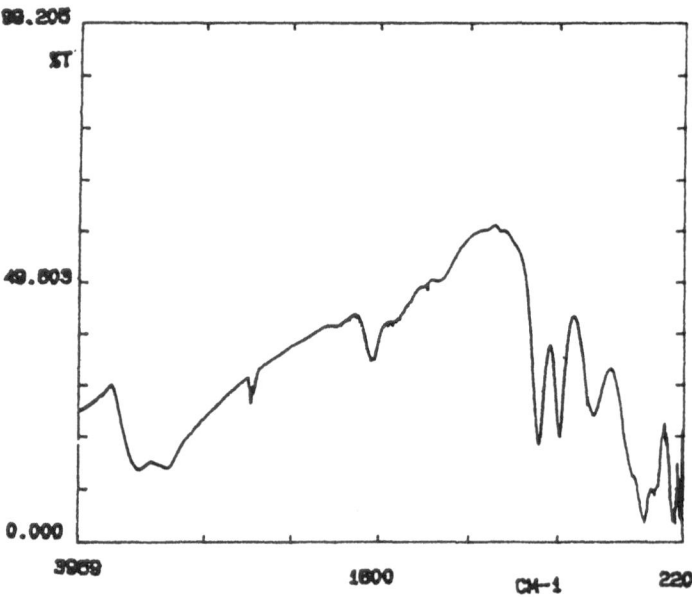

Figure 2.   FT-IR spectrum of goethite.

TABLE 3

Analytical figures for the surface-modified goethites

| MODIFYING AGENT | PERCENTAGES | | | MOLAR RATIOS |
|---|---|---|---|---|
| (t w/w original solution) | C | H | N | C/N |
| LQA(1%) | 1.60 | 2.74 | ND | [57] |
| LQA(10%) | 5.57 | 1.62 | 0.20 | 32[57] |
| LQA(20%) | 11.00 | 4.13 | 0.40 | 32[57] |
| TMB(5%) | 0.59 | 2.37 | 0.18 | 4[ 4] |
| ODTB(5%) | 1.02 | 1.41 | ND | [21] |
| DDB(5%) | 3.40 | 1.57 | 0.17 | 24[26] |
| A336(5%) | 3.49 | 1.15 | 0.29 | 14[25] |
| TOB(5%) | 5.28 | 3.10 | 0.23 | 27[32] |
| TDB (5%) | 1.62 | 2.85 | ND | [48] |

TABLE 4

Surface area data for the modified geothites

| MODIFYING AGENT | SURFACE AREA $/m^2 \, g^{-1}$ |
|---|---|
| LQA(1%) | 70.96 |
| LQA(10%) | 51.37 |
| LQA(20%) | 29.55 |
| ODTB | 51.08 |
| DDB | 44.51 |
| A336 | 49.73 |
| TOB | 26.05 |
| TDB | 32.49 |

The surface area of unmodified goethite = 59.88 $m^2 \, g^{-1}$

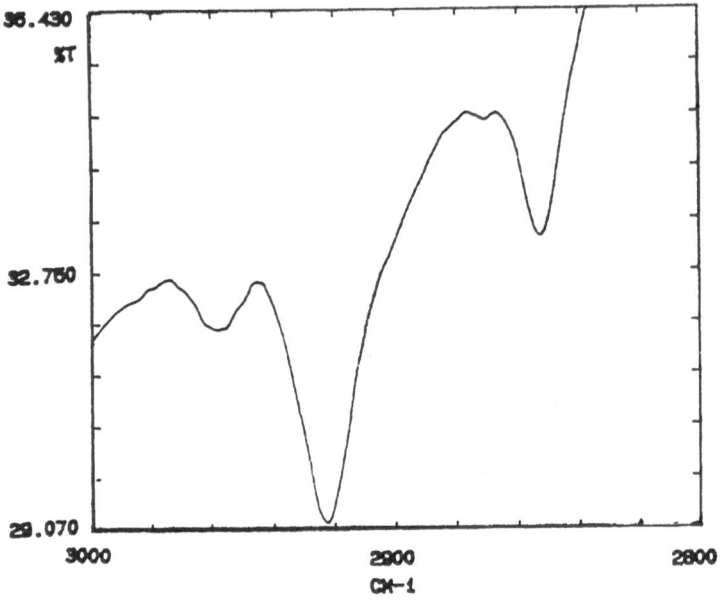

Figure 3.   FT-IR difference spectrum of a modified-
goethite compared to unmodified goethite.

303

Figure 4.

A

(*650)

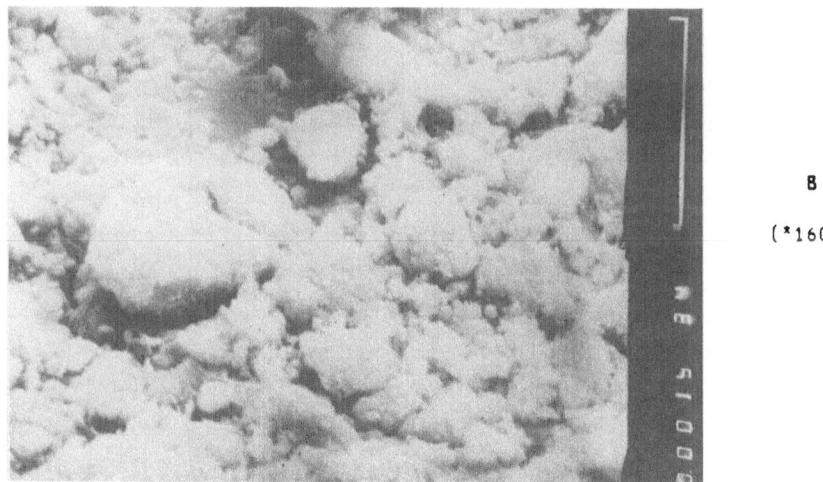

B

(*1600)

Potassium perrhenate (Alfa Chemicals) was dissolved in simulated pond water [17] and diluted to various concentrations. Simulated pond water was composed of NaOH (0.005 M), $Na_2CO_3$ (0.0028 M) and $NaNO_3$ (0.01 M). The Beer-Lambert Law was obeyed over the concentration range 0.1 to 10 m mol $dm^{-1}$.

## RESULTS AND DISCUSSION

The isoelectric point of goethite is 6.5 [18] which means that above this pH the surface is essentially negative and able to adsorb positively charged coatings. This negative charge is due to $OH^-$ groups and it has been shown that goethite has 5.5 OH groups on the surface per 100 $A^2$ [19]. The figures in Table 3 indicate that such positively charged coatings have indeed been formed on the surface of the goethite. Water insoluble quaternary ammonium compounds were used to ensure that the coatings would not redissolve off the surface into the aqueous solution containing the perrhenate to be extracted. The electron micrographs in Figure 4 show the appearance of the goethite before and after the modification by LQA. The surface covering by the quaternary amine, LQA, can be seen and the particle size has been reduced from 25 nm to 9 nm. It also shows the surface of the modified-goethite to be hazy whilst the unmodified goethite has a clearly defined surface. The idealised form of this modified-goethite is represented in Figure 5. All the coatings (except LQA 1%) decreased the effective surface area of the goethite (see Table 4). This together with the electron micrographs indicate that LQA formed layers across the grains which effectively blocked cracks and pores and hence decreased the surface area. Figure 2 displays the FT-IR spectrum of unmodified goethite. The band at 3660 $cm^{-1}$ was due [20] to unperturbed surface OH-groups whilst the band at 3160 $cm^{-1}$ was produced by the bulk OH-groups. At 630 $cm^{-1}$ and 400 $cm^{-1}$ there are bands due to Fe-OH bonds stretching approximately parallel to the a and c planes [21]. Figure 3 is typical of the FT-IR difference spectra obtained with the modified and unmodified surfaces. There are bands at 2954 $cm^{-1}$, 2922 $cm^{-1}$ and 2852 $cm^{-1}$ corresponding to the stretching of the C-H bonds [22]. The bending modes of deformation for C-H bonds occur at 1467 $cm^{-1}$, 1454 $cm^{-1}$ and 1409 $cm^{-1}$ on the FT-IR spectrum. This confirmed that the quaternary ammonium compounds have indeed coated the goethite for all the reagents used except TMB.

TABLE 5

Equilibrium total capacity figures for the perrhenate anion extracted
onto the modified goethites

| MODIFYING AGENT | NUMBER OF CARBON CHAINS (C) | NUMBER OF CARBON ATOMS IN CHAIN | CAPACITY /mmol $Kg^{-1}$ |
|---|---|---|---|
| TMB(5%) | 0 | - | 0 |
| ODTB(5%) | 1 | 13 | 9.7 |
| LQA(1%) | 2 | 18 | 1.7 |
| LQA(10%) | 2 | 18 | 19.2 |
| LQA(20%) | 2 | 18 | 22.5 |
| DDB(5%) | 2 | 12 | 58.3 |
| A336(5%) | 3 | 8 | 42.1 |
| TOB(5%) | 4 | 8 | 60.1 |
| TDB(5%) | 4 | 12 | 10.0 |

TABLE 6

Kinetic data for the extraction of perrhenate by various modified-
goethites

| TIME /min | CAPACITY /mmol $Kg^{-1}$ | | | | |
|---|---|---|---|---|---|
| | LQA(10%) | ODTB | DDB | A336 | TOB |
| 0 | 0 | 0 | 0 | 0 | 0 |
| 1 | | 5.4 | 56.7 | 59.1 | 61.0 |
| 5 | 16.8 | 8.1 | 61.0 | 59.8 | 60.5 |
| 10 | 19.2 | 7.8 | 60.5 | 59.8 | 59.7 |
| 20 | 17.3 | 9.8 | 62.1 | 60.8 | 59.5 |
| 40 | 18.7 | 12.1 | 56.2 | 61.1 | 57.6 |
| 300 | | 5.8 | 57.6 | 58.4 | 51.6 |
| 1400 | | 11.2 | 65.8 | 54.3 | 53.5 |

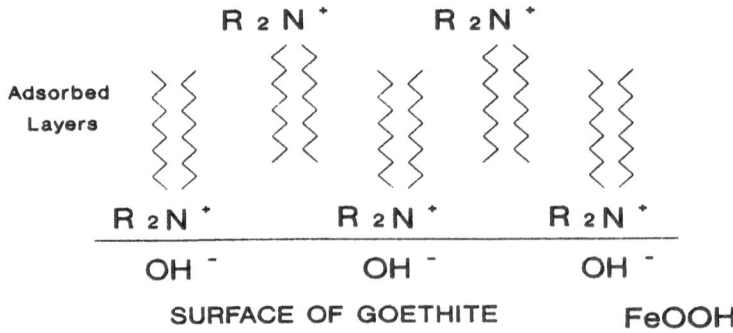

Figure 5.   The idealised surface of the modified goethite
            (bilayer formation).

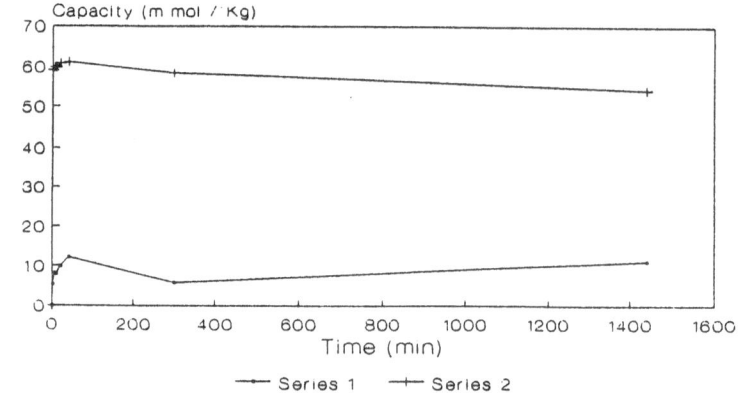

Figure 6.   Kinetic data for the extraction of $ReO_4^-$ by ODTB
            (Series 1) and A336 (Series 2).

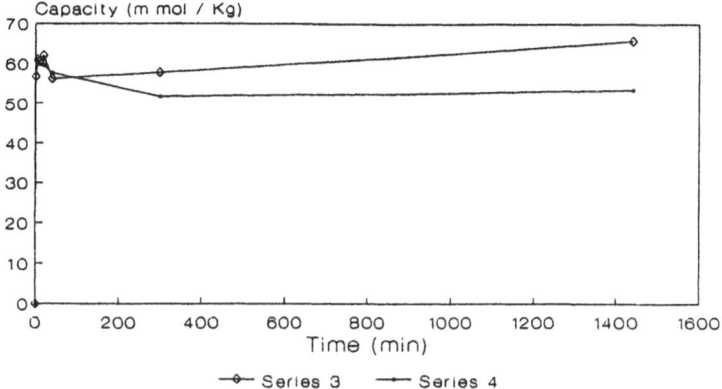

Figure 7. Kinetic data for the extraction of $ReO_4^-$ by
DDB (Series 3) and TOB (Series 4).

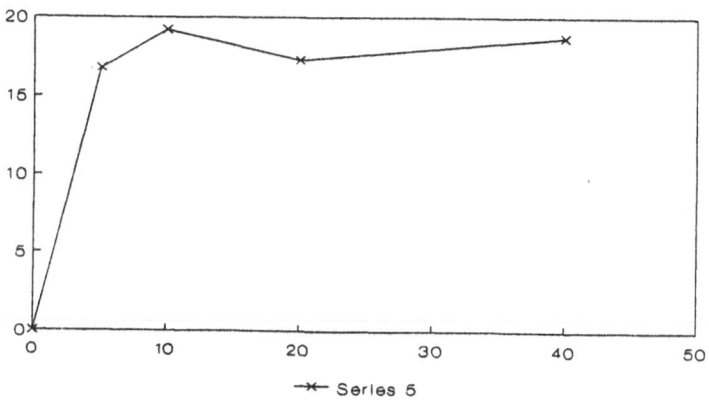

Figure 8. Kinetic data for the extraction of $ReO_4^-$ by
LQA(10%) (Series 5).

## Extraction of Perrhenate from Alkaline Solutions

Unmodified goethite did not extract perrhenate at pH 11.3 from any solution which contained NaOH (0.005 M), $Na_2CO_3$ (0.0028 M) and $NaNO_3$ (0.01 M) and may be regarded [17] as simulated pond water. Consequently the removal of the $ReO_4^-$ anion discussed below was due to the coating of the quaternary amine. The capacitites of the surface-modified goethites were related to the availability of quaternary ammonium groups for ion exchange.

Equilibrium data for the modified-goethites are displayed in Table 5. It shows the dependence of capacity on the carbon chain length and the number of carbon chains. It has been found [9] that long adsorbed carbon chains (e.g. >16 carbons) cohere to each other by van der Waal's forces which reduce the adsorption of additional molecules in between the hydrocarbon chains. Taking the modified-goethites that had extracted $ReO_4^-$ and using the electron micrograph the M alpha emission lines of Rhenium were detected proving directly that these modified-goethites did remove $ReO_4^-$

## Kinetics of Anion Exchange

The kinetics of anion exchange are often reported [23] to be slow. The results for the kinetics of the modifed-goethites are shown in Table 6 and Figures 6-8 and it can be seen that the time taken for half of the reactive sites to be occupied was under 1 minute which is acceptable for a commercial process.

## ACKNOWLEDGEMENTS

This work has been supported by the U.K. Department of the Environment as part of their Radioactive Waste Management Programme. The results may be used in the formulation of government policy but at this stage do not necessarily represent govenment policy.

## REFERENCES

1. Amphlett, C.B., "Inorganic Ion Exchangers", Elsevier, Amsterdam, 1964

2. Abe, M. and M. Tsuki., J. Radioanalytical and Nuclear Chemistry, (1986) 283.

3.  Keen, J.H., Chemistry and Industry, (1977) 579.

4.  Hooper, E.W., Phillips, B.A., Dagnall, S.P. and N.P. Monckton "An Assessment of the Application of Inorganic Ion Exchangers to the Treatment of Intermediate Level Wastes" (1984) H.M.S.O. for the Department of the Environment.

5.  Parida, K.M., Janungo, S.B. and Sant, B.R., Electrochim. Acta, (1981) 26, 435.

6.  Anderson, M.A. and Rubin A.J. (Editors) "Adsorption of Inorganics at Solid-Liquid Interfaces" (1981) Ann Arbor Science Publishers Inc., pp 51-91.

7.  Jenkins, I.C., Actinides Rev., (1969) 1, 187.

8.  Parfitt, G.D. and Rochester, C.H. (Editors) "Adsorption from Solution at the Solid Liquid Interface" (1983) Academic Press, New York.

9.  Busscher H.J. et al., Colloid and Polymer Sci., (1987) 265, 711.

10. Durand-Piena, G., Lafuma, F. and Audebert, R., J. Colloid and Interface Science, (1987) 119(2), 474.

11. Gregory, J., J. Colloid and Interface Science, (1973) 42, 448.

12. Manning, G.S., "Charge Compensation on Absorption onto Metal Oxides", Quarterly Revs. Biophysics, (1978) 11, 179.

13. Jung, R.F., James, R.O. and Healy, T.W., J. Colloid and Interface Sci., (1987) 118(2), 463.

14. Wilson, A.D., Nicholson, J.W. and Prosser, H.J., Surface Coatings., Elsevier, London (1987).

15. Heller, W., Pure Appl. Chem., (1966) 12, 249.

16. Brauer, G., "Handbook of Preparative Inorganic Chemistry", Academic Press, New York 1962.

17. Dyer, A., Keir, D., Hudson, M.J. and Leung, B.K.O., Chem. Commun., (1984) 1457.

18. Park, G.A., Chem. Rev., (1963) 65, 177.

19. McCaffarty, E. and Zettlemoyer A.C., Disc. Faraday Soc., (1971) 52, 239.

20. Rochester, C.H. and Topham, S.A., J. Chem. Soc. Faraday Trans 1, (1979) 75, 591.

21. Cambier, P., Clay Miner., (1986) 21 191.

22. Kemp, W., "Organic Spectroscopy" MacMillan Publishers Ltd., London 1975.

23. Ryan, J.L. and Wheelwright, E.J., "Peaceful Uses of Atomic Energy" Proceedings of the Second International Conference, Geneva, (United Nations, New York), 1958, 17, 137.

24. Taube, H., Angew Chem., (1983) 23, 329.

25. Kershaw, M.R. and Prue, J.E., Trans. Faraday, Soc., (1967) 63, 1198.

SOLID PHASE EXTRACTION AND CHROMATOGRAPHY OF METAL IONS USING FREE AND
IMMOBILISED BIOMIMETIC CHELATING AGENTS

S. SRIJARANAI, N. RYAN, N. MITCHELL and J.D. GLENNON

Chemistry Department, University College Cork (Ireland)

SUMMARY

Biomimetic chelating agents, incorporating the hydroxamic acid functional
group, can be used for trace metal ion analysis and preconcentration. Fe(III)
and V(V) can be separated by reverse-phase HPLC, following injection into a
mobile phase containing N-methylpropionohydroxamic acid. The immobilisation of
this chelating group onto silica particles yields a solid phase of use in the
extraction and analysis of a series of important transition metals.

INTRODUCTION

In nature there are many examples of specific or selective metal binding

ligands. Some of these are high molecular weight metalloproteins for example

tyrosinase for Cu while others are lower molecular weight ligands for example

valinomycin for $K^+$. Analytical chemists have not been slow to apply these

chelating properties in such areas as spectrophotometry and electroanalysis. In

this paper, the analytical applications of biomimetic ligands inspired by

microbial siderophores will be described. Siderophores are Fe(III) sequestering

compounds with high stability constants such as desferoxamine shown in Figure 1

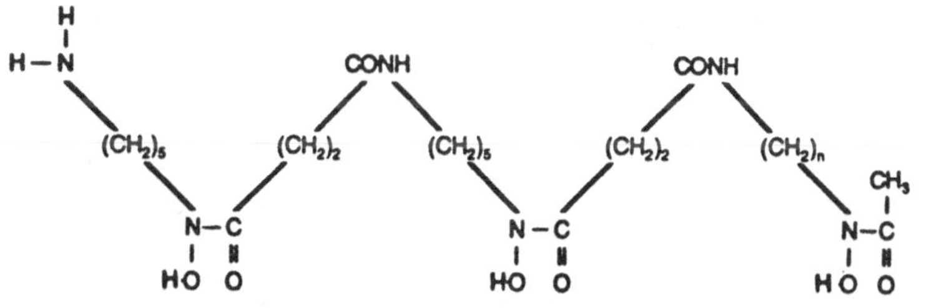

Fig. 1 Structure of the naturally occurring Fe(III)-chelating compound,
desferoxamine.

The simplest biomimetic ligand of this class is acetohydroxamic acid,

$CH_3CONHOH$, which has been used successfully for the analysis of Fe(III) by flow

injection analysis (ref. 1). More recently, N-methylfurohydroxamic acid

chelates of Zr(IV), Hf(IV) and Fe(III) have been separated by high performance liquid chromatography on a polymeric column (ref. 2). Investigations have been carried out in our laboratory into the use of alternative biomimetic chelating agents such as N-methylpropionohydroxamic acid (NMPHA) for metal ion separations.

There are a variety of strategies adopted for the preconcentration of trace metals from aqueous solution, including solvent evaporation, electrodeposition and liquid-liquid extraction (ref. 3). Ion exchange materials can be used for the uptake of metal ions or of charged metal complexes (ref. 4). Uncharged metal chelates can be preconcentrated by a partition mechanism on modern bonded phase materials such as octadecylsilica (ref. 5). An alternative to these approaches is to use chelating solid phases. A variety of chelating solid phases have now been described for applications extending to metal chelate affinity chromatography, chiral ligand exchange chromatography and metal ion chromatography (ref. 6-8).

Desferoxamine (DFA), N-methylpropionohydroxamic acid (NMPHA) and propionohydroxamic acid (PHA) can be readily chemically coupled to silica (ref. 9). Useful attributes of these materials for the solid phase extraction and chromatographic analysis of metal ions will also be illustrated.


## EXPERIMENTAL METHODS

### Materials

The free ligand M-methylpropionohydroxamic acid was synthesised by the reaction of the ester with N-methylhydroxlamine. The biomimetic chelating solid phase, PHA-Si used was prepared from silica (40 micron) as previously reported (ref. 9). The chemicals used for mobile phase preparation (acetic acid, citric acid, sodium hydroxide) were of analytical grade, obtained from BDH chemicals. HPLC grade acetonitrile was obtained from Labscan (Dublin).

Metal ion stock solutions of Fe(III), Al(III), Cd(II), Co(II), Cu(II), and Ni(II) were prepared from the nitrate salts, Au(III) from the chloride, Zn(II) from the sulphate and V(V) from $NH_4VO_3$. Metal ion complexation capacities were studied using a batch experiment, where 25 ml sample solution of metal ion was adjusted to the desired pH and shaken with 0.1 g of chelating phase at room temperature for 15 min. Atomic absorption spectroscopy was used to assess the degree of metal ion uptake, except for V(V) where a spectrophotometric assay was used.


### Instrumentation

The chromatograph consisted of a Waters Model 510 HPLC pump, a Rheodyne 7125 (20 ul) injector and a Waters model 440 fixed wavelength or model 481 variable

wavelength detector.   Chromatographic columns used were the PLRP-S 5um PS-DVB
column (15cm x 4.6mm i.d.) from Polymer Laboratories and a laboratory packed PHA-
Si (40um) column (10cm x 4.0mm i.d.).   Atomic absorption analysis were carried
out on a Pye Unicam SP190 spectrophotometer, and a Shimadzu UV260
spectrophotometer for V(V) assays.


**RESULTS AND DISCUSSION**

<u>NMPHA as Mobile Phase Additive</u> <u>for HPLC Analysis of Metal Ions</u>

   Initial studies using octadecylsilica columns for the separation of metal ions
using acetohydroxamic acid in situ resulted in low retention and poor peak
characteristics as found by other workers (ref. 2).   Improvements in retention
using NMPHA chelates were expected due to increased hydrophobicity while a
changeover to the use of a polymeric column and the higher chelate stability of
N-substituted complexes were expected to improve peak shape and detectibility.   A
typical chromatogram obtained by HPLC with visible detection for a mixture of
Fe(III) and $VO_3^-$ is shown in Fig. 2.   Factors which control the capacity factors
are the pH and the percentage organic modifier.

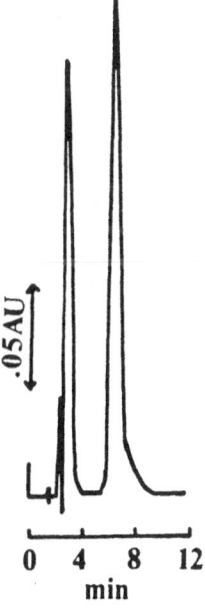

Fig. 2   Separation of Fe(III) and V(V) on a PS-DVB column using a mobile phase of
98:2 water-acetonitrile containing 5mM NMPHA and 0.02M acetate buffer (pH 5.6) ;
detection at 405 nm ; retention data: Fe(III) 7.0 min., V(V) 3.1 min.

Solid Phase Extraction of Metal Ions on PHA-Si

Metal complexation capacities, in units of mmol of metal ion per gram PHA-Si, were determined using atomic absorption spectroscopy for the series of metals shown in Fig. 3. It is clear that pH can be used to control the uptake and elution of metal ions for preconcentration and separation. Particular attention was paid to the uptake of Fe(III) where the pH dependency found and optical properties observed were typical of Fe(III) complexation by siderophores. EDTA solutions readily eluted most metal ions from the chelating phase with the exception of Au(III) and V(V).

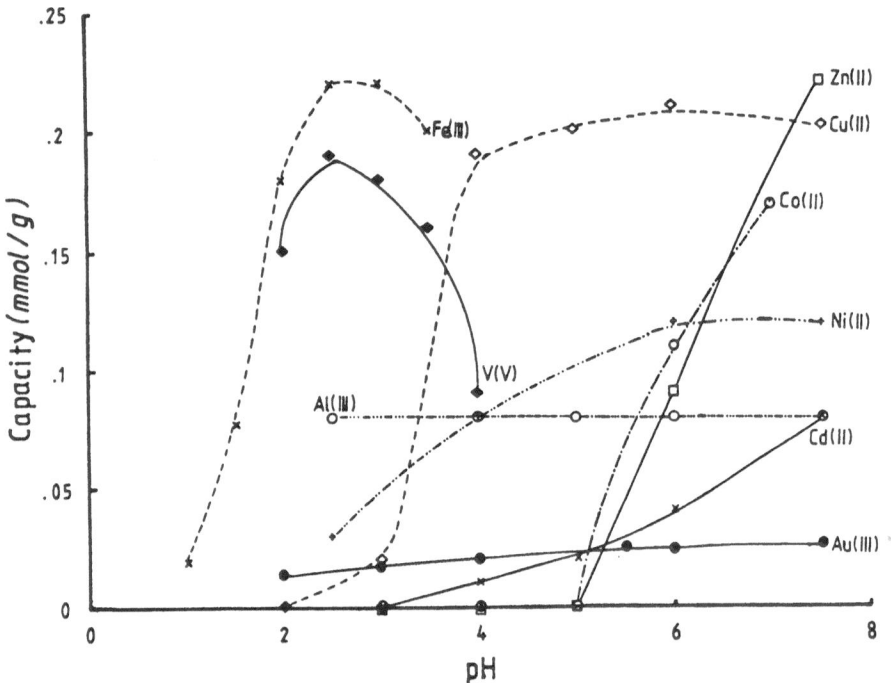

Fig. 3  Metal ion uptake as a function of pH on PHA-Si

Chromatographic Behaviour of Cu(II) on PHA-Si

Preliminary studies were carried out on the usefulness of the PHA-Si as a chromatographic stationary phase for the separation of metals ions in particular Cu(II). The mechanism of retention of metal ions on chelating phases is poorly understood, despite some good demonstrations of metal separations (ref. 8). The PHA-Si was packed into a stainless steel column and mobile phase composition and pH varied. The incorporation of citrate into the mobile phase provided the necessary competition to the chelating phase for Cu(II) when the pH chosen fell within the range of hydroxamate complexation. The effect of pH on retention of Cu(II) and a typical chromatogram is shown in Fig. 4.

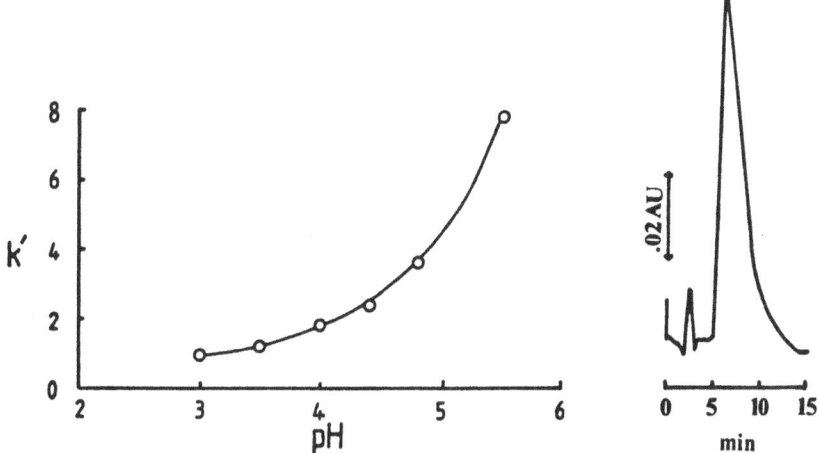

Fig. 4 (a) Capacity factor as a function of pH for Cu(II) on PHA-Si Column (b) Chromatogram obtained on injection of Cu(II), mobile phase 1mM citrate in 0.02M acetate buffer (pH 4.0), flow rate 0.5 ml/min and detection at 280nm.

**REFERENCES**

1. A.T. Senior and J.D. Glennon, Analytica Chimica Acta, 196 (1987) 333-336
2. M.D. Palmieri and J.S. Fritz, Anal. Chem. 59 (1987) 2226-2231
3. D.E. Leyden and W. Wegscheider, Anal. Chem. 53 (1981) 1059A-1065A
4. C. Sarzanini, E. Mentasti, V. Porta and M.C. Gennaro, Anal. Chem. 59 (1987) 484-486
5. G. Schwedt and U. Sicker, LoborPraxis 7 (1983) 816
6. N. Ramadan and J. Porath, J. Chromatogr. 321 (1985) 81-91
7. S. Lam, Chiral ligand exchange chromatography, in: W.J. Lough (Ed.), Chiral Liquid Chromatography, Blackie, Glasgow and London, 1989, pp. 83-101
8. K.H. Faltynski and J.R. Jezorek, Chromatographia 22 (1986) 5-12
9. J.D. Glennon and S. Srijaranai, Analyst (1990) in press

# Part 6

# INDUSTRIAL APPLICATIONS

# ECONOMIC PROCESS FOR SOFTENING PRODUCED WATER USED IN SECONDARY OIL RECOVERY

F. X. McGARVEY and R. GONZALEZ
Sybron Chemicals Inc. Birmingham Road, PO Box 66,
Birmingham, NEW JERSEY, 08011, U.S.A.

## Introduction

Usually natural oil reservoirs are sufficiently pressurized to allow the recovery of about one-third of the crude oil in the deposit. A secondary recovery is possible for another third of the oil using various ways to pressurize the strata. One way to accomplish this uses hot water and high pressure steam. The water used in this process either comes from surface sources or from "produced water" which comes from the oil field itself.

The steam generators used in the flooding operation are high pressure once through boilers which require a very soft water to insure that the equipment does not scale-up. A typical feed water is shown in Table 1. This water analysis is given for a synthetic water used for the studies described in this paper. The values for hardness and alkalinity are not unlike waters available at Bakersfield, California where very large floods are now under way.

The ion exchange process to soften the water has evolved for some twenty-five years. The first units contained primary and secondary columns as shown in Figure 1 [1,2,3]. The early units contained strongly acidic cation exchange resins regenerated with salt. As the available waters had an increase in solids this method failed due to excessive hardness leakage. Work was started using weakly acidic cation exchange resins. Due to the high affinity of these resins for calcium and magnesium, it was found that per million hardness in waters that had more than 10,000 ppm of total dissolved solids. The characteristics of a typical weakly acidic cation exchange resin are shown in Table 2.

## Process

The initial process employed the weakly acidic cation exchange resins in the sodium form. This was accomplished by conversion of the resins to the sodium form with caustic after the resin was exhausted with hardness, and stripped with acid. Oil field workers prefer hydrochloric acid for this step. When sulfuric acid is used, it is necessary to use very dilute acid (0.5-0.7%) to avoid the precipitation of calcium sulfate.

The process employs full regeneration of the resin with acid and base. This means that a resin with a capacity of three equivalents per liter will require an equivalent amount of acid to strip the bed and an equivalent amount of caustic to neutralize the acid prior to the service cycle. In engineering units a cubic foot of hydrogen form resin initially would require about 7 lbs HCl/cu ft and 6.4 lbs NaOH/cu ft for the conditioning step. In addition, the resin will swell about 50% in the conversion from the hydrogen to the sodium form. This requires equipment design to accommodate the volume change.

The cost and availability of caustic in the oil fields made the softening of water by weak acid processes a costly operation. Since the weakly acidic cation exchangers are used to dealkalize water, it was believed that alkalinity in the water can be substituted for the caustic in cases where alkalinity content exceeded the hardness content. The reactions for this process follow:

$$R(COO)_2Ca + HCl \rightarrow 2RCOOH + CaCl_2 \tag{1}$$

$$RCOOH + NaHCO_3 \rightarrow RCOONa + CO_2 + H_2O \tag{2}$$

$$RCOONa + Ca^{++} \rightarrow R(COO)_2Ca + Na^+ \tag{3}$$

The first reaction requires an amount of acid equivalent to the total capacity of the ion exchange resin. The second reaction needs an amount of alkalinity sufficient to neutralize the sites. If there is an excess of hardness over alkalinity, the excess hardness will leak through. For most waters in the oil fields, there is a large excess of sodium alkalinity over hardness as shown in the second equation. The third reaction goes to completion and good hardness reduction can be achieved.

The generation of carbon dioxide as shown in the second reaction is a matter for concern since the acidic water would be expected to have a corrosive effect on the metal surfaces in the boiler. Since the installations usually have several sets of units in different phases of conditioning and exhausting, the problem is not a serious one. The individual primary and polisher units usually operate separately also to keep the carbon dioxide neutralized.

### Experimental

The proper operating conditions require an experimental program to establish the proper amount of acid to be used for stripping. Typical data for the water shown in Table 1 are summarized in the following graphs. The tests were performed in 25 mm columns with 0.75 meter bed depths. Flows were controlled at three bed volumes per hour for acid stripping and service flow was eight bed volumes per hour. Hardness was determined by standard Versenate method and alkalinity and acidity by titration to methyl orange endpoint.

Figure 2 shows the regenerant profile when the acid quantity was held at 64 gms HCl/liter. Since the capacity was 110 grams HCl/liter, it is not surprising to find a serious insufficiency of acid. This results in the leakage of hardness as shown in Figure 3.

Figure 4 summarizes a corrective action where 160 grams HCl/liter was used to strip the resin bed. As expected, a very substantial acid peak is measured with a low pH in the effluent. Figure 5 shows the exhaustion cycle. As expected, an excellent hardness capacity was achieved at low leakage.

The next step was to run some tests at 134 gms HCl/liter. The results are shown in Figure 6 for the regenerant profile. The slow rinse contains most of the excess acid. Figure 7 shows the fast rinse profile which contains considerable hardness which would not be acceptable as feed to the polisher. After this the service cycle shows a very good run with reduction of hardness and high pH as shown in Figure 8.

It was decided to go down a little further to 112 gms HCl/liter. The regeneration profile is shown in Figure 9. This was quite similar to the 134 gms/liter run. The fast rinse again had a substantial volume reaching 16.5 bed volumes. The subsequent service cycle as shown in Figure 11 was quite good. The fast rinse on these runs near the optimum regeneration point was quite long indicating considerable hardness stripping still going on from the bed.

The capacity values and rinse volumes for the various runs are summarized in Table 3 and they are also plotted in Figure 12. This graph shows a sharp drop in fast rinse values near the optimum acid regeneration point. It becomes apparent that the acid regeneration is inefficient near the end point as determined by the free acid in the spent regenerant. When the fast rinse is started, there is some acidity and hardness leakage in the fast rinse. When an excess of acid is used, the fast rinse is quite low as shown in the figure.

## Discussion

The studies show that an optimum acid strip can be measured through a study of the spent regenerant profiles. The actual operation is quite sensitive to conditions. While not examined experimentally it should be possible to soften flood waters so long as the alkalinity exceeds the hardness. How great an excess has not been established, but should certainly be greater than 2 to 1 if a reasonable throughput is to be achieved. Acid usage can be expected to be about 120 grams HCl/liter of resin.

The economic gain from the alkalinity neutralization step can be quite substantial since an equivalent amount of caustic would be required to neutralize the bed. This would amount to 132 grams NaOH per liter. The generation of carbon dioxide has not been too excessive as shown in the spent regenerant profiles.Possibly a plant could deal with this by phasing the operation of several sets of columns feeding the polisher.

## Economics

An appreciation of the savings involved in this process can be gained by a typical plant calculation assuming a bed containing ten cubic meters of carboxylic cation exchange resin.

Assuming a cycle of service and regeneration per day, the unit would consume 1200 kg of 100% HCl. By the conventional process it would require the acid plus 132 kg of NaOH. The unit would treat 25,000 m³ of water. The daily cost for caustic would be 494 dollars (US) and for acid 317 dollars (US). The alkalinity neutralization would reduce the daily cost by about 60%.

Table 1

Analysis of Waters Used for Brackish Water Softening

| Component | Concentration, meq/liter |
|---|---|
| Total Hardness | 6.89 |
| Calcium | 1.80 |
| Magnesium | 4.60 |
| Alkalinity (methyl orange endpoint) | 42.8 |
| Chlorides | 117.6 |
| pH | 9.4 |

Table 2

Summary for Rinse Values for
Weakly Acidic Cation Softening of Brackish Water

| Run No. | Regenerant Level gms HCl/liter | Regenerant Volume BV | Slow Rinse BV | Fast Rinse BV | Service Volume BV |
|---|---|---|---|---|---|
| 2 | 64 | 1 | 2.5 | 13 | 170 |
| 3 | 160 | 3.25 | 1.0 | 2.5 | 262 |
| 4 | 134 | 2.5 | 1.0 | 4 | 240 |
| 5 | 112 | 2.1 | 1.0 | 17.5 | 265 |

Table 3

Summary of Capacity Values for
Weakly Acidic Cation Softening of Brackish Water

| Run No. | Regenerant Level gms HCl/liter | Regenerant BV | Slow Rinse BV | Fast Rinse BV | Service BV | Service Capacity eq/liter |
|---|---|---|---|---|---|---|
| 2 | 64 | 1.25 | 2.75 | 13 | 160 | 1.09 |
| 3 | 160 | 3.3 | 0.75 | 4.7 | 268 | 1.92 |
| 4 | 134 | 2.5 | 1.0 | 6.5 | 240 | 1.83 |
| 5 | 112 | 2.25 | 1.0 | 16.5 | 265 | 1.80 |

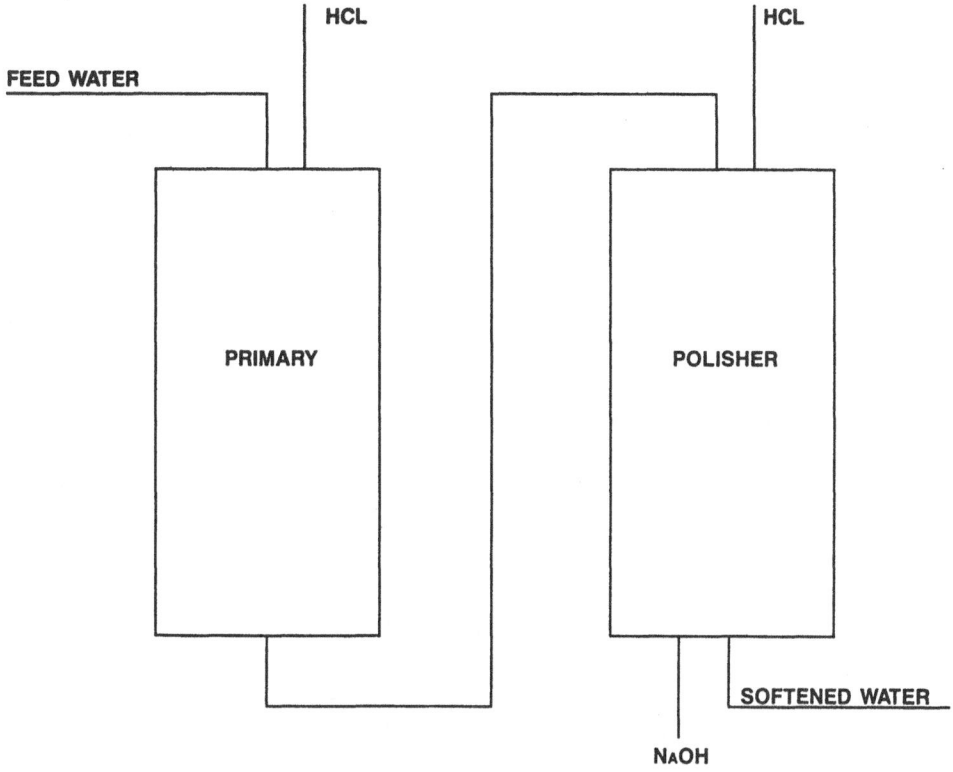

Figure 1 - **PROCESS FLOW**

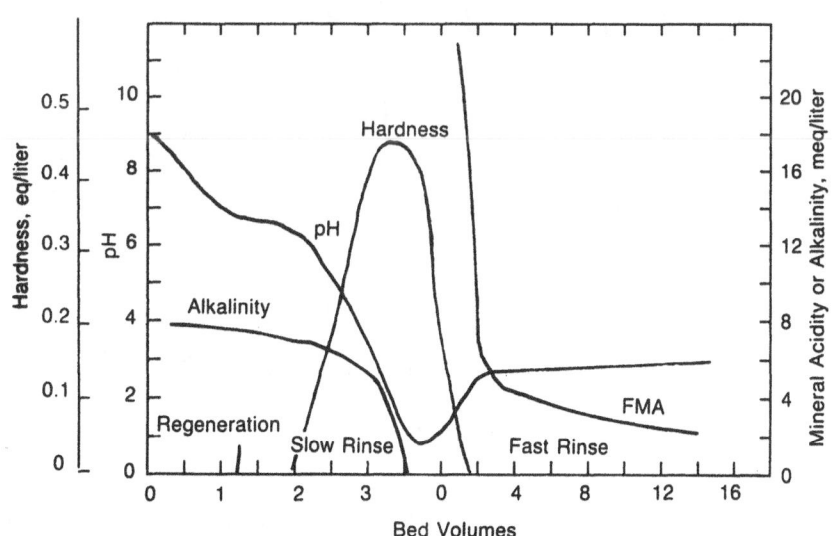

Figure 2 - Regenerant Profile for Cycle 2 - 64 grams HC1/liter

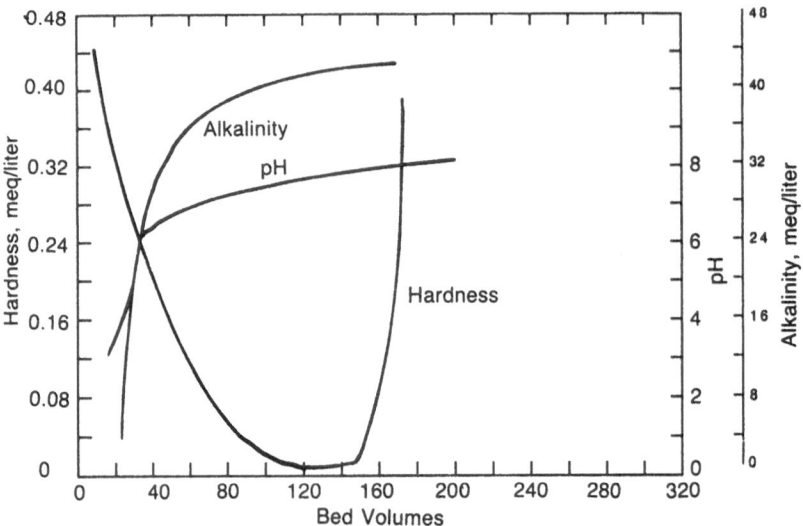

Figure 3 - Service Cycle on Run 2 - 64 gm HC1/liter

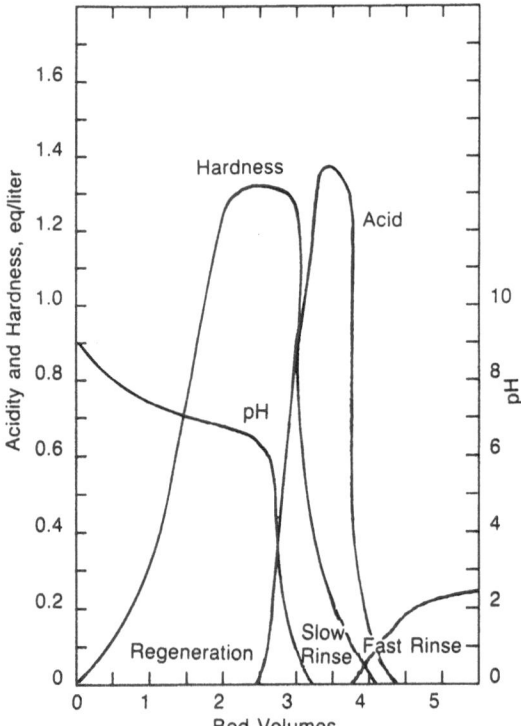

Figure 4 - Regeneration Profile - 160 gms HC1/liter

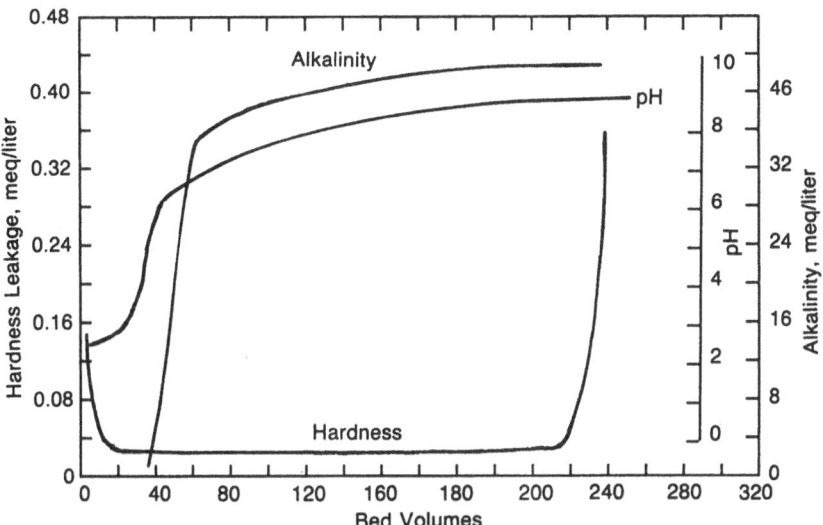

Figure 5 - Service Cycle After Regeneration - 160 gms HCL/liter

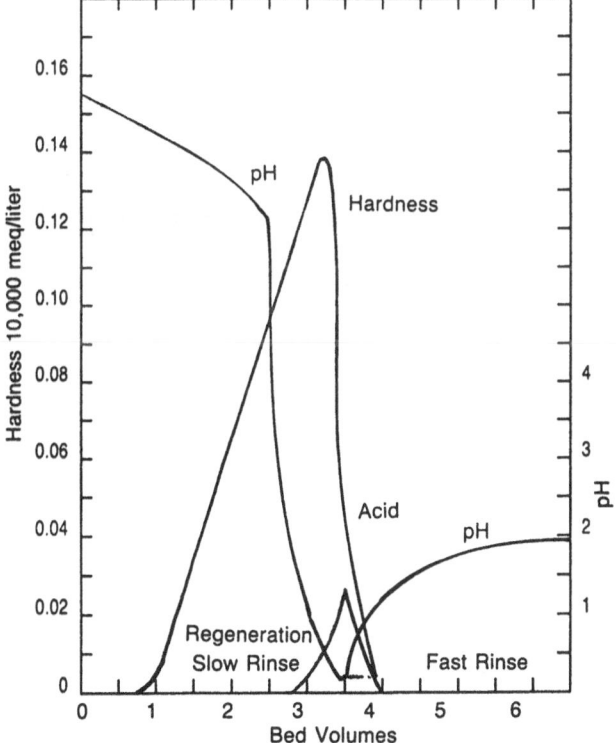

Figure 6 - Regeneration Profile of Cycle 4 - 134 gms HC1 gms/liter

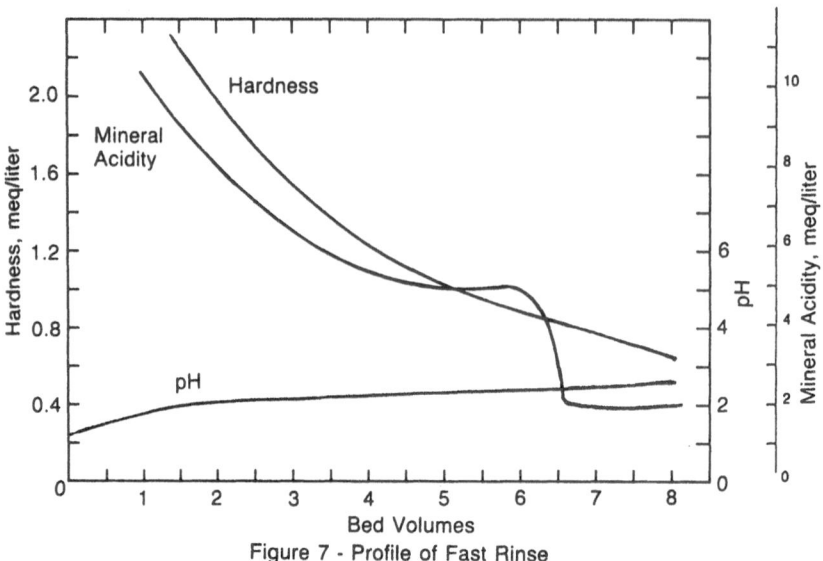

Figure 7 - Profile of Fast Rinse

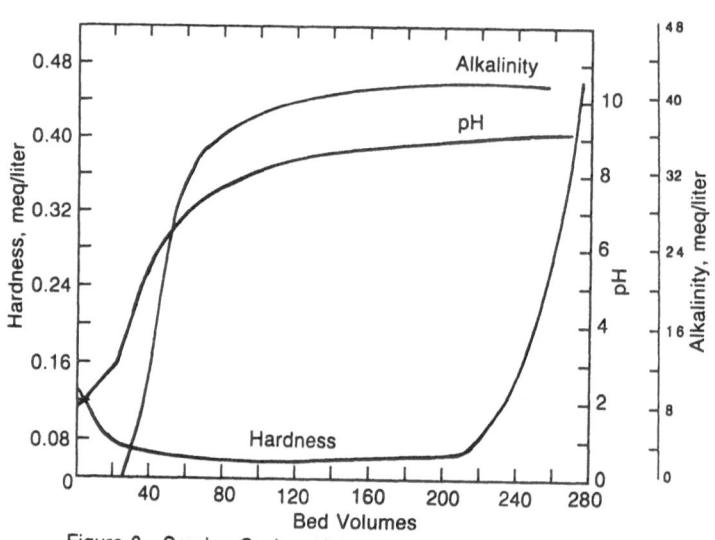

Figure 8 - Service Cycle - 134 gms HC1/liter

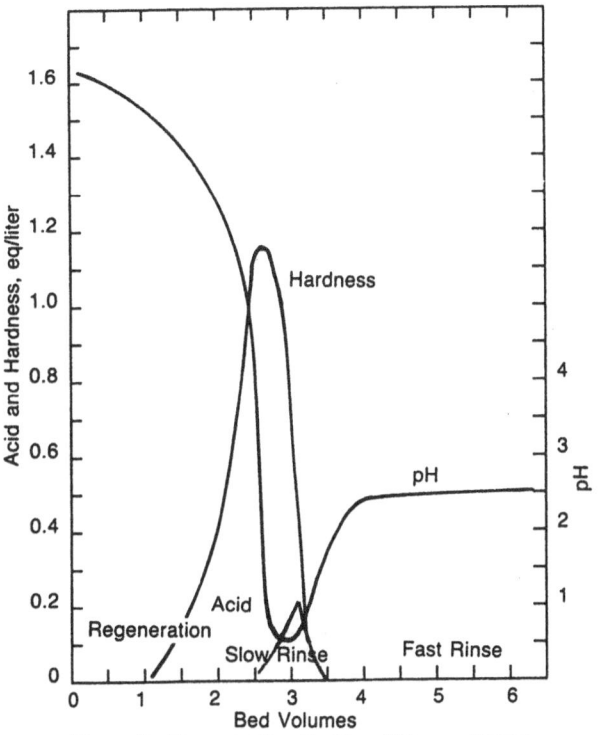

Figure 9 - Regeneration Profile - 112 gms HC1/liter

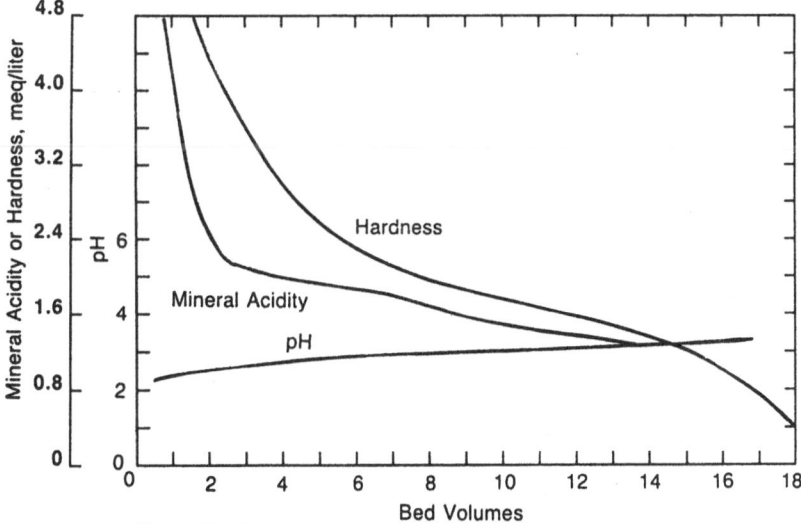

Figure 10 - Fast Rinse - Synthetic Water - 112 gms HC1/ft³

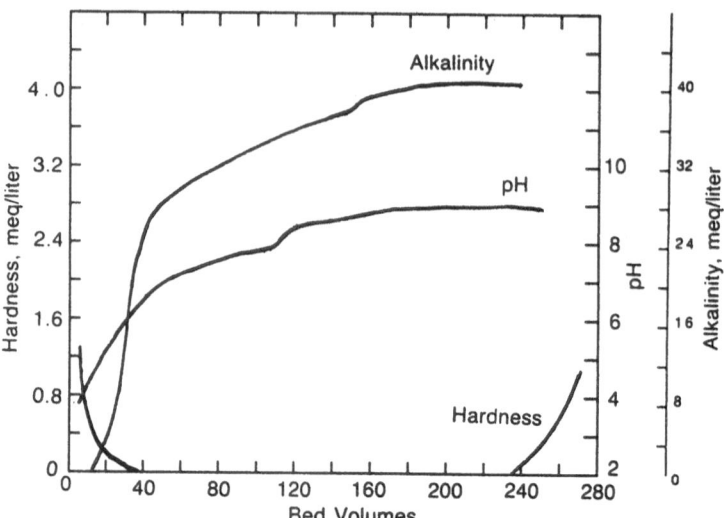

Figure 11 - Service Cycle - Synthetic Water - 112 gms HC1/ft³

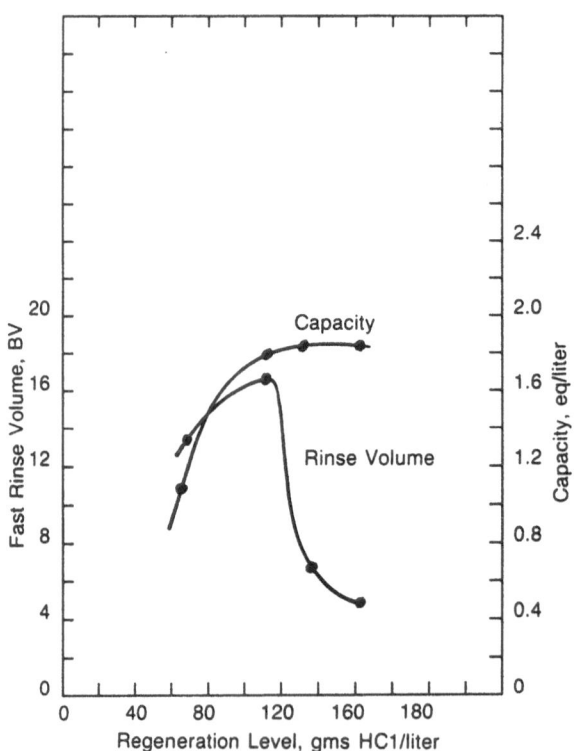

Figure 12 - Rinse and Capacity Values as a
Function of Regeneration Level

## Conclusion

Weakly acidic cation exchange resins are effective materials for softening produce water used in secondary oil recovery. The alkalinity neutralization step can save considerable money, but it is necessary to study the regenerant profile so that the proper regeneration schedule can be measured.

## References

1.  McGarvey, F., Ziarkowski, S. and Gottlieb, M., Water Conditioning for Oil Well Flooding. 18th Annual Liberty Bell Corrosion Course, September 30, 1980, Philadelphia, Pennsylvania, USA.

2.  McGarvey, F. and Gottlieb, M., Column Studies with Selective Ion Exchangers. Energy Progress, December 1981.

3.  McGarvey, F., Hauser, E. and Ziarkowski, S., Ion Exchange Studies on the Softening of High Solids Waters for Enhanced Oil Recovery. 20th Annual Liberty Bell Corrosion Course, September 1982, Philadelphia, Pennsylvania, USA.

# ON-LINE MONITORING OF TRANSITION METALS IN WASTEWATER

GRAHAM HUMPHREY
Dionex (UK) Limited
Albany Park, Camberley, Surrey GU15 2PL

## ABSTRACT

Ion Chromatography, high performance liquid chromatography
(HPLC), and flow injection techniques have been used for the
on-line determination of anions, cations, metals, organics,
and other chemical constituents in a variety of process
streams.  These on-line techniques have been used in several
industries to provide the real time chemical information required
for statistical process control and to identify process upsets
as they occur.

   Ion Chromatography has the distinct advantage of offering
the determination of multiple components in a single analysis,
as well as multiple sample point analysis. These capabilities
have proved extremely beneficial in a number of industries.

## INTRODUCTION

The discharge regulations for many species in wastewater are
becoming more stringent, and penalties for discharging
concentrations above the limits are rising.  The National
River Authority is conducting more spot checks to verify the
accurate reporting of discharges.  Many industries are faced
with the need for continuous monitoring tools for rapid
identification of pollutants     in the waste treatment plant
effluent.  In many cases, staying within the discharge limits
requires optimum performance of the waste treatment plant.
Since common waste treatment processes involve   the
precipitation of heavy metals, the operation for such a process plant
requires the addition of treatment chemicals.

The concentration of treatment chemicals required for optimum performance has been shown to vary with the concentration of metals in the incoming wastewater.

On-line monitoring of the incoming metal concentration can be utilized to control the addition of treatment chemicals. Such a control system can ensure that sufficient chemicals are present during peak treatment periods, as well as decreasing the consumption of chemicals during periods where high concentrations are not required.

Ion chromatography has a significant advantage over other analysis techniques for the on-line determination of metals. Other techniques available, such as Atomic Absorption, require the use of flammable gases in a process environment. Ion Chromatography is easily automated and can provide a profile of the metal content in a short period of time. (**figure 1**)

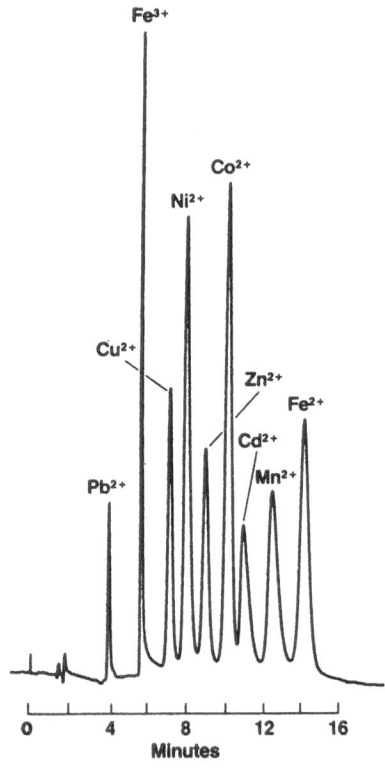

Figure 1          detection of nine Transition Metal ions.

## MATERIALS AND METHODS

The results were obtained, using the series 8100 on-line
process (wastewater) analyser.  The series 8100 process
analyser is the second generation of on-line systems from
Dionex and incorporates features and improvements based on
over four years of operational experience with on-line Ion
Chromatography.

The sample line was immersed in the top of the tube
settler (**figure 2**) and gravity fed to a diaphragm pump to
provide continuous representative samples to the 8100
analyser.  The samples were acidified to pH 2 in order to get
to a "total" metal analysis.

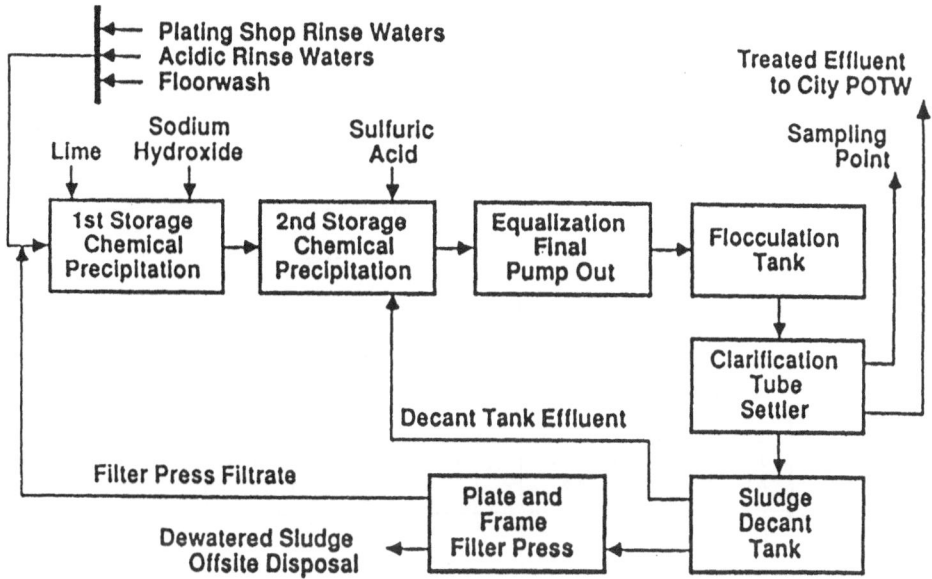

Figure 2 Waste Treatment Block Diagram

After acidification,the samples were loaded on a fixed volume
loop and injected onto a chromatographic column for the
separation of the metal ion complexes formed with the
pyridine-2,6-dicarboxylic acid eluent.  This was then followed
by a post column reaction step by adding a mixture of
4-(2-pyridylazo)resorcinol, ammonium hydroxide and acetic
acid.  Subsequent detection was performed with a visible
detector at a wavelength of 520nm.

The series 8100 process (wastewater) analyser is completely
controlled by the process 300 computer control and data
handling system, which provides control over automatic
sampling, sample preparation and the analysis.  It also offers
a variety of data output functions including trending reports.
If desired, the process 300 computer may be located remotely
in an office or control room.


## RESULTS

For a comparison, samples were sent to an EPA approved
laboratory for the determination of transition metals using
Method 3020, where samples are acid digested and analysed by
Graphite Furnace Atomic Absorption, calibrated between 0.1,
0.5, and 1ppm of each metal.  The metals generally present at
trace levels in the effluent samples we compared were: copper,
nickel, zinc and iron at typical levels below 0·5ppm.

    The results shown in figure 3 are comparable, considering
the use of a different set of calibration standards and two
dramatically different analytical procedures.

|  | Cu | Ni | Zn | Fe |
|---|---|---|---|---|
| Sample 1 |  |  |  |  |
| IC | .077 | .200 | .078 | .082 |
| AA | .10 | .17 | <.1 | 1.6 |
| Sample 2 |  |  |  |  |
| IC | .072 | .057 | .106 | .300 |
| AA | .12 | <.1 | .15 | 2.1 |
| Sample 3 |  |  |  |  |
| IC | .095 | .169 | .029 | .034 |
| AA | <.1 | <.1 | <.1 | .19 |
| Sample 4 |  |  |  |  |
| IC | .003 | – | .024 | .090 |
| AA | <.1 | <.1 | <.1 | .19 |

Figure 3 Comparison of Results IC Verses AA

The only large discrepancy appears in the iron values
obtained.  The lower concentrations by the IC method are
indicative of slower reaction kinetics for dissolution of iron
 hydroxide as compared to the other metals.  Another
possibility could be due to the pressure of strong iron
 cyanide complexes.

## CONCLUSION

Whilst Ion Chromatography may not yield exactly equivalent
results to atomic Apsorbtion with all samples, it can clearly
be utilized as          an  early warning monitoring
technique.  Ion chromatography is also more adaptable    to on-
line operation since there is no need for flammable gases, and
multiple metal determinations may be performed on a single
injection.

Additional benefits may include increased knowledge of
production or operational variations which could directly have
an effect on product yield and quality.

I acknowledge the help given to me in the preparation of
this paper by Karen Haak and Gary Lee of Dionex Corporation.

## REFERENCES

1.   LEE, G., On-line Monitoring of Transition Metals in
     Wastewater.  Dionex Corporation, Sunnyvale, 1985
     Pitttsburgh Conference.

2.   HAAK, K., Field Experience with on-line Ion
     Chromatography, American Laboratory, 1987.

# SOME PROBLEMS IN THE USE OF CHELATING RESINS FOR ENVIRONMENTAL PROTECTION FROM TOXIC METALS

J. R. MILLAR, D. PETRUZZELLI, & G. TIRAVANTI
C.N.R., Istituto di Ricerca sulle Acque, Bari, Italy

## ABSTRACT

No truly specific chelating resin has yet been reported. Even high selectivity for a particular metal is in most cases achievable only under fairly circumscribed conditions, e.g. of pH, ionic strength, nature of the co-ion(s) present, temperature, or absence of other too-strongly competing metal species. Attempts to improve the capability of known selective materials have led to the laboratory synthesis of sorbents many of which would be incapable of economic manufacture in the quantities required for the uses proposed. However, some commercial applications are discussed.

## INTRODUCTION

"The difficult we do immediately; the impossible takes a little longer," [1]
UNOFFICIAL SLOGAN OF THE U.S. ARMED FORCES

Unfortunately, in the words of Matthew Arnold [2] , "Miracles do not happen." Some idea of the difficulty of designing and making a sorbent specific for a given metallic ion can be gauged from the fact that the search was commenced over half a century ago for such a material, and while at the last count [3] (see Table 1) over half a hundred selective resins are commercially available as a result, only a few are in fact commercially viable, and none even approaches the original criterion. At the present time, a number of toxic metals identified in the environment, either of natural origin or more frequently arising from human activities of various kinds, might be regarded as candidates for removal by even imperfectly specific sorbents. Were it not that interference occurs from other (not necessarily similar) metal ions, or that the task of pH control on the influent solution is too great, one or another of the known selective resins might be used. Neither of these factors, by definition, should affect a truly specific resin; again, by definition, to regenerate such a resin would be at best very difficult, and at worst, impossible.

TABLE 1
Main Commercial Chelating & Complexing Resins

| Type | Manufacturer | Resin Name | Notes |
|---|---|---|---|
| Iminodiacetic acid | Bayer | Lewatit TP 207, 208 | |
| | Bio-Rad | Chelex-100 | |
| | Dow Chemical Co. | Dowex A-1 | a |
| | | Dowex XF-4045 | |
| | Mitsubishi | Diaion CR-10 | |
| | Purolite Intnl | Purolite S 930 | |
| | Rohm & Haas | Amberlite IRC-718 | |
| | do. | Duolite ES 466 | b |
| | do. | Imac GT-73 | b |
| | Unitika | Unicellex UR-10 | |
| | [U.S.S.R]. | ANKB-1, -10, -35, -50 | c |
| | VEB Farbenfabriken. | Wofatit MC-50 | |
| Isothiuronium | Ayalon | Srafion NMRR | |
| | | Monivex | |
| | Bayer | Lewatit TP 214 | |
| | Ionac | Ionac SR-3, SRXL | |
| | do. (Resindion) | Relite MAC-3 | d |
| | Rohm & Haas | Duolite ES 345 | b |
| | Sumitomo | Sumichelate Q-10 | |
| Aminophosphonic | Purolite Intnl. | Purolite S 940, 950 | |
| | Rohm & Haas | Duolite ES 467 | b |
| | [U.S.S.R.] | ANKF-2G, -3G, -221 | c |
| | | AEF-1, -2, -3 | c |
| Amidoxime | Rohm & Haas | Duolite CS 346 | b |
| Hydroxyoxime | Dow Chemical Co. | Dowex XF-4196 | |
| Picolylamines | Dow Chemical Co. | Dowex XF-4195, -43084 | |
| Boron-specific | Rohm & Haas | Amberlite IRA-743 | e |
| Dithiocarbamate | Nippon Soda | Nisso ALM-525 | |
| Thiol | Bayer | Lewatit OC-1014 | |
| | Purolite Intl. | Purolite S 920 | |
| | Rohm & Haas | Duolite ES 465 | b |
| | do. | IMAC TMR | b |
| Other | Bayer | Lewatit DN KR | f |
| | Rohm & Haas | Duolite C-3 | b, f |

Notes:  a) No longer supplied directly, (see Bio-Rad).
b) Duolite & IMAC are now Rohm & Haas trademarks.
c) From the 1980 official "Katalog Ioniti".
d) Relite (Resindion) is a Sybron trademark.
e) based on N-methylglucamine-aminated CM poly(S-DVB).
f) Phenolsulphonate resin selective for caesium.

Most of these present-day cation-selective resins are based on the chelation principle. The cation is initially held inside the polymeric matrix by ionic forces, and further co-ordinated via electron-donating groups in appropriate steric relationship contained in the polymer. The additional entropy involved in such a complex binding makes the ion less easily liberated from the resulting chelated polymeric complex. From "monomeric" analogues, it is known that the chelate stability constant tends to increase with decrease in the ionization constant of the ionic moiety, so that there is always a competition between the $H^+$ ions in solution and the chelated ions for the binding group, and hence the selectivity for any bound ion will be pH-dependent. Where the cation in the influent solution is complexed by "monomeric" anions, ligand-exchange [4] also plays a part. Although this paper is not directly concerned with anion-selective resins, these also usually function by ligand-exchange, with the hydrophilicity of the resin phase, and the size of the positively-charged cationic groups being determining factors [5].

The toxic metals of the title can be classified in a number of ways. Those capable of acute toxicity have long been acknowledged, but many such as Ni , Cr , Cd , and As have been identified as human carcinogens [6], and in recent decades the harmful somatic effects of minute traces of metals such as Pb or Hg have been thoroughly documented [7,8]. Attention was first drawn to problems arising from sub-toxic levels of Pb nearly twenty years ago [9]; the European Community hopes to have phased out leaded petrol by 1992. Sources of such problems are often manifold, and literally, so in the case of water-borne lead where its classic use in plumbing, as well as sinkers lost by fishermen, the shooting of waterfowl, and airborne roadway dust, all contribute.

Even common metals like Zn or Al may give rise to chronic sub-toxic effects [10], although they are essential to life. In addition to such concerns, a range of metals can occur in radioactive forms as a result of irradiation or from nuclear fission of uranium, thorium, or plutonium, and although the most obvious effect of radiation is carcinogenic, there may also be more subtle problems of teratogenicity to be dealt with. However, even if "organic" was the buzzword of the eighties, a sense of proportion is essential in any environmental concern; there is no substance that is completely innocuous, nor any so toxic that it cannot be used safely.

The contribution of selective sorbents to the reduction of hazard occurs in two main ways. The first is by direct treatment of, e.g., water supplies to remove residual traces of the offending metal, the second in the treatment of waste solutions from which the environment itself may be at risk. The direct treatment of potable water by cation exchange becomes ineffective when the offending levels of toxic metals are very small in comparison with the normal cation content, and the use of a highly selective sorbent as a "polisher" becomes the method of choice. On the other hand, for the treatment of waste solutions at relatively high toxic metal concentrations, conventional techniques can be quite adequate, particularly if the components of the waste may be segregated as individual streams, to avoid a possible competition for the resin groups by two similar ions which leads to unacceptable levels of leakage. Problems arise on the large scale because it is not always economically feasible to adjust the pH of a large bulk of solution, for example the acidification of seawater to facilitate uranium removal, or to ensure that the one macroion that interferes with the separation is in a higher (or lower) oxidation state in the presence of other reducing (or oxidizing) agents.


## DEVELOPMENT OF COMMERCIAL MATERIALS.

Many early chelating resins were made by condensation polymerisation of phenolic or aromatic amines containing known ligand groups. It was, however, difficult to control the (normally quite high) crosslinking of such materials, and the groups giving rise to the chelate structure were often modified either chemically or sterically during reaction or polymerisation to the extent that their complexing ability was impaired [11].

Later workers made use of the ready functionalisation of addition polymers such as poly(styrene-co-DVB) or crosslinked polyacrylic matrices [12]; more recently still, the ability has been exploited of crosslinked poly(glycidyl methacrylate) or poly(vinylbenzylchloride) to alkylate, and thus attach to the (often macroporous) polymer structure, any of a number of pre-formed ligands [13]. The significant effect of the nature of the matrix, and the mode of attachment of the functional groups [14] is by now commonplace, but this was not so when the first commercial chelating resin was introduced.

This resin was Dowex A-1, introduced some thirty-odd years ago [15], an iminodiacetate resin with groups analogous to those in the "monomeric" chelating agent of the same name, or in EDTA. It was exhaustively investigated by Rosset for his doctorate thesis [16], although it has undergone a number of changes in its synthesis since. The importance of the homogeneity of its functional groups was not recognised until some 10 years after its début [17], and the original resin is no longer being made by Dow. Subsequent manufacturers of this type of resin used macroporous matrices to obviate its notoriously poor kinetics [18], and controlled the substitution reaction more closely, so that the amount of singly-reacted (aminoacetate) groups is kept to a minimum and both capacity and selectivity are optimised. Both improvements add an appreciable amount to the manufacturing cost of the resin.

Although iminodiacetic acid itself (IDA) is a broad-spectrum rather than a highly selective chelating agent, the newer IDA-type resins (mainly used for analytical separations) found a commercial application in the removal of small amounts of the alkaline earth metals from solutions containing quite large amounts of alkali metal ions. These resins were in fact used in the sodium form to "soften" brines for electrolytic chlorine production in the older diaphragm cells to levels around 3-5 ppm. of $Ca^{++}$.

Current membrane cells, however require less than 0.02 ppm., and this can be achieved at substantially higher capacity by a similar use of the related aminomethylenephosphonate exchangers. Somewhat similar materials were synthesized by Kennedy & Ficken in 1957 [19], and by Manecke & Heller in 1960 [20], but the earliest equivalent to the present-day resins was first described in 1973 by Szczepaniak [21]. The macroporous version of this material is probably the most successful of the commercially-available chelating resins, with annual sales of the order of 1000 $m^3$. It is currently being investigated in China, Japan, and the Soviet Union for various environmental applications. In the field of toxic metal removal, traces of radiothorium isotopes have been removed from $HNO_3$- rich raffinates of uranium ore concentrates [22], while the efficient removal of by-product uranium from wet process phosphoric acid solutions is also feasible at temperatures up to 60 °C elution of the recovered uranium being effected by the use of a more powerful complexing agent, a solution of ammonium carbonate, at ambient temperatures [23].

Similar techniques are required for other selective resins. The poly(isothiuronium) resin, first reported in 1958 by Cerny & Wichterle [24], investigated from 1967 onwards in Israel [25], and commercially developed by the Ayalon company as Srafion NMRR (1969, gel-type) and later Monivex (1972, macroporous), is a case in point. Complete elution of platinum group metals sorbed from neutral or slightly acidic solutions is virtually impossible; with the macroporous version, loading from highly acid solutions does permit elution [26], though only with a slightly acid solution of the powerful complexing agent thiourea (to prevent hydrolytic formation of competing thiol groups on the resin).

The development of the thiol resins themselves, by 1955 already reported by various workers in Japan [27], the U.S.[28], and Britain [29], was accelerated by the trauma of "Minimata disease" in Japan during the late fifties, when it was found that discharge of methylmercury to coastal waters in Minimata Bay had caused contamination of fish, and consequent neurological disorders to nearly 2000 people, and the deaths of some 400. Many of these thiol resins were good selective sorbents for Hg, whether ionised or covalent, and with their modern macroporous matrices can virtually eliminate methylmercury from concentrations as low as 10 ppb.[30]. They have also been used for laboratory separations of other heavy metals.

Other than the picolylamine resins which have a recognised use in the hydrometallurgy of copper [31], the remaining types of chelating resin have found little or no large-scale application; a recent survey has been made [32] of the treatment of electrolytic & electroless nickel plating baths, and the so-called nickel strike bath; no single resin was found which would work efficiently and economically, although some marginal savings could theoretically be made. Essentially, these resins are of present significance only in analytical applications, where their rather poor reproducibility necessitates regular calibration of the techniques used.

## COST OF CHELATING RESINS

It is impossible to discuss the costs of resins in absolute terms, as this information is very dependent on raw materials costs at the manufacturing site, batch size and production techniques, form and purity of the resin as supplied, shipping cost to user, and expected life of the resin in service, not to mention the suppliers's particular sales policy for the product.

However, if the standard gel-type styrene sulphonic acid resin is taken as 100, approximate relative costs of some other commercial resins are given in Table 2 below:

TABLE 2
Relative costs of commercial resins

| Type of resin | Cost |
| --- | --- |
| Std. gel styrenesulphonate | 100 |
| macroporous styrenesulphonate | 150-180 |
| Std. gel carboxylic resin | 230-240 |
| macroporous carboxylic resin | 260-270 |
| Std. gel Type I quaternary | 260-290 |
| macroporous Type I quaternary | 300-350 |
| macroporous weak-base resin | 330-380 |
| most chelating resins | 600-900 |

This in conjunction with their relatively poor stability in comparison with other cation-removing exchangers means that they have to work hard to earn their keep. While a thiuronium resin scavenging platinum group metals for resale can command a premium price for a one time use, followed by incineration to yield the product directly, [33], a careful cost analysis of Co and Ni recovery from lead smelter residues using the Dow picolylamine chelating resin XFS-4195, reported by Kennedy and co-workers three years ago, showed economics very dependent on the market price of cobalt [34].

In another example, cadmium, a toxic metal present in the wet-process phosphoric acid referred to earlier, and preventing its use directly for the manufacture of fertilizer, is not removed under the acid conditions by a chelating resin, but a process using a conventional macroporous strong-acid cation exchanger in a continuous countercurrent contactor has been described [35]. Even with the cheaper strong-acid cation resin, the overall cost of cadmium removal is sufficiently high that it will only become economic when the legal requirements are more stringent than at present.

## UNUSUAL SELECTIVITY IN STANDARD RESINS

Among the long-established commercial ion-exchangers there are several that show interesting selectivity behaviour which may serve to avoid the use of specialty chelating resins for some applications. In any conventional cation exchanger the electroselectivity effect will ensure that ions of higher charge are preferred to those of lower charge to an increasing and sometimes quite spectacular degree as the concentration of the solution is decreased [36]. In some cases, a change in oxidation state will convert a metal cation to an anion, thus enabling it to be stripped from, or preventing its uptake by, the negatively-charged functional groups. A good example is that of $Cr^{III}$ which can fairly readily be oxidised to chromate or dichromate anion. Carboxylic resins show a special selectivity for metals forming four-ring "-ato" complexes [37], and have been tried as a means of selectively removing $Cr^{III}$ from a mixture with $Fe^{III}$ and $Al^{III}$, with only partial success [38].

The special affinity for Cs of the phenol methylenesulphonate resins, such as the obsolete Permutit Zeo-Karb 315, or Duolite C-3, finds a use in removal of radiocaesium from radwaste solutions [39], and the Mannich-base structure of the phenolic weak-base resin Duolite A-7, and its ability to complex $Cu^{++}$ and also $Fe^{+++}$ sulphates from solution, led Hodgkin and Eibl to design two new resins, christened Sirorez-Cu & Sirorez-Fe, selective for copper [40] and iron [41] respectively. It is ironic that because of the unavailability of the expensive Sirorez material, Duolite A-7 (which, despite having been around for 40 years or so, is still being made by the present trademark owners, Rohm & Haas) has been the resin of choice for the removal of $Fe^{III}$ from its mixture with $Cr^{III}$ and $Al^{III}$, details of which are given in another paper to be presented in this volume [42].

## CONCLUSIONS

It should be stressed that only when commercial or large-scale use is demonstrated to be feasible and believed to be economical will the manufacturer be prepared to investigate the making of an "exotic" resin on a plant scale. Because the synthesis of such resins is (usually) more complex, the intermediates more expensive, and resin life expectance less than for conventional ion exchangers, the capital and (to a lesser extent) the running costs of such a system may rule it out of court. All things change, however; with our present environmental concern, and our new-found ability to put a price on amenity perhaps tomorrow's problems may be able to make better use of today's answers.

## REFERENCES

1) de Calonne, C.A., "si c'est possible, c'est fait; impossible? ça se fera." [ex Michelet's *Histoire de la Révolution Française*, (1847), Vol. 1, part ii, §8, adapted.]

2) Arnold, M(atthew), end of Preface to *Literature and Dogma*, 1883 edn.

3) Millar, J.R., *Design, Synthesis, & Applications of Chelating Ion-Exchange Resins*", I.R.S.A. Internal Rpt. No.126, C.N.R., Italy, (1989)

4) Helfferich, F.G., *J. Amer. Chem. Soc.*, (1962), *84*, 3237,3242.

5) Sengupta, A. & Roy, T., *Ion Exchange for Industry*, ed. M. Streat, Ellis Horwood Ltd., Chichester, 1988, 195-203.

6) see *I.A.R.C. Monographs on the Evaluation of the Carcinogenic Risk of Chemicals to Humans*, Vols. 1-20, WHO, Lyon, (1972-9)

7) Bryce-Smith, D., *Chem. Brit.*, (1975), *11*, 202-6.

8) d'Itri, P.A. & F.M., *Mercury Contamination; a Human Tragedy*, Wiley Interscience, New York, (1977).

9a) Chem. & Ind., (1988), 269, 346.

b) *Aluminium in Food and the Environment*, eds. R. Massey & D. Taylor, Royal Society of Chemistry, Letchworth, (1989)

10) Bryce-Smith, D., *Chem. Brit.*, (1972), *8*, 240.

11) Millar, J.R., *Chem. & Ind.*, *1957*, 606-12.

12) Nickless, G. & Marshall, G.R., *Chromatog. Rev.*, (1964), *6*, 154-190.

13) Lindsay, D., & Sherrington, D.C., *React. Polym.*, (1985), *3*, 327-39.

14) Warshawsky, A., *Israel J. Chem.* (1979), *18*, 318-24.

15a) Morris, L.R., to Dow Chemical Co., U.S. Pat. 2,875,162 (1959).

b) Mock, R.A., Marshall, C.A. & Morris, L.R., to Dow Chemical Co., U.S. Pat. 2,910,445 (1959)

16a) Rosset, R., *Bull. Soc. Chim. France*, *1964*, pp. 1845 *et seq.*. *1966*, (1), pp. 59 *et seq.*.

b) Rosset, R., D.Sc. thesis, Paris, 6 June 1967, C.N.R.S. no.AO 1490.

17) c.f. Hering, R., *Chelatbildende Ionenaustauscher*, Akademie Verlag, Berlin, (1967)

18) Kressman, R.E., *Ion Exchange & its Applications*, (discussion, p. 22), S.C.I. London, 1955.

19) Kennedy, J., & Ficken, G.E., *J. Appl. Chem.*, (1958), *8*, 465-8.

20) Manecke, G., & Heller, H., *Angew. Chemie*, 1960, *72*, pp. 532 *et seq.*

21) Szczepaniak, W., *Chemia Analitycz.*, Warsaw, (1973), *18* (5), 1019-26.

22) Eccles, H., in *Ion Exchange Technology*, eds. D. Naden & M. Streat, Ellis Horwood, Chichester (1984), 698-708.

23) Gonzalez-Luque, S., & Streat, M., in *Ion Exchange Technology*, edited by D. Naden & M. Streat, Ellis Horwood, Chichester (1984), 679-689.

24) Cerny, J., & Wichterle, O., *J. Polymer Sci.*, (1958), *30*, 501-12.

25) Koster, G., & Schmuckler, G., *Anal. Chim. Acta,*(1967), *179*, pp.38 *et seq.*

26) Warshawsky, A. in *Ion Exchange & Sorption Processes in Metallurgy*, *(Critical Reports on Applied Chemistry*, Vol. 19, eds. M. Streat & D. Naden, John Wiley, London (1987), 127-65.

27) Nakamura, Y., *J. Chem. Soc. Japan, Ind. Chem. Sect.*, (1955), *58*, 269-73.

28) Gregor, H.P., Dolar, D., & Hoeschele, G.K., *J. Amer. Chem. Soc.*, (1955), *77*, pp. 3675 *et seq.*

29) Parrish, J.R., *Chem. & Ind.*, *1956*, 137.

30) Warshawsky, A., in *Ion Exchange & Sorption Processes in Metallurgy*, *(Critical Reports on Applied Chemistry, Vol. 19*, eds. M. Streat & D. Naden, John Wiley, London (1987), 166-225.

31) Grinstead, R.R., *J. Metall.*, (1979),*31*, 13-16.

32) Gold, H., Czupryna, G., Levy, R.D., Calmon, C., & Gross, R.L., in *Metals Speciation, Separation & Recovery*, eds. R. Paterson & R. Passino, Lewis Publishers, Chelsea MI, (1987), 619-645.

33) see Reference 26.

34) Kennedy, D.C., Becker, A.P., & Worcester, A.A., in *Metals Speciation, Separation & Recovery*, eds. R. Paterson & R. Passino, Lewis Publishers, Chelsea MI, (1987), 593-613.

35) Booker, N.A. & Streat, M., in *Ion Exchange for Industry*, ed. M. Streat, Ellis Horwood, Chichester (1988), 617-31.

36) Anderson, R.E., in *A.I.Ch.E Symposium Series*, (1975), *71* (152), pp.236 *et seq.*

37) Millar, J.R. & Kressman, T.R.E., Brit. Pat. 812,815 (29 Apr. 1959).

38) Petruzzelli, D., Millar, J.R., & Tiravanti, G., unpublished work.

39) Higgins, I.R. & Messing, A.F., ORNL-2491 (1958), see *Nucl. Sci. Abstr.* (1959), *13*, 68.

40) Hodgkin, J.H., & Eibl, R., *React. Polym.*, (1985), *3*, 83-89.

41) Idem, ibid., (1986), *4*, 285-91.

42) Tiravanti, G., Petruzzelli, D., Liberti, L., & Passino, R., (paper "Specific Resins for Metal Ion Separation; The Cr(III), Fe(III), Al(III) System", this volume 1990).

# THE ANALYSIS OF INDUSTRIAL WATER TREATMENT PRODUCTS USING ION CHROMATOGRAPHY SEPARATION WITH MULTIPLE DETECTION SYSTEMS

D. S. RYDER
Research Department,
Ciba-Geigy Industrial Chemicals,
Tenax Road, Trafford Park, Manchester M17 1WT, U.K.

## SYNOPSIS

Ion chromatography is employed in the Industrial Water Treatment field both for monitoring additive levels in application test rigs and for assessment of the purity of new research products. In order to cover the wide range of compounds generated by research, single column ion chromatography is used applying both ion exchange columns and dynamic ion exchange systems, together with conductivity, refractive index, direct and indirect photometric detection. Sequential triple detection provides increased confidence in both qualitative and quantitative analysis. Examples of applications involving the detection of carboxylic acid monomers, phosphonates, oxo acids of phosphorus, inorganic ions and a substituted triazine are presented.

## INTRODUCTION

The move towards wholly organic corrosion and scale inhibitor additives for water systems as an environmentally acceptable alternative to materials such as nitrite, and chromate based treatment programmes, requires the development and application of new monitoring methods. In addition during research and development stages of new products, analysis methods are required in order to establish purity, modes of operation and to monitor the efficacy of potential products. Ion exchange processes lend themselves well to the largely polar ionisable compounds encountered in the industrial water treatment field. Whether ion exchange columns or dynamic ion exchange, based on ion pairing systems, are employed detection need not be a limiting factor and may be tailored to suit the needs of the separation rather than restricting it. In the Research Department at Manchester we use conductivity, refractive index, direct U.V. and indirect U.V. depending upon the requirements of the application. Some systems enable the use of sequential triple detection which can be a useful investigational tool particularly in method development. The following paper presents some examples of this multiple approach which is required in order to solve the range of applicational problems encountered in the Water Treatment field.

## EXPERIMENTAL

### Instrumentation

A Hewlett Packard 1090 M liquid chromatograph was used together with a work station fitted with a two channel interface (35900A). The following detectors were connected in series (a). L.D.C. Conductomonitor III, a temperature controlled conductivity detector, (b). Hewlett Packard diode array U.V. detector interfaced directly to the work station and (c). E.R.M.A. E.R.C. 7510 differential refractive index detector.

### Materials and Reagents

All materials employed were of reagent grade and were used without further purification. Dionised and filtered water was from an Elgastat U.H.Q. system.

Chromatographic conditions were as described in the individual figures.

### Direct U.V. Detection

Direct U.V. detection of compounds may be enhanced by employing diode array in addition to retention times as a supplementary identifying characteristic, thus increasing confidence in the interpretation of results. Spectral information is frequently used for materials such as aromatics with strong U.V. chromophores but it can also prove effective even with aliphatic materials such as carboxylic acid monomers having peak maxima at wavelengths below 210 nm.

The comparison of U.V. spectra of acrylic acid and acetic acid carried out during chromatography using an ionically suppressed system with U.V. transparent phosphoric acid buffer on an O.D.S. column illustrates this ability, see Figure 1.

Diode array detection gives a large amount of extra information via its spectral analysis but it is somewhat less sensitive than conventional U.V. detection. However, additives with an inherently high U.V. absorption such as the substituted triazine ferrous metal corrosion inhibitor shown in Figure 2 may be detected at levels well below those used in application. This was chromatographed using a dynamic ion exchange system employing ion pairing on an O.D.S. column.

Direct U.V. may be misleading in that some compounds have inherently much greater U.V. sensitivity than others, e.g. the unsaturated acrylic acid is approximately one hundred times more sensitive than the saturated acetic acid. Thus what may appear on the chromatogram as apparently a 50 : 50 mixture of these two acids would actually be a 99 : 1 mixture. This would obviously cause great difficulties if the compounds in question were unknowns. Indirect U.V. may be used to reduce this effect.

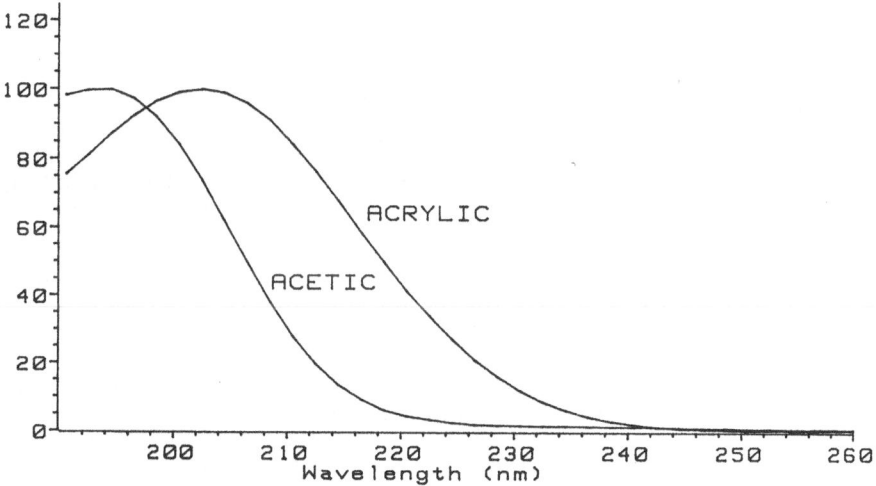

Figure 1. Comparison of U.V. spectra of chromatographed acids on hypersil ODS with 0.5% phosphoric acid.

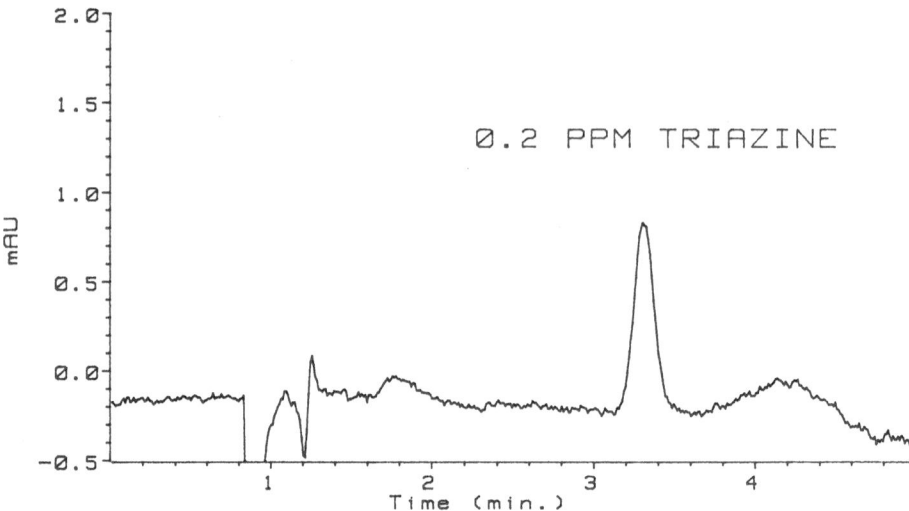

Figure 2. Detection of 0.2 ppm of a substituted triazine ferrous metal corrosion inhibitor on a test rig water. 10 UL injection on hypersil ODS using 60 : 40 water/methanol containing 0.4% tetra butyl ammonium hydroxide and 0.05% phosphoric acid direct UV detection at 216 nM.

**Indirect U.V.**

Indirect U.V. detection as described by Small et al[1] allows the sensitive detection of solutes with little or no U.V. absorption by incorporation of a strong U.V. absorber into the eluent. This often also acts as the sole buffering electrolyte as in the case of phthalate or hydroxy benzoate buffers. A negative vacancy peak appears where non U.V. absorbing ions elute from the column, these having replaced, due to ion exchange equivalence, the U.V. absorbing eluent ions. The negative peaks produced are usually electrically inverted to produce a 'normal' chromatogram. Using the HP 1090 this may be achieved by reversing the settings of the reference and sample wavelengths. It is also important to use a wide band width setting for the sample and a narrow band width for the reference in order to minimise the noise levels. As detection relies on the equivalent reduction in U.V. absorption of the phthalate ion and is not directly dependent on the nature of the analyte ions the responses for different compounds are more nearly universal. This is seen in the comparison of organic acids separated on a Vydac 302 1C column with a 2 mM phthalate buffer at pH 4.9, see Figure 3.

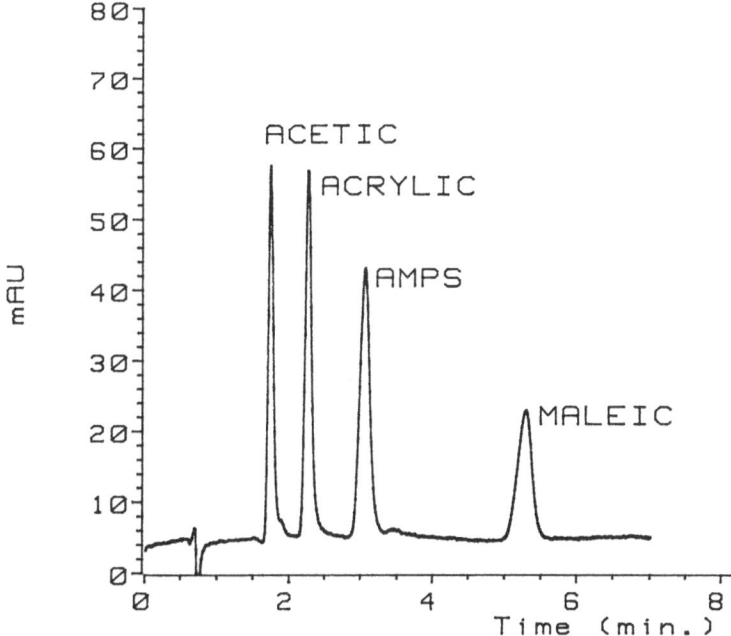

Figure 3. Carboxylic acid monomers at equi molar concentration using indirect U.V. detection. Vydac 302 IC column and phthalate buffer pH 4.9 detection : sample 550 nM, reference 270 nM. (AMPS) = 2-acrylamido 2-methyl propane sulphonic acid.

## Conductivity

In order to assess the purity of research materials it is necessary to determine levels of the oxidation states of phosphoric acid i.e. $PO_4-$, $PO_3-$, $PO_2-$ in polymers made from phosphite and hypophosphite. These ions are best separated as reported previously[2] with a weak acid at pH's giving rise to low dissociation e.g. succinic acid on a Vydac 302 IC column. This system is well suited to conductivity detection yielding a stable low noise baseline with high sensitivity to the analytes, see Figure 4. The detection limit for the three ions $PO_4-$, $PO_3-$ and $PO_2-$ are 0.2 ppm for a 100 ul injection. Good quantitative linearity is achieved at least from 5 ppm to 50 ppm. As the U.V. peak maximum of succinic acid is 210 nm it is not suited for either direct or indirect U.V.

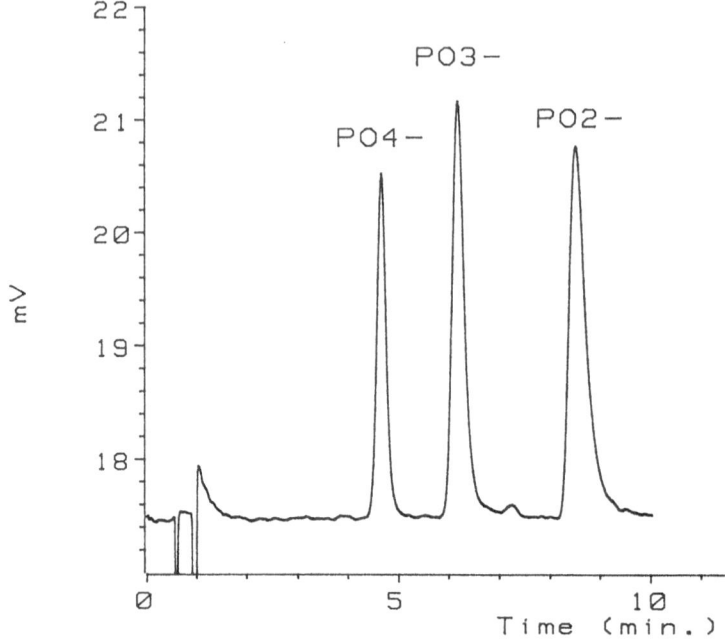

Figure 4. The separation of the oxo phosphorus anions using conductivity detection 100 μl of 10 ppm of each anion on Vydac 302 IC with 0.02M aqueous succinic acid.

## Refractive index

Anti - corrosion additives such as multifunctional phosphonates present particular chromatographic difficulties as they exhibit very little U.V. absorbance even at low wavelengths and are strongly retained on ion exchange columns whilst being very poorly retained on reverse phase columns. If a strong buffer such as nitric acid is used to elute from an anion exchange column such as the Ionosphere T.M.A. the background conductance is too high for adequate detection using conductivity and as nitric acid has a significant U.V. absorbance at low wavelengths direct U.V. may not be used and only poor sensitivity is obtained with indirect U.V. However, refractive index responses are favourable and application levels such as 5 ppm may be adequately detected as shown in Figure 5 allowing the monitoring of phosphonate additives in application test rigs.

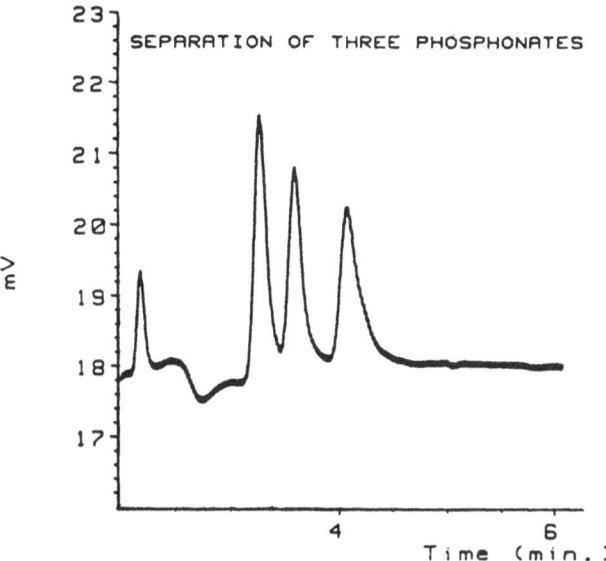

Figure 5. Detection of three phosphonate corrosion inhibitors in rig water at 10 ppm using refractive index detection. 200 µL injection onto ionosphere **TMA** with aqueous nitric acid at 2 mL/litre.

## Sequential Triple Detection

A system employing phthalate buffer at pH 4.9 together with a Vydac 302 1C column enables sequential detection using conductivity, indirect U.V., and refractive index. It is important to site the refractive index detector last in the series as this normally has the greatest dead volume. There is little detectable peak broadening from the first, conductivity, to the last refractive index detector and a delay of only 6 seconds at a flow rate of 2 ml per minute is seen. Each of the detectors gives a similar sensitivity in this system to the inorganic ions which need to be determined in water treatment compounds or rig waters e.g. chloride, nitrate, nitrite and sulphate, see Figure 6.

Organic acids are less well detected by conductivity in this system and refractive index is prone to drifting baselines. However, the relative response of each analyte compared to a standard such as $SO_4-$ is different for each detector. This effect may be used as a characteristic for identity confirmation. It may be considered that two compounds are unlikely to have identical responses for three different modes of detection. In addition triple detection is a useful tool in method development where there is often an uncertainty as to whether a 'missing' solute has not been eluted or whether it has not been detected by the detector in use. The selection of the optimum detector often also assists in reducing the effects of interference from unwanted peaks.

FIGURE 6. Separation of inorganic anions at 25 ppm each, (1) Chloride, (2) Nitrite, (3) Nitrate and (4) Sulphate. On a Vydac 302 IC column with phthalate buffer at pH 4.9.

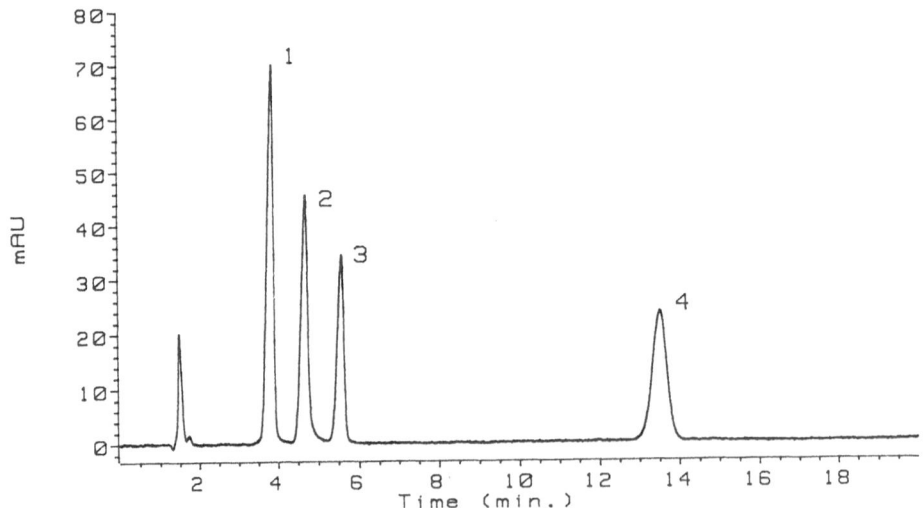

Figure 6(a) Indirect UV

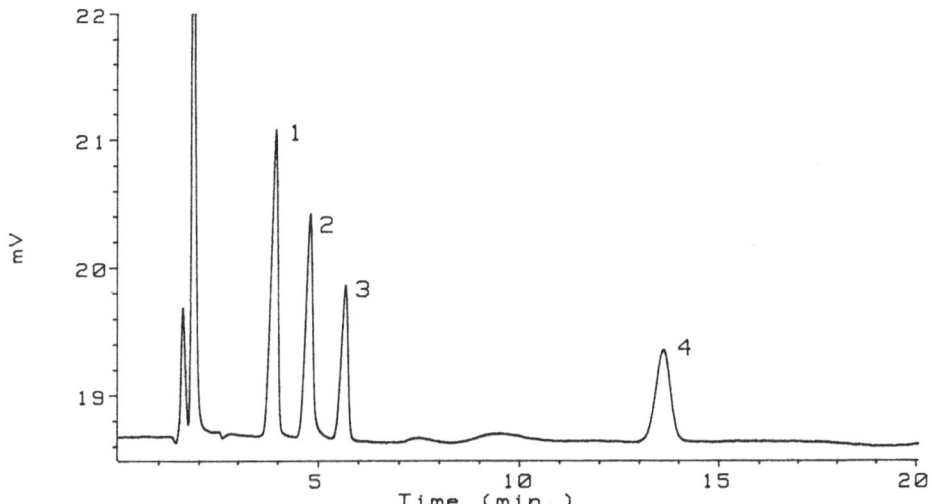

**Figure 6(b) Refractive index**

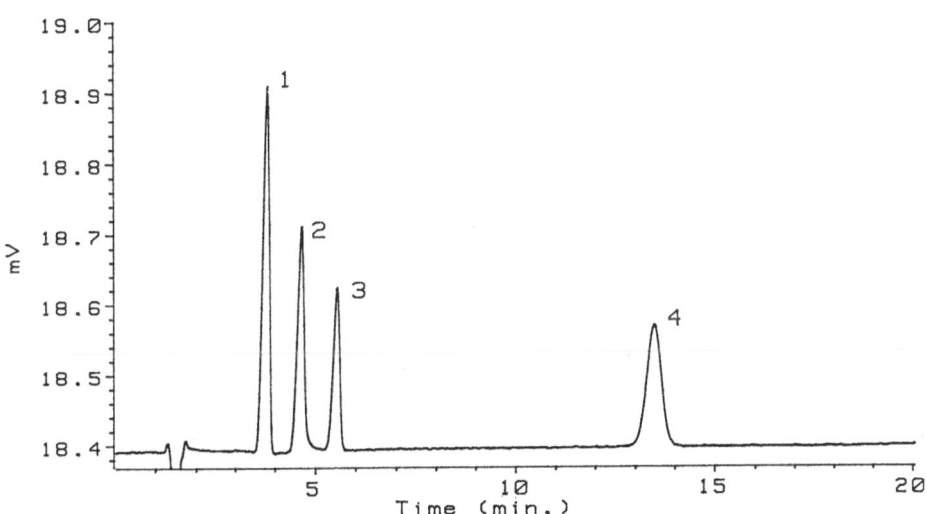

**Figure 6(c)  Conductivity.**

## CONCLUSION

A flexible approach using single column ion chromatography employing a selection of columns and detection systems is suited to the wide range of compounds encountered in the industrial water treatment field. Increased confidence in the results obtained is achieved when a combination of up to three detectors is used sequentially.

## REFERENCES

1.  SMALL, H. AND MILLER, T.E., ANAL. CHEM., 1982, 54, 462

2.  RYDER, D.S., J. CHROMATOGRAPHY, 1986, 354, 438

# INDEX OF CONTRIBUTORS

# SUBJECT INDEX